Library of
Davidson College

TAXONOMY AND ECOLOGY

THE SYSTEMATICS ASSOCIATION PUBLICATIONS

1. BIBLIOGRAPHY OF KEY WORKS FOR THE IDENTIFICATION OF THE BRITISH FAUNA AND FLORA
3rd edition (1967)
Edited by G. J. KERRICH, R. D. MEIKLE *and* NORMAN TEBBLE

2. THE SPECIES CONCEPT IN PALAEONTOLOGY (1965)
Edited by P. C. SYLVESTER-BRADLEY, B.Sc., F.G.S.

3. FUNCTION AND TAXONOMIC IMPORTANCE (1959)
Edited by A. J. CAIN, M.A., D.Phil., F.L.S.

4. TAXONOMY AND GEOGRAPHY (1962)
Edited by DAVID NICHOLS, M.A., D.Phil.

5. SPECIATION IN THE SEA (1963)
Edited by J. P. HARDING *and* NORMAN TEBBLE

6. PHENETIC AND PHYLOGENETIC CLASSIFICATION (1964)
Edited by V. H. HEYWOOD, Ph.D., D.Sc. *and* J. MCNEILL, B.Sc., Ph.D.

7. ASPECTS OF TETHYAN BIOGEOGRAPHY (1967)
Edited by C. G. ADAMS *and* D. V. AGER

8. THE SOIL ECOSYSTEM (1969) *Edited by* J. G. SHEALS

LONDON. Published by the Association

SYSTEMATICS ASSOCIATION SPECIAL VOLUMES

1. THE NEW SYSTEMATICS (1940)
Edited by JULIAN HUXLEY (Reprinted 1971)

2. CHEMOTAXONOMY AND SEROTAXONOMY (1968)*
Edited by J. G. HAWKES

3. DATA PROCESSING IN BIOLOGY AND GEOLOGY (1971)*
Edited by J. L. CUTBILL

4. SCANNING ELECTRON MICROSCOPY (1971)*
Edited by V. H. HEYWOOD

* Published by Academic Press for the Systematics Association

THE SYSTEMATICS ASSOCIATION
SPECIAL VOLUME NO. 5

TAXONOMY AND ECOLOGY

Proceedings of an International Symposium held at the
Department of Botany, University of Reading

Edited by

V. H. HEYWOOD

Department of Botany, University of Reading, England

1973

Published for the
SYSTEMATICS ASSOCIATION
by
ACADEMIC PRESS · LONDON · NEW YORK

ACADEMIC PRESS INC. (LONDON) LTD.
24/28 Oval Road
London NW1 7DX

United States Edition published by
ACADEMIC PRESS INC.
111 Fifth Avenue
New York, New York 10003

Copyright © 1973 by
THE SYSTEMATICS ASSOCIATION

Second Printing 1975

All Rights Reserved

No part of this book may be reproduced in any form by photostat, microfilm, or any other means, without written permission from the publishers

Library of Congress Catalog Card Number: 73-7039
ISBN: 0 12 346960 0

PRINTED IN GREAT BRITAIN BY
T. AND A. CONSTABLE LTD., EDINBURGH

List of Contributors

BAKER, H. G. and BAKER, I., *Department of Botany, University of California, Berkeley, California, U.S.A.*
BARKMAN, J. J., *Biologisch Station, Kampsweg 27, Wijster (Dr.), Netherlands.*
BERRIE, A. D., *Department of Zoology, University of Reading, Reading, England.*
EHRENDORFER, F., *Institute of Botany, University of Vienna, Austria.*
EISIKOWITCH, D., *Department of Botany, Tel-Aviv University, Israel.*
GUINOCHET, M., *Laboratoire de Taxonomie Végétale Expérimentale et numérique de l'Université de Paris-Sud associé au C.N.R.S., 91 Orsay, France.*
HAWKSWORTH, D. L., *Commonwealth Mycological Institute, Ferry Lane, Kew, Richmond, Surrey TW9 3AF, England.*
HEDBERG, O., *Institute of Systematic Botany, University of Uppsala, Sweden, P.O. Box 541, 75121 Uppsala 7, Sweden.*
HEYWOOD, V. H., *Department of Botany, Plant Science Laboratories, The University, Whiteknights, Reading RG6 2AS, England.*
JANZEN, D. H., *Department of Zoology, University of Michigan, Ann Arbor, Michigan, U.S.A.*
JONES, D. A., *Department of Genetics, University of Birmingham, Birmingham B15 2TT, England.*
LEWIS, D. H., *Department of Botany, University of Sheffield, Sheffield S10 2TN, England.*
MEEUSE, A. D. J., *Hugo de Vries-Laboratorium Voor Bijzondere Plantkunde, Universiteit van Amsterdam, Plantage Middenlaan 2a, Amsterdam, Holland.*
MORLEY, B., *Department of Agriculture and Fisheries, National Botanic Gardens, Glasnevin, Dublin 9, Ireland.*
SNAYDON, R. W., *Agricultural Botany Department, The University of Reading, Reading, England.*
THOMPSON, P. A., *Royal Botanic Gardens, Wakehurst Place, Ardingley, Haywards Heath, Sussex RH17 6TN, England.*

* Present address: *Unit of Genetics, Department of Plant Biology, University of Hull, England.*

Preface

During the past 25 years the Systematics Association has sponsored a series of symposia covering most of the major new developments in systematics and taxonomy. It has also paid particular attention to relationships with other fields of study which have a bearing on systematics, such as biochemistry, geography, data processing and scanning electron microscopy. For its symposium at the University of Reading on 11-14 September 1972 the Council of the Association decided to take a critical look at taxonomy and its sister science ecology and at the numerous ways, both obvious and less obvious, in which they interact.

Both taxonomy and ecology are regarded as sciences of synthesis in that they draw upon various subdisciplines and techniques for their information. Since these latter are largely the same, there is a major overlap between taxonomy and ecology. This, added to the fact that they are both to a large extent dependent upon each other, makes the relationships between them highly complex and multi-dimensional.

In planning the programme, it was decided to explore a number of basic themes which were made the basis of the symposium sessions—Ecological Factors and Speciation, Biocoenology and Taxonomy, Relationships, and Practical Role of Ecological Data in Taxonomy. Many of the topics proposed by the invited speakers were so interdisciplinary, embracing plant/animal relationships of various kinds, that we decided to label two of the sessions as simply relationships.

By the very nature of the subject, a complete coverage of all relevant topics was not possible or even attempted but the papers presented here should give a good idea of the main ways in which the relationships between taxonomy and ecology are developing at various levels, from the population to the family and class, from the microhabitat to the ecosystem. It was not the purpose of this symposium to discover great new general principles or to reach new levels of synthesis but rather to widen the dialogue between workers in related, although in practice often isolated, disciplines. That there was a need for greater attempts at communication and mutual understanding, not only between taxonomists and ecologists but also between different kinds of practitioners within each field, became very evident during the lively discussions that took place in the meetings and at the last general session. Some of the greatest debates took place over the interpretation and usage of such basic concepts as species and population.

The symposium was attended by biologists from 15 different countries. Its success was made possible by the hard work and co-operation of many people. In particular I should like to thank Mrs. Abigail Gillett and other members of the Department of Botany at Reading for their invaluable help in making the arrangements for the meeting. The University of Reading offered a reception for the participants on 12 September and we are grateful for this and for their hospitality in allowing the meeting to be held at the University. The Royal Society kindly provided a travel grant to assist several of the overseas speakers attend the symposium.

My colleague, Dr. David Moore, has helped me in many ways in planning the meeting and I am most grateful for his co-operation. Finally, I should like to acknowledge the expert assistance of the Production Department of Academic Press in the preparation of this volume for press.

July, 1973 V. H. HEYWOOD

Contents

		PAGE
THE SYSTEMATICS ASSOCIATION PUBLICATIONS		ii
LIST OF CONTRIBUTORS		v
PREFACE		vii

1. Ecological Factors, Genetic Variation and Speciation in Plants .. 1
   R. W. SNAYDON

2. Ecological Factors and Species Delimitation in the Lichens .. 31
   D. L. HAWKSWORTH

3. Adaptive Evolution in a Tropical-Alpine Environment .. 71
   O. HEDBERG

4. Seed Germination in Relation to Ecological and Geographical Distribution .. 93
   P. A. THOMPSON

5. Phytosociologie et Systématique .. 121
   M. GUINOCHET

6. Taxonomy of Cryptogams and Cryptogam Communities .. 141
   J. J. BARKMAN

7. The Relevance of Symbiosis to Taxonomy and Ecology, with Particular Reference to Mutualistic Symbioses and the Exploitation of Marginal Habitats .. 151
   D. H. LEWIS

8. Snails, Schistosomes and Systematics: some Problems Concerning the Genus *Bulinus* .. 173
   A. D. BERRIE

9. Anthecology, Floral Morphology and Angiosperm Evolution .. 189
   A. D. J. MEEUSE

Contents

10 Comments on Host-Specificity of Tropical Herbivores and its Relevance to Species Richness 201
    D. H. JANZEN

11 Co-evolution and Cyanogenesis 213
    D. A. JONES

12 Some Anthecological Aspects of the Evolution of Nectar-Producing Flowers, Particularly Amino Acid Production in Nectar 243
    H. G. BAKER and I. BAKER

13 Ecological Factors of Importance to *Columnea* Taxonomy .. 265
    B. MORLEY

14 Mode of Pollination as a Consequence of Ecological Factors .. 283
    D. EISIKOWITCH

15 Co-evolution of Plant Hosts and their Parasites as a Taxonomic Tool 289
    A. D. J. MEEUSE

16 Adaptive Significance of Major Taxonomic Characters and Morphological Trends in Angiosperms 317
    F. EHRENDORFER

17 Ecological Data in Practical Taxonomy 329
    V. H. HEYWOOD

AUTHOR INDEX 349

SUBJECT INDEX 358

1 | Ecological Factors, Genetic Variation and Speciation in Plants

R. W. SNAYDON

Agricultural Botany Department, University of Reading, England

Abstract: The effects of ecological factors upon patterns of morphological variation among plants are considered at the level of the individual, the population, the species and above the species. The implications of these effects are discussed in relation to the practical value of taxonomic classifications in plant ecology.

The effects of environmental and biotic factors upon both phenotypic and genotypic variation between individuals are considered. Although this variation is largely disregarded by taxonomists, it has considerable ecological and evolutionary importance; it therefore deserves further attention.

Morphological and physiological differences between populations are frequently associated with environmental differences, and appear to be adaptive. These differences have been studied extensively be genecologists and biosystematists, but have been shunned by taxonomists. Current needs in ecology and genecology make taxonomic treatment more necessary, and recent advances in multivariate analysis make it more feasible.

The effects of ecological factors on plant speciation are considered. Attention is drawn to the importance of ecological factors in species divergence. The importance of reproductive isolation mechanisms, as prerequisites for adaptive divergence, is questioned.

The ecological value of taxonomic classifications would probably be increased if functional characters were used more widely in plant taxonomy.

INTRODUCTION

Taxonomy and ecology are both synthetic disciplines, but have very different objectives. Taxonomy is concerned with the classification, description and naming of variation among organisms, whilst ecology is concerned with the study of interactions between organisms and their environment, and between organisms. At first sight, such diverse objectives seem to exclude the possibility of appreciable interrelationship between the two disciplines, but closer inspection shows that the two disciplines interrelate at a number of points (Constance, 1953; McMillan, 1954; Kruckeberg, 1969; Selander, 1969).

Taxonomic classifications, descriptions and nomenclature are used by a wide variety of biologists, and it is now widely accepted (Davis and Heywood, 1963) that a major aim of taxonomy is to produce a "natural" or "general purpose" classification, which can be used to make inductive generalizations (Gilmour, 1951), not only by taxonomists and phylogenists, but also by geneticists, ecologists, physiologists and biochemists. Ecologists are particularly dependent upon taxonomic nomenclature for purposes of communication, but taxonomic classifications are most useful when they allow inductive generalization. The theoretically ideal taxonomic classification, for the ecologist, would therefore be one with the maximum content of ecological information and the maximum predictive power in ecology. In such a classification, taxonomic groupings would closely reflect ecological differences. Such an ideal classification would obviously be "special purpose"; in practice it would be more convenient to have a "general purpose" classification, where some ecological utility was sacrificed to gain utility in other disciplines.

Ecology is obviously dependent upon taxonomy for a suitable nomenclature and classification but, conversely, taxonomy is dependent upon other disciplines, including ecology, for a suitable supply of taxonomically useful information. There is therefore interdependence between taxonomy and ecology, and a need for cooperation.

Interrelationships between taxonomy and ecology are likely to be most frequent where the attention of taxonomists and ecologists is focussed on the same objects. The taxonomic hierarchy:

individual → population → species → genus → family → order

and the ecological hierarchy of level or organization (Odum, 1971):

individual → population → (species →) community → ecosystem

share common ground at the level of the individual, the population, and perhaps the species; interrelations between the two disciplines might therefore be expected to be greatest at these points, though some points of contact may also occur at higher levels in both hierarchies. The relative importance of the various levels and the methods of study differ, of course, in the two disciplines.

We might consider the interrelationships between taxonomy and ecology from practical or operational viewpoints; this raises such questions as:

(i) Do taxonomists, working in herbaria, sufficiently consider the needs of ecologists working in field conditions?

(ii) Can ecological inferences be drawn from taxonomic classifications? Do taxonomic groupings, which are largely morphologically defined, reflect ecological similarities, which are largely physiologically determined?

(iii) Is the type of ecological data, normally collected by ecologists, useful to the taxonomist? Are habitat data, for example, valid taxonomic data?

(iv) To what extent are techniques of sampling, data analysis, and interpretation common to both taxonomy and ecology? How analogous is the ecological classification of vegetation to classification in taxonomy?

Alternatively, the interrelationships between the two disciplines may be considered from theoretical viewpoints, and in particular the effects of ecological factors upon the variation that is actually or potentially important in taxonomy. These effects of ecological factors are likely to be most apparent, or more easily interpreted, where divergence has been most recent, i.e. at the level of the individual, population and species.

This contribution considers the effects of ecological factors, both environmental and biotic, upon the individual, population and species, and considers the practical implications of these effects in ecology and taxonomy. Most of the examples are drawn from plant taxonomy and ecology; many of the conclusions may not necessarily be relevant to animal taxonomy and ecology.

ECOLOGICAL FACTORS AND THE INDIVIDUAL

Ecological factors have two-fold effects upon the individual: the direct effects upon the phenotype, and the indirect effects upon the genotype.

Individuals, alive in the field, pressed on herbarium sheets, or pickled in museum jars, are the result both of their genetic constitution and of environmental influences during their development. The relative magnitude of environmental and genetic effects depend upon the character considered, the taxon under consideration, and the environmental conditions investigated.

Ecological factors also have an indirect effect upon the genotype of the individual, since the genetic structure of populations is the outcome of long-term interactions between genetic processes, that generate and maintain genetic variation, and environmental factors, which mould that genetic variation.

1. Ecological Factors and the Phenotype

(a) *The Phenotype and the Physical Environment.* Characters that are greatly modified by environmental conditions have been regarded traditionally as "bad" taxonomic characters; environmental modifications seem to have been regarded as "noise" which prevents the recognition of the genetic constitution of the individual. To what extent is this attitude still valid?

Attitudes to phenotypic variation in genetics have changed considerably since the early interests in phenotypically stable characters gave way to studies of the more numerous characters that are modified by environmental conditions. It is now apparent that any distinction between genetic and environmental influences upon the phenotype is artificial and rather meaningless (Dobzhansky, 1970). For example, the capacity of phenotypes to respond to environmental influences is itself under genetic control (for a recent review see Bradshaw, 1965), and there is some evidence that environmental modifications may become genetically "fixed" (Waddington, 1953).

Physiological studies of environmental modification, e.g. Björkman and Holmgren (1963), and Whitehead (1963), indicate that it is not random or biologically insignificant, but is frequently adaptive, and of evolutionary and ecological importance (Heslop-Harrison, 1964; Bradshaw, 1965; Cook, 1968). Thoday (1953) and others have pointed out that phenotypic plasticity and genetic differentiation are alternative strategies in adaptation; each has advantages and disadvantages, and both strategies occur widely. Examples are known where phenotypic plasticity in response to a particular environmental factor occurs in one species, and genetic differentiation occurs in a closely related species, e.g. the response of *Lysimachia* species to light intensity (Turesson, 1922). In many populations, both strategies may operate simultaneously in relation to a given environmental variable, i.e. both genetically fixed forms and phenocopies may occur mixed in a single population (Turesson, 1922; Nelson, 1965); the balance of the two mechanisms may vary between populations, and also between species (Marshall and Jain, 1968; Jain, 1969). In view of these considerations it is perhaps not surprising that, *inter alia*, Wells (1969) has made a detailed criticism of the "misleading distinction between environmentally induced phenotypic modification and genetically fixed ecological races".

The status of phenotypic variation has apparently changed in genetics and ecology in the past few decades; this may be a suitable time to re-evaluate the status of phenotypic variation in taxonomy (Cook, 1968).

(b) *The Phenotype and the Biotic Environment.* Interactions with other organisms frequently have large effects upon the phenotypic expression of plant characters. Competition, in particular, greatly influences the phenotype (Donald, 1963); Harper (1961, 1967) and Bradshaw (1965) have discussed the adaptive significance of this phenotypic plasticity.

The reciprocal effects of parasite and host species on the phenotype of each other seem to have received less attention, though Wilkins (1963) showed that the effects of host species upon the phenotype of the semi-parasite *Euphrasia* were similar in magnitude to the effects of contrasting soil type upon several non-parasitic genera (Marsden-Jones and Turrill, 1938). It is difficult, as Wilkins points out, to assess the adaptive significance of the reciprocal modifications of host and parasite.

(c) *General Considerations.* As a result of recent studies in ecology, physiology and genetics it is now apparent that phenotypic variation is of considerable ecological and evolutionary importance. This may therefore be an opportune time to reconsider the taxonomic status of phenotypic variation.

From an ecological viewpoint, there are sound practical reasons why characters that are subject to environmentally induced variation should be considered in taxonomic descriptions and classifications. Firstly, the ecologist must identify taxa in the field, where individuals are subject to phenotypic modifications. He uses non-variable characters when possible, but must often resort to variable characters; he therefore requires information on the extent of environmental modification. More detailed information on the relationship between phenotypic characteristics and environmental conditions may allow the ecologist to draw valid ecological conclusions; Cook (1968), for example, states that "plastic responses are specific to particular environmental influences and it is rather shocking that so little information on phenotypic modification is present in formal taxonomic works". Secondly, the taxonomic rejection of environmentally variable characters has focussed attention on more stable floral characters, rather than on the more plastic vegetative characters. Since most species flower for only a limited period each year, the field ecologist is at a severe disadvantage. Adequate keys, based upon vegetative characters, are only available for a few taxonomic groups, e.g. the grasses (Hubbard, 1968), so that each field ecologist relies upon highly individual, intuitive keys, developed after long experience and rarely formalized. Thirdly, the emphasis upon environmentally stable characters in plant taxonomy has tended, although not deliberately, to exclude ecologically important characters. The resulting classifications therefore have a smaller content of ecological information, and less predictive value in ecology, than might be the case if plastic but more functional characters were considered.

In view of our apparent inability to distinguish clearly between genetic and environmental effects, and in view of the ecological and physiological importance of environmentally induced phenotypic variation, we must face the important philosophical or theoretical question: should the aim of taxonomy be to classify abstract genotypes or concrete phenotypes? The question of course begs other more practical questions, such as whether the classification of genotypes is physically possible. So far the taxonomist has escaped the dilemma by recognizing only characters that are environmentally stable, and the phenotype closely reflects the genotype, but in doing so he has discarded large numbers of characters that are ecologically and physiologically important. Greater use of characters that are environmentally modified would produce more ecologically useful descriptions, and might also lead to a more "general purpose" classification at the lower end of taxonomic hierarchies, but would probably not greatly change classifications at the upper end of taxonomic hierarchies.

2. Ecological Factors and the Genotype

Considerable genetic variation occurs within most plant populations. In the past, this variation has generally been considered as random variation, the product of genetic processes within the population, which has escaped natural selection. There, is, however, increasing evidence that much of the variation is the result of ecological processes within population. Disruptive selection, due to environmental variation in both space and time, maintains genetic variation in many plant populations, and frequency-dependent interactions between genotypes may also maintain genetic variation. Some of the variation within populations, therefore, appears to have short-term (ecological), or long-term (evolutionary) adaptive significance.

Characters that are genetically variable within populations, i.e. have wide variation within samples, have traditionally received little attention from taxonomists, and have generally been regarded as "bad" taxonomic characters (Davis and Heywood, 1963). If this variation has ecological and evolutionary significance, there may be good reason to give it further taxonomic consideration.

(a) *The Genotype and the Physical Environment*. Genetic variation within populations is determined by interactions between genetic processes that generate variation, and ecological factors, both environmental and biotic, that mould this variation.

It would be reasonably easy to predict the outcome of these interactions if the environmental conditions were stable in time and uniform in space, but environments are rarely stable or uniform (Lewontin, 1969). Levene (1953), Dempster (1955), Levins (1962), Maynard Smith (1962) and Levins and MacArthur (1966) have considered the probable effects of environmental heterogeneity on the genetic structure of populations, and conclude that genetic variability should be greater in heterogeneous environments. There is some evidence of this in the field (Dobzhansky, 1962; Carson and Head, 1964; Stalker, 1964; Van Valen, 1965), though it is usually difficult to estimate environmental heterogeneity in the "wild", and the adaptive significance of the characters measured is uncertain. Evidence from more closely defined conditions (Fig. 1) indicates that the amount of genetic variation within populations of *Anthoxanthum odoratum* L. was also closely correlated with environmental heterogeneity, measured both directly as soil pH (Fig. 1a), or indirectly by plant performance (Fig. 1b). Many of the characters measured in the latter example are apparently of adaptive significance (Snaydon and Davies, 1972); we might therefore infer that the genetic variation within populations has adaptive significance, but the crucial test would be whether particular genotypes are associated with particular microenvironments. Few such studies have been made, though the studies by Harberd (1961a) and Smith (1965) indicate that the microdistribution of genodemes of *Festuca rubra* L. may be associated with small-scale variation in the environment. Similar evidence also comes from studies of

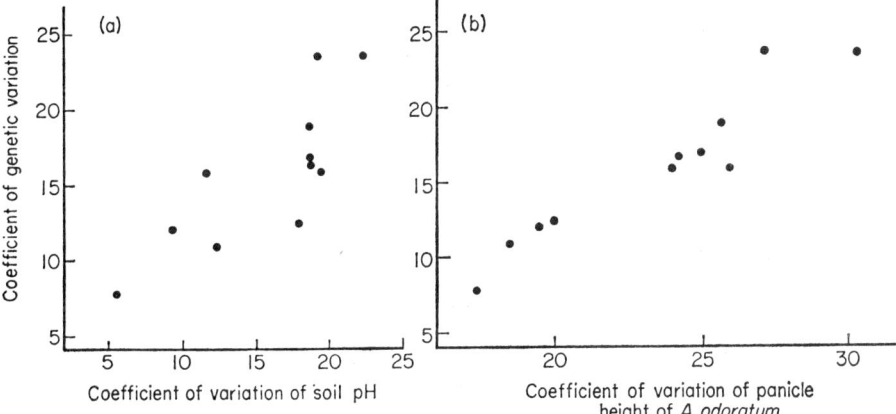

FIG. 1. Relationships between genetic variation within populations of *Anthoxanthum odoratum*, collected from the Park Grass Experiment at Rothamsted, and measures of spatial variation in the environments from which they were collected. The genetic variation of each population is expressed as the mean coefficient of genotypic variation, based on 8 genotypes per population, and 10 morphological characters. The plants were grown in a replicated spaced plants trial in uniform garden conditions. (a) Genetic variation in relation to spatial variation in soil pH. (b) Genetic variation in relation to spatial variation in the height of *A. odoratum*.

population differentiation at the sharp boundaries between contrasting environments (Jain and Bradshaw, 1966; Antonovics and Bradshaw, 1970). Populations may differ, in some cases, over distances of less than 100 mm (Fig. 2); the differences closely follow the sharp changes in environmental conditions, and are apparently adaptive.

Changes in environmental conditions with time also affect the genetic variability within populations. The genetic structure of both plant and animal species may change rapidly in time (Dobzhansky, 1947; Charles, 1964; Brougham and Harris, 1967). Genetic variation within populations is normally greater in variable environments, both in the laboratory (Beardmore, 1961; Beardmore

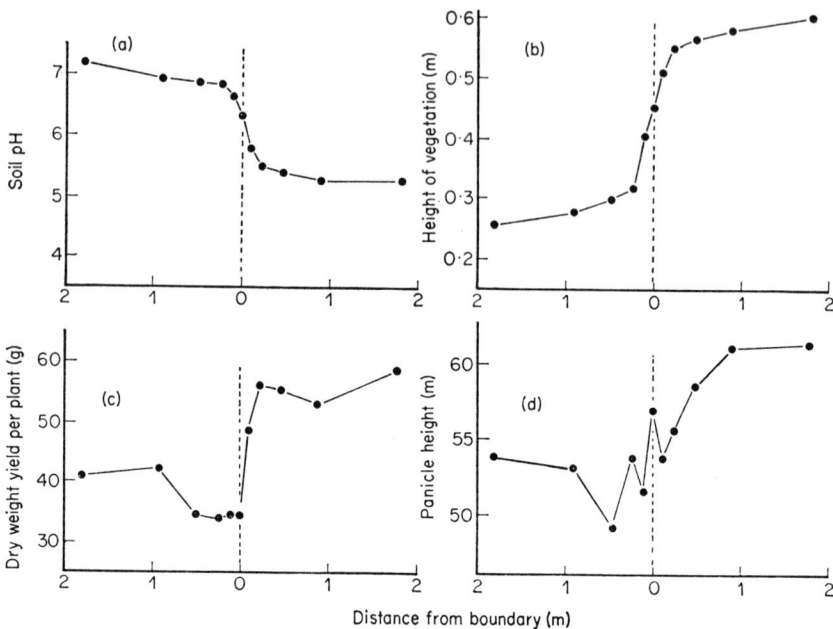

FIG. 2. Spatial changes in environmental conditions at the boundary between two plots of the Park Grass Experiment at Rothamsted (a and b), and associated morphological differences between populations of *Anthoxanthum odoratum* collected at the boundary (c and d). (a) Soil pH. (b) Height of vegetation. (c) Dry matter yield of *A. odoratum* plants in a spaced plant trial. (d) Panicle height of *A. odoratum* in a spaced plant trial.

and Levine, 1963) and in the field (Fig. 3); this genetic variation is apparently important for the adaptibility of populations (Lewontin, 1957).

(b) *The Genotype and the Biotic Environment.* An increasing amount of evidence, much of it obtained in the last decade, indicates that some of the variation within plant and animal populations is the result of biotic interactions. Clarke (1962) and Ford (1971) have reviewed the evidence that polymorphism in several animal species is due to density-dependent predation. Suneson (1949) has argued that some of the variation within plant populations is due to density-dependent effects of pathogens. The results of Beardmore (1963), Workman and Allard (1964), Allard and Adams (1969) and Schultz and Usanis (1969) indicate that density-dependent effects, involving competition and synergism between genotypes within populations, may maintain genetic variation within plant populations. These biotic interactions imply the evolution of co-adaptation within and between species. Pimental *et al.* (1965) and Seaton and Antonovics

(1967) have demonstrated that this may occur rapidly in experimental populations, though others have been unable to repeat their results.

(c) *General Considerations.* Genetic variation within populations is apparently determined by complex interactions between genetic and ecological factors, where the environmental factors also partly determine genetic factors, such as breeding system, mutation rate and dominance. The effect of ecological factors, both environmental and biotic, upon genetic variation within populations is probably greater than had previously been assumed. Conversely, the

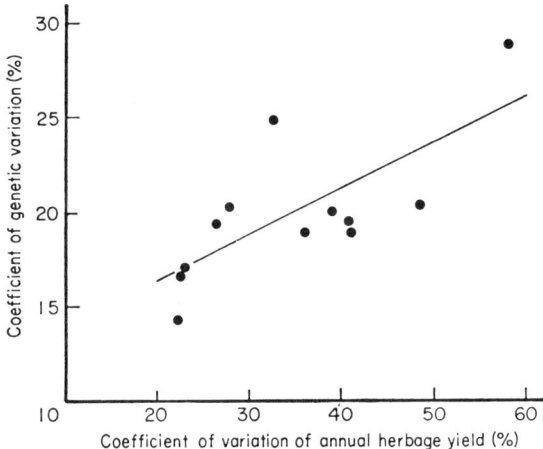

FIG. 3. The relationship between the genetic variation within populations of *Anthoxanthum odoratum*, collected from plots of the Park Grass Experiment at Rothamsted, and year-to-year variation in the yield of herbage harvested from those plots over a 30-year period. The genetic variation of each population is expressed as the mean coefficient of genotypic variation, based on 6 morphological characters of 10 genotypes grown in a spaced plant trial.

effect of genetic factors is probably less than had been assumed; for example, there seems to be less difference in genetic variability within populations of inbreeding and outbreeding species, collected in similar areas (Kannenberg and Allard, 1967), than is usually expected.

Genetic variation within populations has usually been studied at the morphological level, but it may also be studied at the physiological and biochemical level, as recent studies of allozyme and other biochemical polymorphisms have demonstrated (Fairbrother, 1968; Scandalios, 1969; Turner, 1970).

Much genetic variation within populations, whether morphologically or biochemically expressed, is apparently of adaptive significance and is of interest to ecologists, physiologists and biochemists, as well as to geneticists and students of evolution. Since this variation is of importance to a diversity of biological disciplines, and since the declared purpose of taxonomy is to describe, name and classify variation in a manner useful to these disciplines, then it seems that such variation should be recognized taxonomically and not excluded by the ground rules of taxonomy.

The definition of taxonomic groupings at the species level must depend, to a quite large extent, upon an understanding of variation at the intra-specific level; it can therefore be argued (p. 14) that the description and analysis of intra-population variation is, to some extent, an essential prerequisite for the taxonomic treatment of variation at the species level and above.

ECOLOGICAL FACTORS AND THE POPULATION
1. Ecological Factors and Intra-specific Variation

Heslop-Harrison (1964) and Hiesey and Milner (1965) have reviewed the vast amount of literature on genetic variation between plant populations and its relationship to environmental factors. The ecological importance of this ecotypic variation between populations is now generally recognized, but there is less agreement about the taxonomic treatment of the variation. Thorpe (1940), Turrill (1946), Heywood (1959) and Davis and Heywood (1963), among others, have discussed the taxonomic classification of this variation, but the present situation is still much as Heywood (1959) described it: "The tendency has been to side-track the question of its possible taxonomic recognition on the grounds of impracticability, lack of sufficiently wide data, etc.". There are certainly theoretical and practical problems in describing and classifying intra-specific variation, but they are by no means insuperable and, as Heywood (1959) states, "It cannot be emphasised too strongly that it is the taxonomist's problem, not the genecologist's or anyone else's". Twenty years ago, ecologists were criticized by a taxonomist (Constance, 1953) for their lack of appreciation of ecotypic diversity within plant species; such a failing among ecologists was perhaps not surprising, since little attempt had been made to treat intra-specific variation taxonomically. Most ecologists now recognize the ecological importance of ecotypic diversity within species, thanks to the efforts of genecologists, but they are still hamstrung by the almost total absence of any taxonomic description, nomenclature or classification.

Intra-specific variation mainly concerns taxonomically "bad" characters, i.e. characters that:

(i) are subject to wide variation within samples,
(ii) have high intrinsic genetic variability,
(iii) are susceptible to environmental modification,

(iv) are not consistently correlated with other characters that mark the already defined taxon (Davis and Heywood, 1963).

In view of the current state of knowledge in population genetics and ecological genetics, and in view of declared objectives in taxonomy, criteria (i), (ii) and (iii) should perhaps be reconsidered. Davis and Heywood (1963) have already questioned the validity of (iv) because it leads to circular argument.

There are other apparent reasons for the taxonomists' neglect of intra-specific variation. Firstly, much of the variation is physiological or biochemical, though many of these differences have morphological expressions, and most studies of ecotypic differences have dealt with morphological, rather than with physiological or biochemical characters. Secondly, the slow progress in the taxonomic treatment of intra-specific variation has been partly the result of an emphasis on classification, rather than on description. It is generally recognized that intra-specific variation is multidimensional, and that the unidimensional classifications common in taxonomy are unlikely to deal satisfactorily with this variation. The ecologist, however, does not necessarily need a unidimensional classification at the intra-specific level; he is more interested in descriptions that relate intra-specific variation to environmental conditions. The study by Lewis (1969) and Wilkins and Lewis (1969) of leaf shape in *Geranium* is an excellent example of how intra-specific morphological variation may be treated in an ecologically useful manner.

Many of the problems faced by the taxonomist in the description and classification of intra-specific variation are likely to be similar to those faced by the ecologist in the description and classification of vegetation. Several recent reviews have drawn attention to these similarities (Crovello, 1970; Goodall, 1970), and the area is obviously one where strong interrelationships exist between taxonomy and ecology (see Barkman and Guinochet, this volume). The problems of sampling loom large in both disciplines.

2. Sampling Intra-specific Variation

Sampling is probably one of the major problems preventing the objective delimitation of species and classification of intra-specific variation, yet it has received scant attention in the taxonomic literature. The problems of sampling in taxonomy are to some extent analogous to the ecological problems of sampling vegetation prior to the classification of plant communities. The problems exist at two levels: (i) the sample unit, i.e. individuals and populations in taxonomy, and quadrats and stands in ecology, and (ii) characters, i.e. morphological attributes in taxonomy, and floristic composition in ecology. The analogy, like all analogies, is not perfect. For example, the sample unit in ecology is ill-defined, and has variously been defined as the quadrat or stand (Greig–Smith, 1964), on the other hand the sample unit in taxonomy, the individual, is reasonably well defined, though it too may be ill-defined in vegetatively reproducing plants (Harberd, 1961a), and perhaps in parthenogenetic and apomictic species.

(a) *Sampling the Units of Classification.* Ecologists have recognized the need for suitable objective sampling of vegetation for several decades (Greig–Smith, 1964, Lambert and Dale, 1964); the literature on vegetation sampling was already large when Goodall (1962) reviewed it. By contrast, sampling in taxonomy has hardly been considered. Michener (1970), in passing, referred to the need for "very large and random" samples in taxonomy. McMillan (1954) was more explicit about the requirement for suitable sampling in taxonomy. Techniques for sampling the variation within and between populations have been considered by genecologists (Harberd, 1957, 1961b; Wilkins, 1959; Clausen, 1960; Bradshaw, 1962); something approaching random sampling is probably used within populations, but the selection of sample sites, i.e. between population sampling, is highly subjective. This subjective sampling is admissible in most genecological studies, where there is no necessity to sample fully the variation within the species, but it is not admissible in taxonomic studies, where the object is to define, as objectively as possible, the pattern of variation, and the limits of the species. It is difficult to determine to what extent the samples of taxonomic material in herbaria and museums are representative, though Brower (1963) found incidental evidence of selective sampling among museum collections of butterflies.

Early classification of vegetation was bedevilled by the use of subjective sampling, and much of the controversy as to whether plant associations were discrete or continuous centred around the differences in sampling methods used. Experience in ecology has shown that more extensive and objective sampling tends to substantiate the continuous nature of variation, e.g. McVean and Ratcliffe (1962). One wonders to what extent the use of extensive and objective sampling in taxonomy will also lead to less well defined species, though there are good reasons for believing that species are more clearly defined than plant communities. I see no reason why taxonomists should fear any increased fuzziness at the edges of species, that might follow such developments.

(b) *Sampling the Characters.* Compared with the scant attention given to techniques of sampling taxonomic units, sampling of characters has perhaps received too much attention in taxonomy; the debate between phylogenists and pheneticists on the use of various characters will be familiar to taxonomists. In contrast to this, there has been comparatively little debate on the selection of "characters" in the classification of vegetation. The problem is quite similar in both disciplines, however. For example, the selection of taxonomic characters on the basis that they are "consistently correlated with other characters that typify the already defined taxon", is closely analogous to the use of so-called faithful or "characteristic" species in phytosociology; both practices have been criticized, the former by Davis and Heywood (1963) and the latter by Greig-Smith (1964), because they are based on circular argument. Similarly, the use of relatively few floral characters in taxonomy is, to some extent, analogous to the use of dominant species in vegetation classification; both are likely to make groupings appear more discrete than if a larger number of characters were used. In both disciplines the "characters" may be expressed qualitatively or quantitatively, and the choice may greatly influence the resulting classification (Lambert and Dale, 1964).

(c) *General Considerations.* Sampling of both the units of classification and attributes may have large effects upon the subsequent classification of vegetation (Lambert and Dale, 1964) and taxonomic variation, yet taxonomists have given little attention to sampling the units of classification, and ecologists have probably given too little attention to sampling attributes. The methods of sampling individuals and populations for taxonomic purposes have been highly subjective; some collectors have been noted for their tendency to collect "aberrant" plants (W. T. Stearn, personal communication), but there is probably a greater tendency to collect "typical" samples. More objective sampling techniques whether random or systematic, are needed in taxonomy.

3. Classification of Intra-specific Variation

An increasing number of statistical techniques has been used in classification, by both taxonomists and ecologists, in the past few decades (Lambert and Dale, 1964; Sokal and Sneath, 1964; Williams and Dale, 1964; Crovello, 1970; Goodall, 1970). The two disciplines have already benefited from this common interest (Goodall, 1970), and will probably continue to do so (Crovello, 1970).

Since statistical techniques have been used in the classification of vegetation for several decades longer than in taxonomy (Goodall, 1970), the pattern of development in classifying vegetation may indicate possible future trends in taxonomy. One of the more interesting recent trends in the description and classification of vegetation has been the growing agreement on the relative value of classification and ordination techniques (Goodall, 1970) It is now recognized that both techniques have advantages and disadvantages, and have rather different uses. Both classification and ordination techniques may be applied to the same set of data, regardless of whether it shows continuous or discontinuous variation (Greig-Smith, 1964; Anderson, 1965; Goodall, 1966). The fact that continuous variation can be classified is particularly relevant to the taxonomic classification of intra-specific variation. It seems, to an ecologist, that the ground rules of taxonomy at the moment are framed so as to stress the apparent discontinuity between species, by excluding those characters that show continuous variation (see pp. 5, 9 and 13); whether these rules were devised simply to stress discontinuity, or to maintain the identity of currently acceptable taxa is uncertain, but it tends to have both effects. If classification of continuous variation is possible, there is no longer a need to exclude continuously variable characters. The way is then open to increase the number and type of characters used in classification, and so to increase the general utility of classifications, especially to functional biology, including ecology. Already, in the classification of vegetation, the use of a few "characters", such as dominant species and "faithful" species, has given way to the use of many "characters", whilst the use of qualitative (presence/absence) data has tended to give way to quantitative data. Similar changes have occurred in so-called numerical taxonomy; to what extent will these changes be assimilated in taxonomy?

4. General Considerations

There is little doubt that much ecologically important variation occurs below the species level. Much of this variation is of importance not only to the ecologists but also to the physiologist, biochemist and geneticist, and it can therefore be argued that the variation should be described, and perhaps named and classified taxonomically. It can also be argued that the description of this variation at the intra-specific level is essential for the taxonomic treatment of variation at the species level. For example, it is difficult to imagine how the limits of species can be defined without reasonably detailed samples, though the limits may be defined on the basis of less intensive sampling than would be necessary to fully describe and classify intra-specific variation.

At the moment an integrated approach to intra- and inter-specific variation is limited by the fact that most intra-specific variation involves taxonomically "bad" characters that are not used taxonomically at the species level and above. Already there is some hope that these artificial and rather dubious criteria of "good" and "bad" characters are being discarded; this should lead to a more open approach to the taxonomic treatment of intra-specific variation.

As an ecologist, I see the problems of describing and classifying intra-specific variation as both one of the most challenging, and potentially one of the most rewarding areas of interrelationship between taxonomy and ecology. It is challenging in that it requires the use of new techniques, both in biology and biometry, and requires the use of many resources by both taxonomists and ecologists. It is also challenging to the taxonomist in that it calls into question many long-held dogmas. Ehrlich and Holm (1962) decried the "hardening of the concepts" that they saw in community ecology; since then, partly as a result of the application of computer techniques to community analysis, those concepts have noticeably softened in ecology. Perhaps the criticism can now be levelled more aptly at taxonomy. Indeed, taxonomy is in greater danger of such conservatism because of pressures for a stable system of nomenclature, imposed by other biologists. Ecologists have probably been the most vocal of all biologists in their demand for a stable nomenclature so that, as an ecologist, I find myself in the paradoxical position of requiring both stable nomenclature and changes in the ground rules of taxonomy. Such a paradox could, of course, be resolved by separating nomenclature and classification (Amadon, 1966); this is unlikely to occur in the short term (Johnson, 1968). In the meanwhile, I think most ecologists would be willing to suffer some nomenclatural instability to gain greater ecological information in taxonomic classifications.

ECOLOGICAL FACTORS AND THE SPECIES

1. Ecological Factors and Speciation

Ecological factors have a dual role in speciation. The role of ecological factors in maintaining reproductive isolation between species has received widespread attention (Mayr, 1942; Stebbins, 1950; Dobzhansky, 1970); by comparison, the more important role of ecological factors in species divergence has perhaps been underrated. It is easy to forget that reproductive isolation simply maintains, at any one time, the *status quo*: some other mechanism must be invoked to account for the divergence between species.

(a) *Ecological Factors and Species Divergence*. One reason why the role of ecological factors in species divergence has tended to be underrated is that the emergence of a particular species cannot be observed by one person. Indirect evidence of the role of ecological factors in species divergence has been obtained by extrapolation from intra-specific differentiation, by studies of the relationship between species diversity and environmental diversity, and by ecological studies of taxonomically related species.

There is considerable evidence that ecological factors, both environmental and biotic, have moulded and continue to mould intra-specific variation (see previous section). It seems to be generally accepted that morphological and physiological variation between species is an extension of that within species, and has been moulded by similar processes (Stebbins, 1950; Dobzhansky, 1970). The most important process is probably natural selection; but what is the role of genetic drift, as opposed to selection, in species divergence? Many plant species have been taxonomically established on the basis of apparently non-adaptive floral characters, and this has tended to emphasize the role of genetic drift in speciation, though recent studies (Stebbins, 1970; Baker, Ehrendorfer, Eisikowitch and Meeuse, this volume) indicate that floral characters may also be adaptive, and subject to selection involving both biotic and environmental factors. The role of selection in species divergence seems to be more apparent in animals, where functional characters have been used widely in taxonomy, and Mayr (1963) concluded that "it appears that random fixation is of negligible evolutionary importance". Ehrlich and Raven (1969) go further and conclude that "selection is both the primary cohesive and disruptive force in evolution". If selection is really of such importance in species divergence, as well as in population differentiation, then ecological factors obviously play an important role in species divergence.

Studies of the correlation between species diversity and environmental diversity have not produced any clear evidence of the role of ecological factors in species divergence. Some correlation between species diversity and environmental diversity exists (e.g. Johnson and Raven, 1970), but is complicated by difficulties of quantifying diversity, disentangling the effects of migration as opposed to speciation, and considering the effects of past environmental changes.

Ecological and genecological studies of taxonomically related species (e.g. Clausen *et al.*, 1940; Kruckeberg, 1954; Harper *et al.*, 1961; Lewis, 1966; Harper, 1967; Jain, 1969) have provided some evidence of the probable role of environmental factors in their divergence. Similarly, studies of the relationships between hosts and parasites, and prey and predators (e.g. Gillett, 1962; Booth, Janzen, Jones and Meeuse in this volume) indicate that biotic factors may also be important in speciation. The evidence is, however, circumstantial; unfortunately we cannot reconstruct the past history of species, and relate this to the changing patterns of environmental conditions that they have withstood. We must beware of constructing facile explanations of past speciation on the basis of the present ecological and taxonomic status of species.

The processes of speciation seem most likely to be unravelled by studies of the process in action, i.e. by studies of intra-specific variation, and of incipient species, but correlated studies of the taxonomy and ecology of more divergent species are likely to provide additional information, and prove a fruitful meeting-place for the two disciplines.

(b) *How Essential is Isolation?* Ecological factors may be important in species divergence, but to what extent are isolating mechanisms necessary to maintain those differences? Does the morphological and physiological divergence between species lead directly or indirectly to reproductive isolation?

Let us first consider to what extent reproductive isolation may be essential. There are presumably limits to the amount of morphological and physiological diversity that can be maintained within a panmictic population or species; above that limit the number of viable offspring would be too small for the continued survival of the population or species as a whole. This is essentially a physiological/genetic *raison d'être* for isolation. A very wide range of physiological diversity is maintained in many plant species, apparently without the need for isolation mechanisms. In some cases the differences between populations in response to soil and climatic factors are as great as those between species that are widely diverse ecologically (Hiesey and Milner, 1965; Davies and Snaydon, 1973), yet these populations usually are fully interfertile. The limit to the range of diversity that can be maintained within these species is obviously very wide. In addition, many related species are reproductively compatible and may produce ecologically successful offspring, provided there are suitable environmental conditions (Anderson, 1949). Hybridization has also been an important factor in the evolution of many plant species.

Although there may be no intrinsic isolating mechanisms that prevent crossing and viable seed production between certain ecologically and physiologically diverse populations and species; there may be external isolating mechanisms that reduce crossing in the field, e.g. spatial or temporal isolation. In these situations, the evolution and continuity of differences between species and populations will depend upon the relative magnitude of gene exchange, and disruptive selection (Jain and Bradshaw, 1966); both of these processes will be largely determined by ecological factors.

A number of recent studies indicate that gene exchange, even in the absence of isolating mechanisms, may be much smaller than was previously assumed, and that selection pressures may be larger than was previously assumed (Jain and Bradshaw, 1966). Physiologically and morphologically diverse populations may therefore maintain their identity over very short distances, provided environmental conditions change sharply in space. For example, populations of *Anthoxanthum odoratum* change sharply at the boundary between plots of the Park Grass Experiment, Rothamsted, where environmental conditions change very sharply (Fig. 2, a and b). Populations differ significantly over distances of only 100 mm at the boundary (Fig. 2, c and d). Such sharp differences might be due either to limited gene exchange between the population types on each side of the boundary, or to intense selection against deviant genotypes on each side of the boundary. Some indication of the magnitude of gene exchange was obtained by comparing the progeny, collected as seed in the field, with their parents in spaced plant trials, on the basis of eleven morphological characters (Fig. 4). Appreciable gene exchange was apparent up to 0·5 m on the down-wind (left-hand) side of the boundary, and detectable amounts of gene exchange occurred up to 2 m down-wind of the boundary. This result is essentially in agreement with Griffiths' finding (1950) that 10% "contamination" occurred over a distance of 5 m in *Lolium perenne* L.

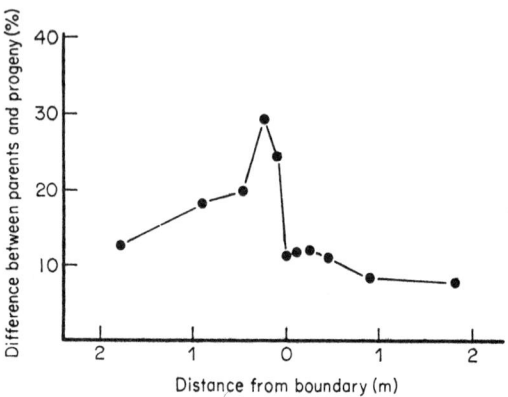

FIG. 4. The mean difference (%) between tiller (parent) populations of *Anthoxanthum odoratum* and their seed (progeny) populations, collected at the boundary between two plots of the Park Grass Experiment at Rothamsted. Each difference is the mean for eleven morphological characters measured for each of 8 genotypes grown from tillers, and 16 grown from seed.

The difference between parents and progeny near the border (Fig. 4), together with the fact that there were no significant differences in seed production from crosses between and within the population (Snaydon, unpublished), indicates that there was no intrinsic isolation barrier between the populations. The only exception was the population collected along a strip, 100 mm wide, precisely at the boundary between the two plots. This population flowered five days earlier than the populations on each side of it (Fig. 5a), and the period between stigma and anther exertion was appreciably less than in other populations (Fig. 5b). The population therefore appeared to be temporally isolated from adjacent populations; this is confirmed by the small difference between parents and progeny (Fig. 4), in spite of significant differences between the population and populations on the windward side of the boundary (Fig. 2). A similar phenomenon was encountered at another very different boundary (Snaydon, unpublished). In this outbreeding, wind-pollinated species, spatial isolation apparently occurs over distances of about 2 m, and selection apparently outweighs the effects of gene exchange over distances of less than 0·5 m. Similar sharp differentiation without intrinsic isolation has been found within other plant species (Jain and Bradshaw, 1966), but also between some ecologically contrasting, but interfertile species that grow closely intermingled in mosaic environments, yet maintain their separate identities (Heslop–Harrison, 1956; Briggs, 1962). Other apparently "good" taxonomic species appear to have no

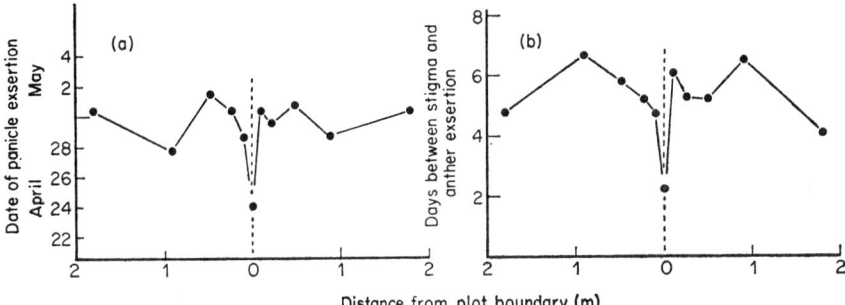

FIG. 5. The date of flowering (a), and number of days between stigma exertion and anther exertion (b), of populations of *Anthoxanthum odoratum* collected at the boundary between two plots of the Park Grass Experiment at Rothamsted.

intrinsic isolation mechanism, and may hybridize when the environment is modified (Anderson, 1949). It is rather surprising that isolating mechanisms have not developed in these cases, and also within the numerous ecologically and geographically wide-ranging species, in view of the fact that isolation mechanisms evolve rapidly in response to selection (Thoday and Gibson, 1962; Paterniani, 1969; Dobzhansky and Pavlovsky, 1971). Lewis (1966) concludes that "complete barriers, when they occur, are more often accidental by-products of evolution than the results of selection for reproduction isolation", and Ehrlich and Raven (1969) conclude that incompatibility is "a common but not universal result, not a cause, of the process, of speciation".

It would seem that isolation mechanisms, and intrinsic isolation in particular, have received too much attention; more attention should be given to the processes that lead to species divergence, rather than those that maintain it. There seems, in either case, to be ample scope for further collaborative studies by ecologists and taxonomists of the processes of speciation.

2. The Species and the Ecologist

The ecologist's attitude to species is equivocal, and the ecological status of species has been questioned, explicitly or implicitly, at intervals during the last few decades (e.g. Salisbury, 1940; Constance, 1953; Ehrlich and Holm, 1962; McMillan, 1969). The basic problem is that ecologists study interactions between organisms and between organisms and their environment, which occur largely at the level of the individual, but also at the level of the population and community; these interactions do not occur at the level of the species (Ehrlich and Holm, 1962), except in rare cases. The ecologist's attention is therefore centred on the individual, the population and the community (McMillan, 1969); to him the species is an abstraction. The ecological value of the species depends upon the amount of ecological information that it conveys. If the species is ecologically homogeneous, and ecologically distinct from taxonomically related species, then it is of value to the ecologist. Such species do exist, and are useful as "indicator species"; they may indicate a particular set of environmental conditions, or a particular plant association (Greig-Smith, 1964), either throughout their whole range, or over a part of it. On the other hand, many plant species are ecologically heterogeneous (Salisbury 1940; Davidson, 1952; Constance, 1953; McMillan, 1960, 1969), and then "ecotypes, rather than species, are the best indicators of ecological conditions" (Hanson and Churchill, 1961), though groups of species may also serve this function. Ecological heterogeneity within plant species has led ecologists to question not only the ecological value of the species (McMillan, 1969), but also the biological discreteness of species (Ehrlich and Holm, 1962). Greig-Smith's (1964) comment on the classification of vegetation was that "the danger of classification is that the techniques involved may result in over-emphasis of discontinuity"; the same might also be said of taxonomic classifications.

3. Changing Concepts of the Species

Taxonomic concepts of the species have, of course, changed several times during the past 200 years. How have these changes influenced taxonomic classifications and their value to ecologists?

Changes in the species concept from the "Linnean" concept of "divinely created, discrete and true-breeding" species (cf. Davis and Heywood, 1963), through the post-Darwinian phylogenetic concept of species, to the genetic concept of species as "groups of actual or potentially interbreeding populations which are reproductively isolated from other such groups" (Mayr, 1942), have apparently had little effect which is of ecological importance on taxonomic classifications. Indeed, one might even question whether they have had any effect at all on taxonomic classifications, as was stressed by Walters (1963). For example, a random sample of 500 species from the latest British Flora (Clapham, Tutin and Warburg, 1968) shows that approximately 60% of current species are solely attributed to Linnaeus. This is, of course, a very crude measure of changes in classification, since, on the one hand, the rules of priority tend to favour Linnaeus, but, on the other hand, he did not have a complete collection of the British flora. It does indicate, however, that in this particular case the present classification is substantially similar to that of Linnaeus. As far as the ecologist is concerned, the most important change has been the large amount of ecologically useful information written into the description, though probably not used in classification.

The recent extensions of neo-Adansonian or isocratic methods in taxonomy are important to ecologists. They allow a greater use of ecologically important attributes in classification, and are less restricted by criteria of "good" and "bad" characters (see pp. 6 and 10). It promises a more general-purpose taxonomy, and more aptly deserves the epithet "biological species" than does the genetic species, which currently bears the name.

ECOLOGY AND TAXONOMIC HIERARCHIES

In general, the ecological information content of taxonomic hierarchies is not large; taxonomic "distance" between plant genera, families, etc., does not generally reflect differences in ecological behaviour, and the relationship between taxonomic and ecological similarity tends to become more tenuous higher up the taxonomic hierarchy. It would, however, be misleading not to consider some of the cases where such relationships do exist. At the level of the order and above there is still a surprising similarity between taxonomic groupings and ecological behaviour; plants, animals, fungi and bacteria are broadly equivalent to the producer, consumer and decomposer levels in ecosystems. Within each of these groups there are also broadly equivalent taxonomic and ecological groupings; for example, the taxonomic grouping into algae, bryophytes and angiosperms broadly corresponds to ecological differences, and the major taxonomic groupings within the algae broadly correspond to ecological differences.

Below the level of the order there is also considerable similarity between taxonomic and ecological grouping; many plant families are ecologically quite homogeneous, e.g. the Chenopodiaceae, Salicaceae, Ericaceae, Plumbaginaceae, Orobanchaceae and Potamogetonaceae. The similarity is not sufficient, however, to be of wide use in ecology. Even genera are normally too diverse to be of any great ecological value. Species are the highest taxonomic level that is of appreciable ecological value; we have already seen (pp. 10 and 20) that many species are so ecologically heterogeneous that they are of limited use in ecology.

The fact that higher levels of the taxonomic hierarchy are of limited value to the ecologist is relatively unimportant, provided that the lower levels of the hierarchy are ecologically useful. There seems to be no reason why different levels of taxonomic hierarchies might not be defined by different criteria. The higher levels of hierarchies are mainly defined by floral characters, and the resultant classification is mainly of use to phylogenists. This classification would probably not change appreciably if functional vegetative characters were used more widely in classifications. On the other hand, the lower levels of taxonomic hierarchies would probably change appreciably if these characters were used, and would become more useful to ecologists, physiologists and biochemists.

RETROSPECT

Patterns of taxonomic diversity may be studied from evolutionary, genetic and ecological viewpoints, but these three disciplines have not contributed equally to taxonomy, largely because of differences in the historical development of the disciplines. The great interest in evolution, stimulated by Darwin in the middle of the last century, has had an immense and continuing influence on taxonomists, if not on taxonomy; phylogenetic considerations still appear to be important in taxonomy. Genetics has had a significant and increasing influence on taxonomy in the last few decades, especially in the fields of cytotaxonomy (Moore, 1968), and reproductive isolation (Solbrig, 1968). Ecology, physiology and biochemistry have developed, as scientific disciplines, more recently. They have only just begun to have an influence on taxonomy. Biochemical characters have already become a recognized part of the taxonomist's armoury (Turner, 1970). Ecological data can now be found more frequently in Floras, though these data are usually added *a posteriori*, rather than used in the process of classification. Ecologically important, functional characters, whether morphological, physiological or biochemical, are also being used increasingly by numerical taxonomists, though progress has been very uneven. These characters have been used most extensively in microbial taxonomy, where suitable morphological characters have not been available. As a result, the techniques of numerical taxonomy have been more readily accepted in microbiology than in plant and animal taxonomy.

Functional characters have also been used, almost *ab initio*, in animal taxonomy. As a result, the taxonomic treatment of animal variation seems to have been more useful to ecologists than that of plant variation. Is the greater use of functional characters in animal taxonomy due to any greater correlation between phylogeny and ecological adaptation in animals than in plants? There does appear to be greater niche specificity in animals than in plants. This may, in turn, be the result of greater habitat diversity, together with greater mobility and habitat selection, reinforced by greater reproductive isolation, in animals than in plants. Clearly there is scope for a more detailed comparison of the interrelationships between taxonomic diversity and ecological diversity in plants as opposed to animals.

Relationships between patterns of morphological variation in plants, and the ecological factors with which they are associated, are most apparent below the level of the species, where divergence has been most recent (pp. 5–11). The relationships are less apparent further up the taxonomic hierarchy (pp. 20–21). This contrast is, to some extent, due to the fact that different techniques have been used to study variation above and below the species level. Above the species level, variation has been studied by taxonomists, with the emphasis on floral morphology. Below the species level, variation has been studied by genecologists and biosystematists, and the emphasis has tended to be on functional, ecologically important characters. But the contrast is not due solely to the use of different techniques above and below the species level. Divergence between higher taxa presumably took place longer ago. The environmental differences that gave rise to that divergence have long since disappeared, and there has been ample opportunity for subsequent adaptive radiation, convergent evolution (Phillipson, 1961; Went, 1971), and genetic drift. Correlations between phylogenetic relationship and ecological similarity are likely to be less apparent, therefore, at higher levels of taxonomic hierarchies. This, of course, accounts for the lesser value of higher levels of taxonomic hierarchies to ecology. The ecological value of taxonomic classifications, at these lower levels, could be greatly increased if functional characters were more widely considered, and if intra-specific variation was given more detailed treatment. I feel confident that a greater use of functional characters in plant taxonomy would greatly increase the value of lower levels of taxonomic classifications to ecologists, physiologists and biochemists, without seriously upsetting the phylogenetic nature of the higher levels of hierarchies.

Data concerning environmental conditions, ecologically associated taxa, and ecological distribution pose particular problems as potential taxonomic attributes. If these data were used in taxonomic classifications, there would be serious danger of circular argument, since the resultant taxa would be used as indicators of the particular environmental conditions and plant associations, which were used to define them. Such attributes should therefore be used with care; perhaps it would be safest to use them only to validate ecologically the taxa that have already been defined by other attributes.

Taxonomy and ecology have common origins in field-based natural history, but the two disciplines have diverged widely since then. Taxonomy has, until recently, been largely concerned with phylogeny. On the other hand, ecology has been largely concerned with the description and analysis of the *status quo* of organisms in relation to each other, and to their environment. Recent developments in biology have made a *rapprochement* between the two disciplines both more possible and more necessary. Recent developments in genecology/ ecological genetics/biosystematics have, at one and the same time, forged direct links between taxonomy and ecology, and forged common links between those two disciplines and other disciplines, such as genetics and evolutionary studies. They have made the ecologist even more aware of the inadequacies of many taxonomic classifications. At the same time, recent developments in computer technology, and in the application of multivariate statistical techniques to biology, have made it possible to handle the ever-increasing amount of information that these inter-disciplinary studies have produced.

The time now seems ripe for more intercommunication between taxonomy and ecology, but an accord is not likely to come about without some revision of long-held concepts on both sides. It is certainly worthwhile to bear in mind the comment on population biology made by Ehrlich and Holm (1962, p. 2), that "people are inclined to confuse concepts with established facts and then consider it unnecessary to investigate the facts further". Already the facts are beginning to highlight deficiencies in some of our long-held concepts; we need to investigate these deficiencies, rather than turn a blind eye to the facts. I feel sure that taxonomists and ecologists have much to offer each other, as long as each approaches the areas of common interest with an open mind, not heavily blinkered by rigid concepts; I hope that this volume may speed the *rapprochement*.

REFERENCES

ALLARD, R. W. and ADAMS, J. (1969). Population studies in predominantly self-pollinated species. XIII Intergenotype competition and population studies in barley and wheat. *Am. Nat.* **103**, 621–645.

AMADON, D. (1966). Another suggestion for stabilizing nomenclature. *Syst. Zool.* **15**, 54–58.

ANDERSON, E. (1949). "Introgressive Hybridization." Wiley, New York.

ANDERSON, D. J. (1965). Classification and ordination in vegetation science: controversy over a non-existent problem? *J. Ecol.* **53**, 521–526.

ANTONOVICS, J. and BRADSHAW, A. D. (1970). Evolution in closely adjacent plant populations. VIII Clinal pattern at a mine boundary. *Heredity* **25**, 349–62.

BEARDMORE, J. A. (1961). Diurnal temperature fluctuations and genetic variance in *Drosophila* populations. *Nature, Lond.* **189**, 162–163.

BEARDMORE, J. A. (1963). Mutual facilitation and the fitness of polymorphic populations. *Am. Nat.* **97**, 69–74.

BEARDMORE, J. A. and LEVINE, L. (1963). Fitness and environmental variation. I A study of some polymorphic populations of *Drosophila pseudobscura*. *Evolution* **17**, 121–129.

BJÖRKMAN, O. and HOLMGREN, P. (1963). Adaptability of the photosynthetic apparatus to light intensity in ecotypes from exposed and shaded habitats. *Physiologia Pl.* **16**, 889–914.

BRADSHAW, A. D. (1962). The taxonomic problems of local geographical variation in plant species. *In* Systematics Association Publication **4**, 7–16.

BRADSHAW, A. D. (1965). Evolutionary significance of phenotypic plasticity in plants. *Adv. Genet.* **13**, 115–155.

BRIGGS, B. G. (1962). Interspecific hybridization in the *Ranunculus lappaceus* group. *Evolution* **16**, 372–390.

BROUGHAM, R. W. and HARRIS, W. (1967). Rapidity and extent of changes in genotype structure induced by grazing in a ryegrass population. *N.Z. Jl agric. Res.* **10**, 56–65.

BROWER, L. P. (1963). The evolution of sex-limited mimicry in butterflies. *Proc. XVI Int. Congr. Zool.* **4**, 173–179.

CARSON, H. L. and HEAD, W. B. (1964). Structural heterozygosity in marginal populations of nearctic and neotropical species of *Drosophila* in Florida. *Proc. natn. Acad. Sci. U.S.A.* **52**, 427–430.

CHARLES, A. H. (1964). Differential survival of plant types in swards. *J. Br. Grassld Soc.* **19**, 198–204.

CLAPHAM, A. R., TUTIN, T. G. and WARBURG, E. F. (1968). "Excursion Flora of the British Isles" (2nd edition). Cambridge University Press.

CLARKE, B. (1962). Balanced polymorphism and the diversity of sympatric species. *In* Systematics Assoc. Publication **4**, 37–46.

CLAUSEN, J. (1960). Sampling for genecology. *Rep. Scott. Pl. Breed. Stn* **1960**, 69–75.

CLAUSEN, J., KECK, D. D. and HIESEY, W. H. (1940). Experimental studies on the nature of species. I Effect of varied environments on Western North American plants. *Publs. Carnegie Inst.* No. 520.

CONSTANCE, L. (1953). The role of ecology in biosystematics. *Ecology* **34**, 642–649.

COOK, C. D. K. (1968). Phenotypic plasticity with particular reference to three amphibious plant species. In "Modern Methods in Plant Taxonomy" (V. H. Heywood, ed.). Academic Press, London.

CROVELLO, T. J. (1970). Analysis of character variation in ecology and systematics. *Ann. Rev. Ecol. Syst.* **1**, 55–78.

DAVIDSON, J. F. (1952). The use of taxonomy in ecology. *Ecology* **33**, 297–299.

DAVIES, M. S. and SNAYDON, R. W. (1973). Physiological differences among populations of *Anthoxanthum odoratum* collected from the Park Grass Experiment, Rothamsted. I Response to calcium. *J. appl. Ecol.* **10**, 33–45.

DAVIS, P. H. and HEYWOOD, V. H. (1963). "Principles of Angiosperm Taxonomy". Oliver and Boyd, Edinburgh.

DEMPSTER, E. R. (1955). Maintenances of genetic heterogeneity. *Cold Spring Harb. Symp. Quant. Biol.* **20**, 25–32.

DOBZHANSKY, T. L. (1947). Genetics of natural populations. XIV A response of certain gene arrangements in the third chromosome of *Drosophila pseudobscura* to natural selection. *Genetics* **32**, 142–60.

DOBZHANSKY, TH. (1962). Rigid *vs.* flexible chromosome polymorphism in *Drosophila*. *Am. Nat.* **96**, 321-328.

DOBZHANSKY, TH. (1970). "Genetics of the Evolutionary Process." Colombia University Press, New York.

DOBZHANSKY, TH. and PAVLOVSKY, O. (1971). Experimentally created incipient species of *Drosophila*. *Nature, Lond.* **230**, 289-292.

DONALD, C. M. (1963). Competition among crop and pasture plants. *Adv. Agron.* **15**, 1-118.

EHRLICH, P. R. and HOLM, R. W. (1962). Patterns and populations. *Science, N.Y.* **137**, 652-657.

EHRLICH, P. R. and RAVEN, P. H. (1969). Differentiation of populations. *Science, N.Y.* **165**, 1228-1232.

FAIRBROTHER, D. E. (1968). Chemosystematics with emphasis on systematic serology. *In* "Modern Methods in Plant Taxonomy" (V. H. Heywood, ed.). Academic Press, London.

FORD, E. B. (1971). "Ecological Genetics" (3rd edition). Chapman and Hall, London.

GILLETT, J. B. (1962). Pest pressure, an underestimated factor in evolution. *In* Systematic Association Publication **4**, 37-46.

GILMOUR, J. S. L. (1951). The development of taxonomic theory since 1851. *Nature, Lond.* **168**, 400-402.

GOODALL, D. W. (1962). Bibliography of statistical Plant Sociology. *Excerpta Bot.* **4**, Sect. B., 253-322.

GOODALL, D .W. (1966). The nature of the mixed community. *Proc. ecol. Soc. Aust.* **1**, 84-96.

GOODALL, D. W. (1970). Statistical plant ecology. *A. Rev. Ecol. Syst.* **1**, 99-124.

GREIG-SMITH, P. (1964). "Quantitative Plant Ecology" (2nd edition). Butterworths, London.

GRIFFITHS, D. J. (1950). The liability of seed crops of ryegrass (*Lolium perenne*) to contamination by wind-borne pollen. *J. agric. Sci.* **40**, 19-38.

HANSON, H. C. and CHURCHILL, E. D. (1961). "The Plant Community." Reinhold, New York.

HARBERD, D. J. (1957). The within population variance in genecological trials. *New Phytol.* **56**, 269-280.

HARBERD, D. J. (1961a). Observations on population structure and longevity of *Festuca rubra* L. *New. Phytol.* **60**, 184-206.

HARBERD, D. J. (1961b). The case for extensive rather than intensive sampling in genecology. *New Phytol.* **60**, 325-338.

HARPER, J. L. (1961). Approaches to the study of plant competition. *S.E.B. Symposia* **15**, 1-39.

HARPER, J. L. (1967). A Darwinian approach to plant ecology. *J. Ecol.* **55**, 247-270.

HARPER, J. L., CLATWORTHY, J. N., MCNAUGHTON, I. H. and SAGAR, G. R. (1961). The evolution and ecology of closely related species living in the same area. *Evolution* **15**, 209-227.

HESLOP-HARRISON, J. (1956). Some observations on *Dactylorchis incarnata* L. *Proc. Linn. Soc. Lond.* **166**, 51-83.

HESLOP-HARRISON, J. (1964). Forty years of genecology. *Adv. ecol. Res.* **2**, 159-247.

HEYWOOD, V. H. (1959). The taxonomic treatment of ecotypic variation. *In* Systematic Association Publication **3**, 87–113.

HIESEY, W. M. and MILNER, H. W. (1965). Physiology of ecological races and species. *A. Rev. Pl. Physiol.* **16**, 203–216.

HUBBARD, C. E. (1968). "Grasses." Pelican Books, London.

JAIN, S. K. (1969). Comparative ecogenetics of two *Avena* species occurring in central California. *Evolutionary Biol.* **3**, 73–113.

JAIN, S. K. and BRADSHAW, A. D. (1966). Evolutionary divergence among adjacent plant populations. I The evidence and its theoretical analysis. *Heredity* **21**, 407–441.

JOHNSON, L. A. S. (1968). Rainbow's end: the quest for an optimal taxonomy. *Proc. Linn. Soc. N.S.W.* **93**, 8–45.

JOHNSON, M. P. and RAVEN, P. H. (1970). Natural regulation of plant species diversity. *Evolutionary Biol.* **4**, 127–162.

KANNENBERG, L. W. and ALLARD, R. W. (1967). Population studies in predominantly self-pollinating species. VIII Genetic variability in the *Festuca microstachys* complex. *Evolution* **21**, 227–240.

KRUCKEBERG, A. R. (1954). Plant species in relation to serpentine soils. *Ecology* **35**, 267–274.

KRUCKEBERG, A. R. (1969). Ecological aspects of the systematics of plants. *In* "Systematic Biology." Natn. Acad. Sci., Publ. **1692**, 161–203.

LAMBERT, J. M. and DALE, M. B. (1964). The use of statistics in phytosociology. *Adv. ecol. Res.* **2**, 59–99.

LEVENE, H. (1953). Genetic equilibrium when more than one ecological niche is available. *Am. Nat.* **87**, 331–333.

LEVINS, R. (1962). Theory of fitness in a heterogeneous environment. I The fitness set and adaptive function. *Am. Nat.* **96**, 361–378.

LEVINS, R. and MACARTHUR, R. (1966). The maintenance of genetic polymorphism in a spatially heterogeneous environment: variations on a theme by Howard Levene. *Am. Nat.* **100**, 585–589.

LEWIS, H. (1966). Speciation in flowering plants. *Science, N.Y.* **152**, 167–172.

LEWIS, M. C. (1969). Genecological differentiation of leaf morphology in *Geranium sanguineum* L. *New Phytol.* **68**, 481–499.

LEWONTIN, R. C. (1957). The adaptation of populations to varying environments. *Cold Spring Harb. Symp. Quant. Biol.* **22**, 395–408.

LEWONTIN, R. C. (1969). The meaning of stability. *Brookhaven Symp.* **22**, 13–24.

MCMILLAN, C. (1954). Parallelism between ecology and taxonomy. *Ecology* **35**, 92–94.

MCMILLAN, C. (1960). Ecotypes and community function. *Am. Nat.* **94**, 245–255.

MCMILLAN, C. (1969). Discussion. *In* "Systematic Biology." Natn. Acad. Sci., Publ. **1692**, 203–206.

MCVEAN, D. N. and RATCLIFFE, D. A. (1962). "Plant communities of the Scottish Highlands." H.M.S.O., London.

MARSDEN-JONES, E. M. and TURRILL, W. B. (1938). Fifth report on the transplant experiments of the British Ecological Society. *J. Ecol.* **26**, 359–389.

MARSHALL, D. R. and JAIN, S. K. (1968). Phenotypic plasticity of *Avena fatura* and *A. barbata*. *Am. Nat.* **102**, 457–67.

MAYNARD SMITH, J. (1962). Disruptive selection, polymorphism and sympatric speciation. *Nature, Lond.* **195**, 60–62.

MAYR, E. (1942). "Systematics and the Origin of Species." Columbia University Press, New York.
MAYR, E. (1963). "Animal species and Evolution." Harvard University Press, Cambridge, Mass.
MICHENER, C. D. (1970). Diverse approaches to systematics. *Evolutionary Biol.* **4**, 1–37.
MOORE, D. M. (1968). The karotype in taxonomy. *In* "Modern Methods in Plant Taxonomy" (V. H. Heywood, ed.). Academic Press, London.
NELSON, A. P. (1965). Taxonomic and evolutionary significance of lawn races of *Prunella vulgaris. Brittonia* **17**, 160–174.
ODUM, E. P. (1971). "Fundamentals of Ecology" (3rd edition). Saunders, Philadelphia.
PATERNIANI, E. (1969). Selection for reproductive isolation between two populations of maize, *Zea mays* L. *Evolution* **23**, 534–547.
PHILLIPSON, W. R. (1961). Relationship and convergence in Angiosperms. *Phytomorphology* **10**, 367–371.
PIMENTAL, D. FEINBURG, E. H., WOOD, P. W. and HAYES, J. T. (1965). Selection, spatial distribution and the coexistence of competing fly species. *Am. Nat.* **99**, 97–109.
SALISBURY, E. J. (1940). Ecological aspects of plant taxonomy. *In* "The New Systematics" (J. S. Huxley, ed.). Oxford University Press.
SCANDALIOS, J. G. (1969). Genetic control of multiple molecular forms of enzymes in plants: a review. *Biochem. Genet.* **3**, 37–79.
SCHULTZ, W. M. and USANIS, S. A. (1969). Intergenotype competition in plant populations. II Maintenance of allelic polymorphism with frequency dependant selection and mixed selfing and random mating. *Genetics* **61**, 875–891.
SEATON, A. P. C. and ANTONOVICS, J. (1967). Population inter-relationships. I Evolution in mixtures of *Drosophila* mutants. *Heredity* **22**, 19–33.
SELANDER, R. K. (1969). The ecological aspects of the systematics of animals. *In* "Systematic Biology." Natn. Acad. Sci. Publ. **1692**.
SMITH, A. (1965). The assessment of pattern of variation in *Festuca rubra* L. in relation to environmental gradients. *Rep. Scott. Pl. Breed. Stn.* **1965**, 163–169.
SNAYDON, R. W. and DAVIES, M. S. (1972). Rapid population differentiation in a mosaic environment. II Morphological variation in *Anthoxanthum odoratum* L. *Evolution* **26**, 390–405.
SOKAL, R. R. and SNEATH, P. H. A. (1964). "Numerical Taxonomy." Freeman, San Francisco.
SOLBRIG, O. T. (1968). Fertility, sterility and the species problem. *In* "Modern Methods in Plant Taxonomy" (V. H. Heywood, ed.). Academic Press, London.
STALKER, H. D. (1964). Chromosome polymorphism in *Drosophila euronotis. Genetics* **49**, 669–687.
STEBBINS, J. L. (1950). "Variation and Evolution in Plants." Columbia University Press, New York.
STEBBINS, G. L. (1970). Adaptive radiation in the Angiosperms. I. Pollination mechanisms. *A. Rev. Ecol. Syst.* **1**, 307–326.
SUNESON, C. A. (1949). Survival of four barley varieties in a mixture. *Agron. J.* **41**, 459–461.
THODAY, J. M. (1953). Components of fitness. *S.E.B. Symp.* **7**, 96–113.

THODAY, J. M. and GIBSON, J. B. (1962). Isolation by disruptive selection. *Nature, Lond.* **193,** 1164–1166.

THORPE, W. H. (1940). Ecology and the future of systematics. *In* "The New Systematics" (J. S. Huxley, ed.). Oxford University Press.

TURESSON, G. (1922). The genotypic response of the plant species to the habitat. *Hereditas* **3,** 211–350.

TURNER, B. L. (1970). Molecular approach to population problems at the infraspecific level. *In* "Phytochemical Phylogeny" (J. B. Harborne, ed.). Academic Press, London.

TURRILL, W. B. (1946). The ecotype concept. A consideration with appreciation and criticism especially of recent trends. *New Phytol.* **45,** 34–43.

VAN VALEN, L. (1965). Morphological variation and width of ecological niche. *Am. Nat.* **99,** 377–390.

WADDINGTON, C. H. (1953). Genetic assimilation of an acquired character. *Evolution* **7,** 118–126.

WALTERS, S. M. (1963). Methods of classical plant taxonomy. *In* "Chemical Plant Taxonomy" (T. Swain, ed.). Academic Press, London.

WELLS, P. V. (1969). Discussion. *In* "Systematic Biology." Natn. Acad. Sci. Publ. **1692,** 206–211.

WENT, F. W. (1971). Parallel evolution. *Taxon.* **20,** 197–226.

WHITEHEAD, F. (1963). Experimental studies of the effect of wind on plant growth and anatomy. IV Growth substances and adaptive anatomical and morphological changes. *New Phytol.* **62,** 87–90.

WILKINS, D. A. (1959). Sampling for genecology. *Rep. Scot. Plant. Breed. Stn.* **1959,** 92–96.

WILKINS, D. A. (1963). Plasticity and establishment in *Euphrasia*. *Ann. Bot. N.S.,* **27,** 533–552.

WILKINS, D. A. and LEWIS, M. C. (1969). An application of ordination to genecology. *New Phytol.* **68,** 861–871.

WILLIAMS, W. T. and DALE, M. B. (1965). Fundamental problems in numerical taxonomy. *In* "Advances in Botanical Research" (R. D. Preston, ed.), Vol. 2. Academic Press, London. pp. 35–68.

WORKMAN, P. L. and ALLARD, R. W. (1964). Population studies in predominantly self pollinating species. IV Analysis of differential and random viabilities in mixtures of competing pure lines. *Heredity* **19,** 181–189.

2 | Ecological Factors and Species Delimitation in the Lichens

D. L. HAWKSWORTH

Commonwealth Mycological Institute, Kew, Surrey, England

Abstract: The known effects of ecological factors on thallus form, colour and pruinosity, anatomy, the production of soralia and isidia, chemical components and physiology are reviewed with reference to particular examples. Many variations that occur in thallus form, colour and pruinosity and anatomy are often a result of phenotypic plasticity but in some cases appear to have a genotypic origin and require taxonomic recognition. As chemical races in lichens are often based on differences in secondary metabolic products within the medulla which have no obvious adaptive value, careful studies of the ecological requirements of such races are of particular interest, since they may be expected to be related to other genotypic physiological differences.

The host or substrate range of different species varies considerably but species from different taxonomic groups are often adapted to similar precise ecological situations so enabling phytosociological taxa to be characterized. Some of the lichens with the most exacting host requirements are in genera allied to or containing non-lichenized species.

Lichen taxonomists should be aware of the effects ecological factors can have on particular characters and take care to study the variation in them in field situations before erecting new taxa. It is now becoming possible to study intact lichen thalli in laboratory conditions and taxonomists are urged to attempt to use such methods together with transplant experiments in investigating races characteristic of particular ecological situations.

The new combinations *Anaptychia mamillata* (Tayl.) D. Hawksw. and *Cladonia uncialis* subsp. *dicraea* (Ach.) D. Hawksw. are made.

INTRODUCTION

In common with other major plant groups, different lichen species have different ecological requirements and it is not uncommon to find that the individual species within a genus are restricted to quite distinct habitats. Consequently it is reasonable to assume that ecological factors have played an important rôle in their speciation. Although lichens are dual organisms it should be noted that they behave as individual species and show distribution patterns comparable to those seen in vascular plants (see Good, 1964).

Lichens are, however, one of the most difficult groups of plants in which to study speciation. Their slow growth rates make transplant experiments very time-consuming and breeding and genetical studies are currently impossible. Transplant experiments have formed an important aspect of the study of the effects of air pollutants on lichens (see Hawksworth, 1973) and although they have also been used in other ecological studies (e.g. Hale, 1954; Wetmore, 1970) only rarely have they been attempted to resolve taxonomic problems (Richardson, 1967). Growth of intact lichen thalli under controlled conditions on artificial media, soil and moist filter paper in the laboratory is now practicable (Kershaw and Millbank, 1969, 1970; Ahmadjian and Heikkilä, 1970; Galun et al., 1972), and Dibbens (1971) have demonstrated that it is possible to grow certain lichens by sowing macerated parent material on sterilized soil under carefully controlled conditions in a phytotron. These methods clearly have potential applications in taxonomic investigations but at the present time the lichenologist has to base his species concepts on individuals and populations growing in the field. Apart from the important studies of Thomson (1948) and Richardson (1967), experimental taxonomy is largely foreign to the literature of lichenology, remaining an unrealized ideal.

Because of their longevity for most practical purposes well-grown populations of lichen species in the field may be considered to be in a state of equilibrium with their environment. Consequently any study of the relationships of ecological factors to variations seen within and between populations of a species, particularly those leading to well-marked discontinuities that might be expected to have some genetic basis, could be assumed to lead to some understanding of the ecological parameters concerned in speciation.

Species concepts in lichens are currently based on sharp discontinuities in one or several morphological and anatomical and (or) chemical characters, particularly where there is evidence that genotypic differences are involved (e.g. the two entities growing side by side in a uniform ecological situation and retaining their identities), or there are differences in either ecological requirements or geographical distributions, or both.

In many instances while it is apparent that ecological factors are of paramount importance in particular cases of variation it is not easy to determine precisely what these are. It seems probable that not one but several interacting factors often require consideration. Excellent reviews of the physical and chemical characteristics of lichen substrates which require consideration have been prepared by Barkman (1958) and Brodo (1973). In this paper I propose to discuss some examples of types of variation seen in different ecological situations and their relevance to the delimitation of taxa. An understanding of the effects of ecological factors on characters employed in taxonomy is fundamental to the systematic work in any group, but while a number of authors have emphasized this or reviewed parts of this field (e.g. Sernander, 1907; Grummann, 1941; des Abbayes, 1951; Weber, 1962, 1967; Culberson, W. L., 1970; Brodo, 1973) no general survey from the taxonomic standpoint appears to have been produced in recent years.

MORPHOLOGY

Gross morphological characters play the most important part in the delimitation of species within particular genera of lichens. Almborn (1965, p. 454) has stated that ". . . most, perhaps all, good lichen species should be recognized by the trained eye (aided with a lens or a binocular) without a detailed microscopic examination", and this view is shared by many practising taxonomists. Some characters derived from gross morphology may, however, be influenced by a variety of ecological factors.

1. Thallus Form

As lichens are dual organisms the significance of the two components in the determination of thallus form must be considered. In most lichens the fungal partner (mycobiont) provides the gross features but when grown in pure culture isolated mycobionts fail to assume the shapes characteristic of intact thalli (see Ahmadjian, 1967) and so it is evident that the algal component (phycobiont) plays an important rôle in the determination of thallus form. In the genera of filamentous lichens (i.e. *Coenogonium* Ehrenb. ex Nees, *Cystocoleus* Thwaites and *Racodium* Pers.) the fungal hyphae surround the filaments of *Trentepohlia* Mart.* species and the importance of the phycobiont in the determination of thallus form is clear. In some instances, however, the same mycobiont may combine with two quite different algae (often the blue-green alga *Nostoc* Vaucher or the green alga *Myrmecia* Printz) to form morphologically quite dissimilar thalli. *Lobaria amplissima* (Scop.) Forss., for example, is a large foliose lichen with *Myrmecia* as phycobiont but has inclusions of colonies of *Nostoc* (i.e. cephalodia) which under certain environmental conditions grow out to form a minutely fruticose lichen with the same mycobiont but a different phycobiont called *Polychidium umhausense* (Auersw.) Henss. (syn. *Dendriscocaulon umhausense* (Auersw.) Zahlbr., *D. bolacinum* Nyl.) which may become detached from the parent thallus and exist as an independent plant (des Abbayes,

---

* The scientific names of lichens are considered as referring to the mycobionts so that separate names may be employed for the phycobionts.

1951). In *Sticta* (Schreb.) DC., P. W. James (unpublished) has found that in New Zealand the foliose *S. filix* Laur. (with a green phycobiont) arises from a robust fruticose species of *Polychidium* (Ach.) Gray (with *Nostoc* as phycobiont).

In many cases, however, cephalodia do not develop into separate plants that are recognized as distinct species (e.g. *Lobaria pulmonaria* (L.) Hoffm., *Peltigera aphthosa* (L.) Willd., *Placopsis gelida* (L.) Nyl., *Solorina crocea* (L.) Ach.) although they may occur as tubercle-like outgrowths (e.g. *P. venosa* (L.) Baumg.) which can become detached from the thallus (e.g. *P. lepidophora* (Nyl. ex Vain.) Bitt.). Blue-green algae function as nitrogen-fixing agents in those lichens in which they occur, and also have a different carbohydrate metabolism. It seems probable that in cases where different thallus forms are produced by the same mycobiont with different phycobionts, the mycobiont is giving priority to the particular phycobiont best suited to the environmental situation it is in. The nature of the ecological factors that might be involved in such cases is at present unknown and requires further study.

The form of the thallus in lichens is undoubtedly the character most affected by ecological factors. In many cases such variations (morphotypes) are of no taxonomic significance whilst in others there are grounds for believing that they may be genetically controlled and require taxonomic recognition.

Alectoria sarmentosa (Ach.) Ach. is a circumboreal species in the Northern Hemisphere with a few outlying localities in South America (Hawksworth, 1972a). Corticolous specimens are usually pendent, abundantly branched from the base, and have narrow, mainly terete main stems (Fig. 1 A–B). Morphotypes which are clearly very closely related on the basis of anatomical, chemical and other morphological characters, but which have dorsiventrally flattened expanded main stems (Fig. 1 E–F), grow prostrately over low lichen and dwarf shrub vegetation in montane to subarctic regions. This morphotype has been treated as a species, subspecies, variety, form or unworthy of taxonomic recognition by different authors, even within the last decade. When the world distribution is considered, however, some specimens showing intermediate conditions are found to occur in Europe, usually near the tree-line or just below it. No transplant experiments have yet been carried out but the occasional occurrence of the morphotypes usually associated with trees on the ground and on rocks suggest that genotypic and not merely phenotypic factors are involved; this view is also supported by the apparent absence of intermediate morphotypes in North America. This situation is comparable to those in which the subspecies concepts of vascular plant taxonomists has been applied, and has been interpreted as a single species with two subspecies (Hawksworth, 1972a): i.e. subsp. *sarmentosa* (pendent and usually corticolous) and subsp. *vexillifera* (Nyl.)

D. Hawksw. (prostrate and usually terricolous over low vegetation). This type of dorsiventral expansion of the main stems is seen in some other taxa in this genus (Table I), though in most it is less pronounced, and in a few other genera (see Sernander, 1907). This morphotype may be expected to have some adaptive value as it presents a greater area of tissue (and consequently the proportion of algal cells) to the incident sunlight. A terete pendent thallus, in contrast, might be expected to be able to make the best use of the available light coming from all directions.

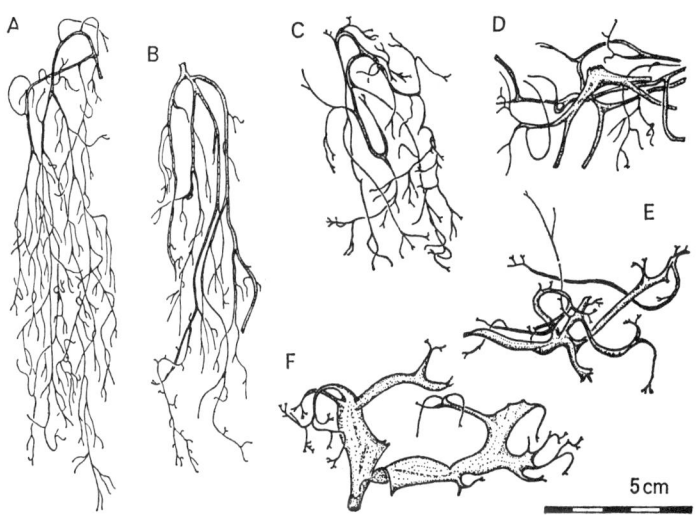

FIG. 1. *Alectoria sarmentosa* (Ach.) Ach. A, Finland, on *Picea abies*; B, U.S.A., Washington, on twigs of *Pseudotsuga*; C, Finland, on *Picea excelsa*; D, France, on horizontal fronds of *Abies*; E, Scotland, over *Rhacomitrium–Empetrum* heath; F, W. Greenland, on the ground. A–D = subsp. *sarmentosa*, E–F = subsp. *vexillifera* (Nyl.) D. Hawksw.

Adaptation to a different ecological situation is seen in the recently recognized *Parmelia britannica* D. Hawksw. & P. James which appears to have been derived from *P. revoluta* Flörke. *P. britannica* is essentially a species of maritime siliceous rocks in western Britain while *P. revoluta* is usually corticolous and much more widespread in Britain (Fig. 2). As *P. revoluta* occasionally occurs on rocks (usually over mosses) and *P. britannica* may occur 3–4 miles from the coast it is clear that genotypic factors are involved in this case. The differences between these two species are summarized in Table II. Maritime influences have clearly been important here and in many other genera which include species restricted to rocky shores or parts of the zonation pattern of rocky shores.

TABLE I. Habits and habitats of some closely related pairs of taxa in *Alectoria* Ach.

	Main stems ± terete			Main stems becoming dorsiventrally compressed and expanded to varying degrees		
A. capillaris (Ach.) Cromb.	usually corticolous	pendent		*A. setacea* (Ach.) Mot.	usually terricolous	prostrate
A. fuscescens Gyeln. var. *fuscescens*	corticolous or saxicolous	pendent		*A. fuscescens* var. *positiva* (Gyeln.) D. Hawksw.	corticolous, saxicolous or terricolous	prostrate
A. intricans (Vain.) Mot.	usually saxicolous	usually pendent		*A. chalybeiformis* (L.) Gray	usually terricolous or saxicolous	prostrate
A. ochroleuca (Hoffm.) Massal. var. *ochroleuca*	terricolous	erect		*A. ochroleuca* var. *ecuadorensis* Zahlbr.	terricolous	becoming prostrate
A. nigricans (Ach.) Nyl. f. *nigricans*	usually terricolous	erect		*A. nigricans* f. *subchalybeiformis* Räs.	terricolous	becoming prostrate
A. sarmentosa (Ach.) Ach. subsp. *sarmentosa*	usually corticolous	pendent		*A. sarmentosa* subsp. *vexillifera* (Nyl.) D. Hawksw.	terricolous	prostrate
A. virens Tayl. var. *virens*	corticolous	pendent		*A. virens* var. *forrestii* D. Hawksw.	saxicolous	prostrate

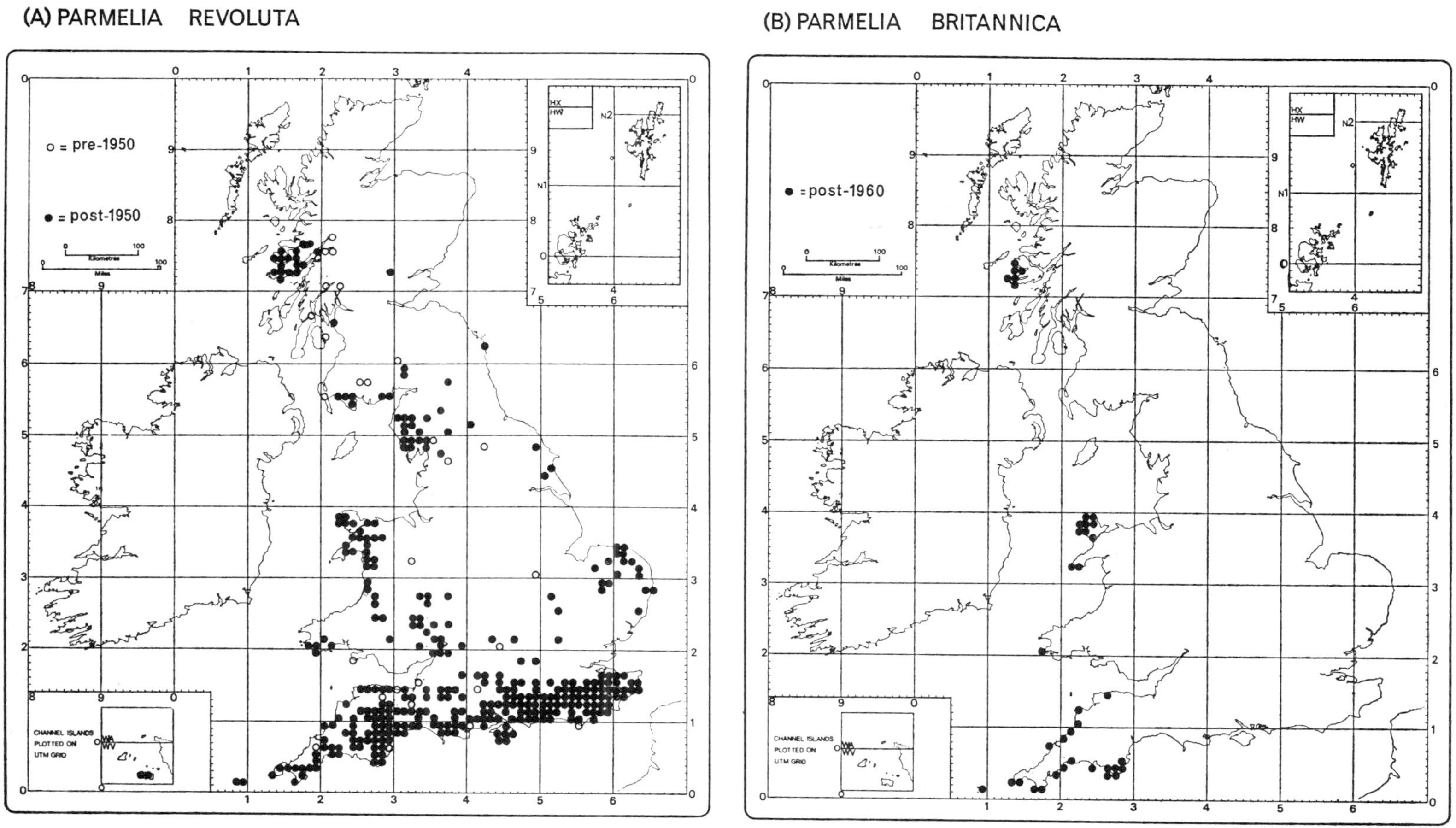

FIG. 2. British distribution of (A) *Parmelia revoluta* Flörke, and (B) *P. britannica* D. Hawksw. & P. James.

TABLE II. Comparison between the ecological requirements and morphological characters separating *Parmelia britannica* D. Hawksw. & P. James and *P. revoluta* Flörke

Parmelia britannica	*Parmelia revoluta*
Soralia always bluish-grey to black	Soralia creamy-white to pale grey but never bluish-grey to black
Lobes strongly sinuate	Lobes less strongly sinuate
Lobes usually 1·5–2 mm wide	Lobes usually 4–5 mm wide
Lobes usually shiny	Lobes usually matt but occasionally slightly shiny
Lobes often becoming convex	Lobes usually flattened
Ends of lobes not usually revolute	Ends of lobes usually revolute
Thallus usually closely adpressed to the substratum	Thallus usually loosely attached to the substratum
On siliceous rocks near the sea	On deciduous trees, or over mosses on rocks (rarely on bare rock), not always near the sea
Western species in the British Isles (Fig. 2B)	Broadly southern and western species in the British Isles (Fig. 2A)

The European *Anaptychia ciliaris* (L.) Körb. is a common species of eutrophicated bark but occasionally occurs on calcareous rocks in inland situations. A distinctive morphotype with consistently narrower and dark brown lobes occurs rarely on maritime rocks in western Britain. This morphotype has usually been treated as an infra-specific taxon within *A. ciliaris* (i.e. f. *melanosticta* (Ach.) Harm., subsp. *melanosticta* (Ach.) Lynge), but intermediates are unknown and in view of its characteristic ecological requirements it has been treated as a distinct species by Trass (1969).*

A narrow-lobed variety of *Xanthoria parietina* (L.) Th. Fr., var. *ectanea* (Ach.) Kickx, which predominates on well-lit maritime rocks had generally been accepted by British authors. Richardson (1967), however, transplanted some specimens of this variety to an inland site near Oxford from coastal rocks in Devon and found that over an eighteen-month period the lobe width increased

---

* Trass (1969) used the name " *A. melanosticta* (Ach.) Trass" (comb. inval., Art. 33) for this morphotype in the rank of species but as *Borrera ciliaris* γ *B. melanosticta* (Ach.) Röhl., *Deutsch. Fl.* **3** (2), 110 (1813), was an infraspecific taxon and not in the rank of species as indicated by Zahlbruckner (1931, p. 716) and an earlier epithet in species rank is available (Kurokawa, 1962, p. 14) the name **Anaptychia mamillata** (Tayl.) D. Hawksw. **comb. nov.** (Basionym: *Parmelia mamillata* Tayl., *Hook. Lond. J. Bot.* **6**, 171, 1847) must be used for this species.

so that the plants could no longer be distinguished from var. *parietina*. Richardson's experiments show that the var. *ectanea* is merely a result of phenotypic plasticity and does not merit any separate taxonomic recognition.

Some lichens assume nodular growth forms, become detached from their substrate, and can be blown about by the wind. This habit is characteristic of the "manna" lichens (i.e. *Lecanora esculenta* (Pallas) Eversm. and related species) but is also occasionally found in foliose lichens. Paulson and Hastings (1914) provide an account of the erratic morphotype of *Parmelia revoluta* (i.e. var. *erratica* (Lindsay) Zahlbr.) in southern England. This morphotype appears to originate from plants which were once attached to pieces of flint, becoming dislodged and assuming an atypical habit as they are rolled over the ground by the wind. This type of variation, when it is evidently a phenotypic modification induced by growth in unstable conditions, is clearly without taxonomic significance.

In the ubiquitous *Lecanora dispersa* (Pers.) Sommerf., a crustose species, several distinctive morphotypes occur and these have been investigated by Laundon (1958) who found that there were some correlations between these and ecological factors (Table III). The morphological variations seen in this case were treated by Laundon as forms because of the minor nature of the morphological differences but some non-British authors recognize some of these morphotypes as distinct species.

Cladonia furcata (Huds.) Schrad. has two distinct morphotypes, one with erect main stems usually about 1 mm in diameter characteristic of peaty soils (subsp. *furcata*), and one with prostrate main stems usually about 2–3 mm in diameter found on highly calcareous soils (subsp. *subrangiformis* (Sandst.) Pišút). In addition to the morphological differences, the morphotype on highly calcareous soils tends to have atranorin in easily detectable amounts more frequently than that from peaty soils, and is probably exclusively European whereas subsp. *furcata* is much more widely distributed. These two taxa maintain their identities when growing in close proximity and as some plants difficult to assign to either subspecies are occasionally found they are most appropriately regarded as subspecies. Differences in soil type are also related to different morphotypes in the species pairs *C. pyxidata* (L.) Hoffm. (peaty soils) and *C. pocillum* (Ach.) O.-J. Rich. (calcareous soils), and *Lecidea oligotropha* Laund. (very acid peat) and *L. uliginosa* (Schrad.) Ach. (mildly acid peat and decaying wood).

Variations in soil texture and stability, shading and available moisture were considered by Weber (1962, p. 311), on the basis of field observations, to be responsible for the morphological variations used to separate *Dermatocarpon dessertorum* Tomin, *D. hepaticum* (Ach.) Th. Fr., *D. lachneum* (Ach.) A. L. Sm.

TABLE III. Comparison between the substrate ranges and morphological types in *Lecanora dispersa* (Pers.) Sommerf. recognized by Laundon (1958)

Taxon	Morphological characters	Substrate(s)
f. *dispersa*	thallus thin, ± even, granular and spreading, not lobed, often evanescent, usually greyish	± smooth siliceous and calcareous rocks, wood, bark, *Armeria*, *Limonium*, bone, leather, iron, etc.
f. *albescens* (Hoffm.) Laund.	thallus with small lobes at the margins, superficial, usually white	± rough calcareous rocks and concrete
f. *verrucosa* (Leight.) Laund.	thallus verrucose, superficial, often in small islets, brilliant white	calcareous rocks near the sea (distinctly maritime)
f. *dissipata* (Nyl.) B. de Lesd.	thallus thin, almost evanescent, not lobed, always sooty-black	calcareous rocks in areas affected by air pollution

and *D. rufescens* (Ach.) Th. Fr. and he consequently interpreted them as morphotypes due to phenotypic plasticity within a single species (*D. lachneum*). Some preliminary transplant experiments carried out by P. W. James (unpublished) indicate that British material named as *Cladonia coniocraea* (Flörke) Spreng. and *C. macilenta* Hoffm. are probably *C. ochrochlora* Flörke and *C. polydactyla* (Flörke) Spreng., respectively, when growing in more humid situations and on rotting logs or other humus-rich substrates. *C. cervicornis* (Ach.) Flot. and *C. verticillata* (Hoffm.) Schaer. may provide a comparable example.

The ecological factors which seem to be particularly important in most of the examples so far discussed appear to arise from the nature of the substrate itself but in some cases less tangible ecological parameters can affect thallus form. Following up some preliminary observations (Kärenlampi, 1964), Kärenlampi and Pelkonen (1971) made a detailed study of the morphological variation seen in *Cladonia uncialis* (L.) Wigg. in eastern Fennoscandinavia. These authors found that plants from the more continental areas of Finland tended to have predominantly tri- and tetrachotomous branching patterns, whereas ones from climatically more maritime regions tended to have predominantly dichotomous branching (Fig. 3). They treated this latter type as var. *dicraea* (Ach.) Kär. & Pelk. which they considered to represent one end of an ecophenoclinodeme. Evidence that genotypic factors are involved in this case stems from the co-existence of both types side by side in the same habitat at some sites. In my view, following the concepts adopted for comparable examples in this paper, *dicraea* should be treated as a subspecies rather than as a variety.* While the type of branching in this species appears to be genetically determined and related to climatic factors, in particularly moist situations in north-western Britain a dilated morphotype occurs within subsp. *dicraea* which seems to be merely a phenotypic response to this factor as intermediates occur in less moist sites (Fig. 4) and there is no evidence to suggest that genotypic factors are involved. Comparable examples of phenotypic morphotypes characteristic of particularly moist situations are seen in some other species of *Cladonia* Hill ex Hill which can assume similarly dilated main stems (e.g. *C. impexa* f. *portentosa* (Duf.) Harm.) or produce unusually large numbers of squamules on the podetia (e.g. *C. squamosa* (Scop.) Hoffm. f. *squamosa*).

Montane and arctic-alpine conditions appear to have resulted in some variations in thallus morphology seen in *Cetraria islandica* (L.) Ach. and its subsp. *crispa* (Ach.) Cromb. (syn. *C. ericetorum* Opiz). Kristinsson (1969) has

---

* **Cladonia uncialis** subsp. **dicraea** (Ach.) D. Hawksw. **stat. nov.** (Basionym: *Baeomyces uncialis* β. *dicraeus* Ach., *Meth. Lich.* : 353, 1803).

made a detailed study of the populations of this species in Iceland and compared his results with some data obtained from European material. In Europe narrow canaliculate lobed plants tend to occur at higher altitudes than broader lobed ones and lack fumarprotocetraric acid (i.e. subsp. *crispa*) but in Iceland the situation is extremely complex and there is no consistent correlation between the

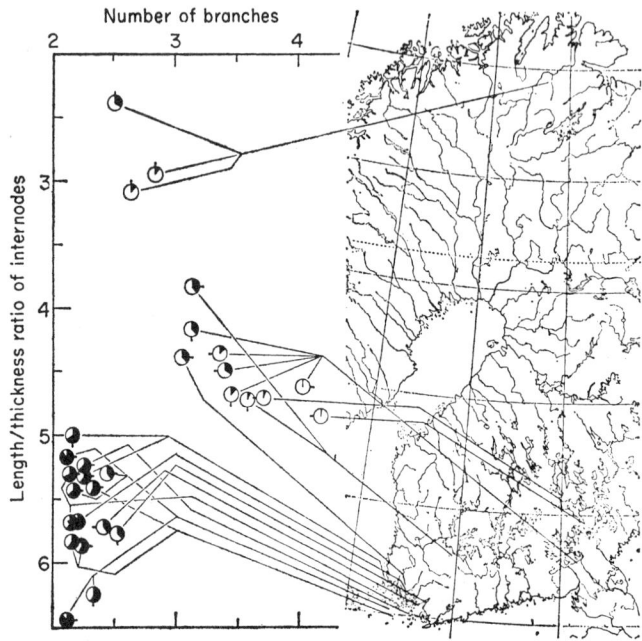

FIG. 3. *Cladonia uncialis* (L.) Wigg. Pictorial scatter diagram of the distribution and correlations of some selected characters and their collection localities in Finland. The white segments of the circles show the proportion of perforated axils and the short line shows the percentage of the area covered by the algal layer (right = over 30%, down = 25–30%, left = 20–25%, up = below 20%). Reproduced with permission from Kärenlampi and Pelkonen (1971, p. 52).

morphotypes and chemotypes. Kristinsson considered that genotypic factors were probably involved in the production of different morphotypes which include a bewildering array of intermediate forms between the classically accepted taxa, because of the occurrence of distinct morphotypes within single local populations. Kristinsson suggests that gene exchange is taking place but is careful to emphasize that proof of this must eventually be derived from cultural studies using single ascospore isolates.

The distinction between fruticose (shrubby) and crustose thallus forms might perhaps be thought of as one of the morphological characters least subject to environmental modifications. Weber (1967), however, found that as the normally crustose *Lecanora calcarea* (L.) Sommerf. (syn. *Aspicilia calcarea* (L.) Mudd) grew over the edges of rocks onto adjacent soil it could assume a fruticose habit. A comparable situation appears to exist in *Lecanora cinerea* (L.) Sommerf. (syn. *Aspicilia cinerea* (L.) Körb.). Such fruticose forms of *Lecanora calcarea* have been described as a distinct genus (*Agrestia* Thoms.) but Weber's studies indicate that in this case only phenotypic modification in response to the change in substrate is taking place. The minute fruticose lichen *Alectoria minuscula* (Nyl. ex Arnold) Degel. has a form characteristic of particularly extreme arctic and antarctic

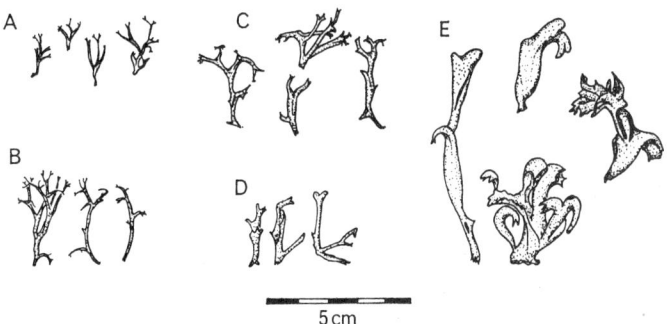

FIG. 4. Specimens of *Cladonia uncialis* subsp. *dicraea* (Ach.) D. Hawksw. from various habitats in different parts of the British Isles. A, Leicestershire, dry heath; B, Fife, hollows in pine plantation on sand-dunes; C, Shetland, shallow acid peat; D, Inverness, recolonized shingle by river; E, Ross-shire very wet peat on rocks.

situations in which the thallus consists of irregular pulvinate crustose masses of secondary branchlets (i.e. f. *congesta* (Zahlbr.) Lamb) so that it may superficially resemble a crustose species.

Environmental factors are also extremely important in considering species delimitation in crustose lichens where variations caused by physical factors, particularly bark and rock types, humidity and attrition have led to unnecessarily large numbers of taxa being proposed in some genera. In considering the yellow species of *Rhizocarpon* Ram. em. Th. Fr. in Europe Runemark (1956, p. 16) noted that in specimens from "extreme" localities the areolae tend to be situated away from each other on a prominent prothallus whilst in material of the same species from more "favourable" sites the areolae tend to be crowded together forming a continuous thallus. Because of this considerable variation in thallus form caused by environmental conditions Runemark emphasizes that caution is required in the use of habit characters in the genus.

Taxonomists working in herbaria with poor material and without the opportunity to examine the variation of populations in the field have sometimes failed to appreciate the extent to which ecological factors can influence thallus form. In a single subgenus of *Acarospora* Massal. (subgen. *Xanthothallia* Magnusson), for example, Weber (1968) considered that 93 taxa, most of which had been proposed as species, could be accommodated within only two species when environmental modifications were taken into account. It seems possible that more species should be accepted in this subgenus, however, and taxonomists must be aware of the dangers of combining taxa without critical examination. Uncritical lumping of species may be as scientifically misleading as the description of meaningless taxa without adequate regard for environmental variables.

Variations in ascocarp shape, arrangement and thallus form have led to the description of numerous infraspecific taxa in some common corticolous crustose lichens (e.g. *Graphina anguina* (Mont.) Müll. Arg., *Graphis scripta* (L.) Ach., *Opegrapha atra* Pers.) which are now treated as of no taxonomic significance and attributed merely to differences in the texture, grain and friability of the bark on which the species are growing. Polymorphism due to variations in the nature of the substrate is particularly well marked in the marine *Arthopyrenia halodytes* (Nyl.) Arnold which has a smooth and sometimes shiny epilithic thallus on hard siliceous rocks, a more or less well developed to completely endolithic thallus on mollusc shells and hard limestones, and perithecia up to three times the normal size when on chalk (Swinscow, 1965; P. W. James, unpublished). Comparable variations also occur in some species of *Porina* Müll. Arg. and *Verrucaria* Schrad. In the marine *V. striatula* Wahlenb. ex Ach., for example, A. Fletcher (unpublished) has found that the degree of development of the blackened ridges on the thallus characteristic of this species are an expression of the degree of exposure.

2. Soralia and isidia

Soralia (which produce soredia) and isidia, vegetative reproductive structures formed on lichen thalli, are extensively used in the taxonomy of lichens. Soralia are, however, extremely rare in pyrenocarpous species and are replaced by specialized structures termed hormocystangia in some gelatinous lichens (e.g. *Lempholemma vesiculiferum* Henss.). Many cases are now known of species pairs, one of which is sorediate or isidiate and usually lacks ascocarps, and the other lacks soredia and isidia and usually has ascocarps. This species-pair ("*Artenpaare*") concept constitutes an important aspect of modern lichen taxonomy and has been discussed in detail by Poelt (1970, 1972) who provides examples. It is suggested that species producing asexual reproductive structures alone have arisen from otherwise identical ascocarp-forming species lacking these structures. As the "primary" (ascocarp-forming) and "secondary" (sorediate or isidiate) species often have either different distribution patterns and (or) ecologies an understanding of any ecological factors tending to promote the formation of (or replacement of ascocarps by) sorediate and isidiate races is of particular interest to the lichen taxonomist. It should also be emphasized, however, that some pairs of species differ in the type of soredial development (e.g. *Physcia ascendens* Bitt. and *P. tenella* (Scop.) DC. em. Bitt., *P. caesia* (Hoffm.) Hampe and *P. wainioi* Räs., *Ramalina obtusata* (Arnold) Bitt. and *R. polymorpha* (Ach.) Ach.).

Du Rietz (1924) pointed out that sorediate taxa appear to be relatively infrequent in arctic regions where, as emphasized by Thomson (1972), thallus fragmentation probably plays an important rôle in the dispersal of macrolichens. The sorediate *Lecanora intricata* var. *soralifera* Suza (syn. *L. soralifera* (Suza) Räs.) and *Ochrolechia androgyna* (Hoffm.) Arnold tend to become rarer in montane and arctic situations where their non-sorediate and commonly fertile counterparts *L. intricata* (Ach.) Ach. and *O. tartarea* (L.) Massal., respectively, are much more frequent. A few sorediate species, however, do extend into arctic regions (e.g. *Lecidea tumida* Massal., *O. geminipara* (Th. Fr.) Vain.). It is also of interest to note that in some species which extend into subarctic and arctic areas from boreal and cool-temperate regions sorediate taxa occur only in the less extreme parts of their ranges (e.g. *Alectoria sarmentosa* var. *sorediosa* (Lång ex Räs.) DR., *Cladonia uncialis* f. *soraligera* (Robb.) Sandst., *Cornicularia muricata* f. *sorediella* (Erichs.) Erichs.).

Species capable of producing soralia tend to form most soredia when growing in particularly humid situations (e.g. *Lecanora expallens* Ach., *Lecidea orosthea* (Ach.) Ach., *Parmelia perlata* (Huds.) Ach., *P. revoluta*) but in most cases this is merely a phenotypic response without taxonomic significance. It is interesting to note, however, that sorediate species are particularly frequent in dry recesses or underhangs of rocks (James, 1970) and dry bark crevices. Such species appear to be adapted so as to utilize atmospheric moisture and some are almost unwettable by water droplets.

The most important factor inhibiting ascocarp production and consequently promoting reproduction by asexual methods in western Europe at the present time is certainly that of air pollution. In many corticolous species in Britain ascocarp production is suppressed where the mean winter sulphur dioxide values exceed about 40 μg/m^3 air (Hawksworth *et al.*, 1973). Such species then either have to rely on asexual methods of reproduction or become relic (i.e. unable to spread into other sites). Some species able to form ascocarps in polluted areas may persist as relics also as in the case of *Caloplaca heppiana* (Müll. Arg.) Zahlbr. in London which now seems unable to colonize new substrates (Laundon, 1967). Air pollutants appear to be able to inhibit the germination of ascospores and the establishment of some species directly or indirectly, either by affecting the pH of the substrate on which they settle or the mechanical action of soot deposits which also favours the premature colonization of alien algae encouraged by the eutrophiated conditions.

De Sloover and LeBlanc (1970) have illustrated how the frequency of ascocarp production along an air pollution gradient declines in several species as the level of air pollution increases. The sorediate *Lecanora conizaeoides* Nyl. ex Cromb. probably arose from an allied esorediate species and greatly extended its range in the mid-nineteenth century becoming common in areas affected by mean winter sulphur dioxide values in the range 55–150 $\mu g/m^3$ air (Hawksworth *et al.*, 1973; Laundon, 1973). This species is, nevertheless, able to form ascocarps but the frequency of these often declines as the centres of polluted areas are approached, and they are completely absent in its innermost sites in London and the industrial areas of the Midlands. The decline in the frequency of ascocarps in this species was studied by Pišút and Jelínková (1971) but their data have been confused by the inclusion of *L. varia* (Hoffm.) Ach. amongst the material they studied to judge from specimens kindly sent to me by Jelínková. The situation in *L. conizaeoides* is, however, probably more complex than might at first appear, as in Scandinavia thalli differing considerably in the degree of soredial production and frequency of ascocarps may grow side by side, indicating that some genetic factors may also be involved. It is likely that in time other sorediate races will arise in polluted areas from normally esorediate ascocarp forming species and a few have already been recognized (e.g. *Lecania erysibe* f. *sorediata* Laund.). It must also be emphasized here that some sorediate lichens tend to produce soralia more rarely towards their innermost limits to air pollution sources (Hawksworth *et al.*, 1973).

In some other species transitions between thalli with abundant ascocarps and few soralia and almost entirely sorediate thalli with no ascocarps occur in areas unaffected by air pollutants (e.g. *Catillaria lightfootii* (Sm.) Oliv., *Lecidea granulosa* (Hoffm.) Ach.), and in *Peltigera spuria* (Ach.) DC. young thalli are sorediate whilst old ones are esorediate and bear ascocarps. The factors controlling soredial production in such cases are generally obscure and in need of detailed investigations.

The ecological factors favouring the production of isidia are less clear. The isidiate counterpart of *Parmeliella plumbea* (Lightf.) Vain., *P. atlantica* Degel., has a more pronounced oceanic distribution in Europe than *P. plumbea* (see Degelius,

1935) but both are characteristically oceanic. In *Pseudevernia furfuracea* (L.) Zopf isidia appear from field observations to be most abundantly produced in plants growing in particularly exposed situations but are still usually readily visible in specimens from sheltered sites. *Usnea fragilescens* Hav. ex Lynge may behave in a similar way. *U. subfloridana* Stirt., however, normally has "isidiate" soralia but when growing in very humid sheltered sites in western Britain their frequency declines and only non-isidiate soralia are usually found on plants from such habitats (i.e. *U. glabrescens* (Nyl. ex Vain.) Vain. auct. angl.). The only experimental work known to me on the factors influencing the production of isidia is that of Thomson (1948) on some species of *Peltigera* Willd. He discovered that in some cases regeneration isidia were formed on breaks in the cortex caused by crushing and cutting with a scalpel after a period of about one year. Experiments of this type in other genera in which isidia are employed in the delimitation of taxa would clearly be very interesting but none appear to have been carried out.

3. Colour and Pruinosity

The colour of the thallus in some lichens is affected by the substrate and (or) the degree of illumination. In many cases this appears to be merely a phenotypic response to a particular environmental situation but in others there are indications that genotypic factors are also involved.

Dark-brown pigments deposited in the outer parts of the cortex of *Alectoria fuscescens* Gyeln. and *A. subcana* (Nyl. ex Stiz.) Gyeln. cause the thalli to be dark brown and fuscous, respectively, when growing in well-lit situations, but they are poorly developed or absent in very shaded sites so that the thalli become pale fuscous and greyish, respectively. In these two species this appears to be merely a phenotypic response to the degree of illumination but in *A. capillaris* (Ach.) Cromb. dark and pale plants may occur side by side in nature indicating that in this case the production of the dark brown pigments is probably genotypically controlled (Hawksworth, 1972a). Dark brown and brown thalli occur in *Parmelia glabratula* (Lamy) Nyl. studied by Laundon (1965). Dark brown plants (subsp. *fuliginosa* (Fr. ex Duby) Laund.) are characteristic of saxicolous habitats whilst brown ones (subsp. *glabratula*) predominate on bark and wood. Occasionally these two subspecies occur side by side on the same substrate and so it is evident that genotypic differences are involved and that they merit taxonomic recognition in the rank of subspecies.

Siple (1938, p. 478) observed that in Antarctica, "Most lichens were decidedly black, whilst others were dark green, grey, brown, or red. Only few types exhibited light colours," and several later workers have supported this view.

Ahmadjian (1970a) suggests that natural selection has operated in favour of species with darkly or heavily pigmented cortices in the Antarctic because such layers above the algal cells would be expected to enhance the chances of survival of the light-sensitive algal cells under Antarctic light conditions. Dark thalli would also be expected to acquire higher temperatures than light coloured thalli under comparable levels of illumination and, as Ahmadjian points out, Galun (1963) found that in the Negev Desert in Israel most lichens have white or greyish-white thalli and that such orange and brown species as occur tend to have a superficial layer of whitish pruina.

Whitish pruina in lichens are caused by superficial deposits of calcium oxalate crystals on the surface of either the thalli and (or) ascocarps. This phenomenon is particularly frequent in species growing on calcareous rocks (e.g. *Caloplaca saxicola* (Hoffm.) Nordin) and was interpreted by Weber (1962) as primarily a response to environmental factors in view of (a) the variation within a single species at one site, (b) the occurrence of pruinose and epruinose types on different rock types in the same region, and (c) because it transcends generic boundaries. Schade (1970) discusses the occurrence of calcium oxalate crystals in some corticolous species of *Parmelia* Ach. Des Abbayes (1951) pointed out that variations in the amount of calcium oxalate crystals on the surface of a single species at one site might be related to the age of the plant but this does not appear to be true for all pruinose species. The studies of Syers *et al.* (1967), however, suggest that the production of calcium oxalate is particularly characteristic of obligately calcicolous species rather than of all lichens occurring on calcareous rocks. The calcium content of the substrate has not usually been thought of as having any marked effects on the gross morphology of lichens but Schade (1966) considered it to be the cause of the morphotype *Cladonia furcata* subsp. *subrangiformis* discussed above (p. 39). The chemical nature and factors affecting the production of pruinia in corticolous pruinose species (e.g. *Physconia pulverulenta* (Schreb.) Poelt) are in need of further investigation.

The chemical composition of the rock can also affect thallus colour. Thalli which are characterized by reddish-orange deposits of iron oxides in their surface layers, which have been termed "oxydated" by Weber (1962), are formed in some species when growing on rocks particularly rich in iron (e.g. *Lecidea lapicida* (Ach.) Ach., *L. lithophila* (Ach.) Ach., *L. macrocarpa* (DC.) Steud.). Colour variants of this type have often been treated taxonomically in the rank of form but field observations indicate that oxydation is usually merely a response to this ecological factor and without taxonomic significance. In *L. atrata* (Ach.) Wahlenb. (syn. *L. dicksonii* auct., non (J. F. Gmelin) Ach.) the thallus is usually oxydated but occasionally grey, but in *Acarospora sinopica*

(Wahlenb. ex Ach.) Körb., *Lecidea silacea* (Ach.) Ach. and *Rhizocarpon oederi* (Web.) Körb., for example, the thalli are consistently oxydated and this is characteristic of the species.

Some extracellular lichen products (see p. 52) are brightly coloured (e.g. parietin, usnic acid, vulpinic acid) and when located in the cortical layers of the thallus provide colours which are readily visible. Taxonomists have consequently often been inclined to recognize morphologically identical plants differing in the colour of their thalli as distinct taxa. Specimens of *Xanthoria parietina* (L.) Th. Fr. are bright orange when growing in well-lit situations but assume a greenish-yellow or yellowish-grey colour (i.e. f. *cinerascens* (Leight.) Berg.) when in shaded habitats. Hill and Woolhouse (1966) studied specimens of this species from three sites and found that material from trees, roofs and maritime rocks had 76·6, 391·5 and 408·1 μg parietin/cm^2 thallus, respectively. All intermediate colour forms exist and considerable variations may occur at a single site (e.g. a horizontal wall-top as compared with its vertical sides) and the resultant colour appears to be merely a phenotypic response in the amount of the anthraquinone parietin produced under different conditions of illumination. The production of the yellow dibenzofuran derivative usnic acid in *Cladonia subtenuis* (des Abb.) Evans (syn. *Cladina subtenuis* (des Abb.) Hale & Culb.) was found by Rundel (1969) to vary in a similar way along a light gradient from 0·17% at 3% of full sunlight to 2·82% at 51% of full sunlight. This type of phenotypic response appears to occur in many other lichens (e.g. *Alectoria sarmentosa*, *A. virens* Tayl., *Cladonia uncialis*, *Evernia prunastri* (L.) Ach., *Teloschistes flavicans* (Sw.) Norm.) and, as in the case of dark brown cortical pigments discussed above (p. 48), may be interpreted as a method of protecting the underlying algal cells from excessive levels of sunlight which might damage them.

In some cases, however, the ability to produce coloured lichen products in the cortical layers appears to be genotypically determined. The crustose lichen *Haematomma ventosum* (L.) Massal. has bright yellowish thalli with usnic acid and greyish thalli lacking this acid which can often be found growing side by side on the same rock in Britain (Hawksworth, 1970). These two chemotypes are not recognized taxonomically as no correlations with ecological or geographical factors appear to occur. It is interesting to note in this species that the chemotype able to form usnic acid varies from yellowish-green or yellowish-grey to bright yellow depending on the degree of illumination in a manner comparable to that seen in *Cladonia subtenuis*. In *C. tenuis* (Flörke) Harm. plants lacking usnic acid (var. *leucophaea* (des Abb.) Ahti) are more strictly oceanic in Europe than those with this acid (var. *tenuis*) and because of their differences in distribution have been treated as varieties by Ahti (1961). In *Haematomma ochroleucum* (Neck.) Laund. (syn. *H. coccineum* Körb.), as in *H. ventosum*, chemotypes with and without usnic acid occur. In *H. ochroleucum*, however, the usnic acid-containing chemotype is usually saxicolous in Britain whilst the usnic acid absent chemotype commonly occurs on both rocks and trees. Because of the different ecological amplitudes of these chemotypes and their ability to grow side by side and maintain their identities on rock Laundon (1970) has treated the usnic acid absent chemotype as constituting a separate variety (i.e. var. *porphyrium* (Pers.) Laund.).

ANATOMY

1. Thallus Anatomy

The anatomy of the thalli in lichens is also subject to modification by ecological factors. In more extreme situations cortical layers in crustose lichens tend to be thicker than in plants from more favourable or shaded habitats. This thickening occurs either by increases in the number of cells or cell layers in the cortex itself or by the deposition of a necrotic layer above the cortex. Such thickenings may be interpreted merely as phenotypic responses designed to reduce the rate of loss of water or water vapour from the algal layer and medulla, reduce the amount of light reaching the algal layer, or provide additional protection against attrition. Galun (1963), for example, found that the Negev Desert material of *Buellia canescens* (Dicks.) de Not. developed a thick amorphous cortical layer not seen in Swedish material of this species, and Looman (1964) reported that the cortex of *Lecanora reptans* Loom. was 100% thicker in exposed situations than in sheltered sites. In *Peltula obscurans* (Nyl.) Gyeln. the degree of cortical development and the type of medulla appear to vary with the pH of the substrate, material from the most calcareous substrates having the least well developed cellular tissues (Wetmore, 1970). In *Rhizocarpon obscuratum* (Ach.) Massal. the black fimbriate prothallus is much better developed on quartz than on the other types of siliceous rocks on which this species occurs. Some species tend to have much better developed thalli when growing on sites frequented by birds (e.g. *Candelariella vitellina* (Hoffm.) Müll. Arg.). Variations of these types are clearly without taxonomic significance and because of the wide variation seen in such characters as cortical thickness the citation of such thicknesses in species descriptions is of limited value unless based on many collections from a wide range of habitats.

In their study of *Xanthoria parietina* referred to above (p. 49) Hill and Woolhouse (1966) also studied the number of algal cells/cm^2 and chlorophyll content ($\mu g/cm^2$) and found marked differences between the collections from maritime rocks and roofs as compared with those from trees. From this it is clear that the density (and in some other cases also the thickness) of the algal layer may be subject to environmentally induced variations.

Species of *Alectoria* Ach. characteristically have a loose arachnoid net of anastomosing hyphae in the medulla and in section often appear hollow centrally. In some instances the medulla may be occupied by a much denser tissue (Grummann, 1941) so that the central cavity is occluded. This phenomenon is particularly marked in some taxa adapted to a prostrate rather than pendent habit (e.g. *A. sarmentosa* subsp. *vexillifera*) which have dorsiventrally compressed and expanded main stems (see Table I, Fig. 1 and Hawksworth, 1972a).

The contacts between the algal cells and fungal hyphae within lichen thalli are also subject to ecological pressures. Ben-Shaul *et al.* (1969) showed that in *Caloplaca aurantia* (Pers.) Hellb. the degree of penetration of the fungal hyphae into the algal cells was greatest in plants from the most xeric sites they investigated. A comparable situation has also been shown to occut in *Lecanora radiosa* (Hoffm.) Schaer. (Galun *et al.*, 1970). Variations can also occur within single thalli as well as between thalli taken from different habitats. James (1970) found that in a single specimen of *L. gangaleoides* Nyl. growing round the lip of an overhanging rock the fungal hyphae were mostly intramembranous (i.e. entering the cell walls of the algal cells) in the part of the thallus on the surface of the lip but mainly intracellular (i.e. forming haustoria which penetrate the algal cells' wall and enter the interior of the cells) in parts of the thallus below the lip. Further work is necessary to determine if the type of fungal-algal contact is subject to ecological variations of this nature in all lichens as so few taxa have yet been investigated from this standpoint.

Only recently has the variation in surface features of lichen thalli as seen with the scanning electron microscope been recognized (Hawksworth, 1969; Peveling, 1970; Richardson, 1971). Ultrastructural surface features might be expected to be modified by ecological factors but so far there is no evidence to suggest that this is the case and differences in them have proved useful at the generic level (Hawksworth, 1969). In one instance minor differences in ultrastructural surface features have been considered possibly to indicate the affinity of a newly recognized species (Hale, 1972).

2. Ascocarps and Ascospores

Ascocarp and ascospore characters are extensively used in the taxonomy of lichens and are often regarded as being among the characters least subject to ecological modifications. The available evidence suggests that this assumption is probably correct but too little attention has perhaps been paid to this aspect of lichenology in the past. The factors governing the persistence or disappearance of thalline margins, thickness of hypothecia and thecia, ascospore pigmentation, size, shape and septation, are all in need of study from the ecological viewpoint before the possibility that they may be affected by environmental factors can be entirely ruled out.

Minor differences in spore colour, shape, size and septation have probably been over-emphasized in many genera of pyrenocarpous lichens by earlier workers. Swinscow (1962, p. 17) observed that "... as experience of pyrenocarpous lichens has accumulated it has become clear that natural variation within species is wider than was previously thought". In *Porina chlorotica* (Ach.) Müll. Arg. slightly longer spores occur in material from particularly shaded sites (f. *tenuifera* (Nyl.) Swinsc.), while specimens from highly calcareous rocks (var. *persicina* (Körb.) Zahlbr.) have spores which are often more rounded at their ends (see Swinscow, 1962).

Chemistry

Lichens produce about one hundred secondary metabolic products unknown in other plant groups. Most of these are orcinol or β-orcinol derivatives and detailed accounts of their chemistry and occurrence are provided by C. F. Culberson (1969, 1970). Culberson and Culberson (1970) discuss their phylogenetic significance and a detailed review of their biosynthesis has been prepared by Huneck (1971). A comprehensive account of the types of chemical variation known in the lichens and their taxonomic treatment falls outside the scope of this paper, but as I have previously emphasized (Hawksworth, 1971) different types of chemical variation require quite different taxonomic treatments. Chemical races characterized by the presence, absence or variations in the concentration of coloured compounds have already been discussed (pp. 49–50) and this section is primarily concerned with colourless compounds deposited on the surfaces of the fungal hyphae in the medulla.

The best example of habitat selection by chemotypes characterized by replacement patterns involving several biogenetically closely related compounds is certainly the *Ramalina siliquosa* (Huds.) A. L. Sm. complex investigated by Culberson and Culberson (1967). These authors found that particular chemotypes were restricted to precise habitats on a single rocky promontory in North Wales (Fig. 5), and a comparable situation has been demonstrated on a single pyramidal boulder in Portugal (Culberson, W. L., 1969a). These chemotypes were treated as distinct species by Culberson (1967) but the rank of variety seems to be most appropriate (Hawksworth, 1972b). A similar example is *R. subfarinacea* (Nyl. ex Cromb.) Nyl. in which, for example, a variety with norstictic acid occurs on maritime rocks but one with norstictic and salazinic acids is found on trees and inland siliceous rocks (Hawksworth, 1968).

In the *Cladonia chlorophaea* (Flörke ex Sommerf.) Spreng. complex in Michigan, Wetherbee (1969) found that although some of the chemical races did not appear to be correlated with particular habitats, grayanic acid and cryptochlorophaeic acid-containing plants had complementary ecologies, occurring mainly on bark and usually on the ground, respectively. These particular chemotypes have compounds that are probably not biogenetically

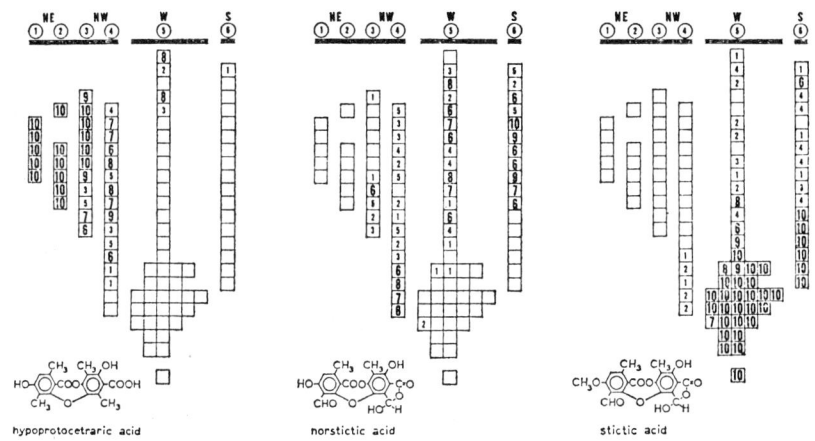

FIG. 5. Habitat selection by chemotypes of *Ramalina siliquosa* (Huds.) A. L. Sm. on a sheltered rocky promontory in North Wales. The numbers in the blocks indicate how many of the 10 plants collected closest to the centre of each 35 × 35 cm block were of the three chemical types that made up 96% of the 980 plants encountered and the line-transects are shown in their topographic position. Hypoprotocetraric acid containing plants dominate the north (most protected) side; stictic acid containing plants dominate the most exposed areas (the bottom of the seaward west and south sides); and norstictic acid containing plants are most abundant in the intermediate zone. Reproduced with permission from Culberson and Culberson (1967, p. 1196).

closely related (Shibata and Chiang, 1965) and in view of some minor morphological differences (Ahti, 1966) are most appropriately treated as distinct species. W. L. Culberson (1969b) found that two chemotypes within the *C. cariosa* (Ach.) Spreng. complex had both different geographical distributions and ecological requirements in the south-eastern United States. *C. polycarpia* Merr. (with norstictic and stictic acids) is primarily confined to the Atlantic coastal plain whilst *C. polycarpoides* Nyl. (with norstictic acid) is absent from most of this region. In an area where both these species occurred, however, Culberson found that although geographically sympatric they were still separated ecologically, *C. polycarpia* occurring on sandy soils and *C. polycarpoides* on clay soils.

Although it is now well established that some chemotypes have become restricted to particular ecological niches it is also equally clear that others have not. No habitat selection was demonstrated by Hale (1963) in the *Cetraria ciliaris* Ach. complex after a study of 16,398 plants, and Graham (1969) came to similar conclusions after investigating 1896 specimens. Similarly, in his study of four chemotypes of *Ramalina montagnei* de Not. in the Virgin Islands, Rundel (1972) was unable to find any evidence for habitat selection in the area he studied. A more complex situation exists in some chemotypes where chemical differences are not related to particular substrates but to slight differences in geographical distribution. In *Pseudevernia furfuracea* (L.) Zopf the chemotype with the depside olivetoric acid (var. *ceratea* (Ach.) D. Hawksw., syn. *P. olivetorina* (Zopf) Zopf) is much more frequent in northern Europe than the chemotype with the corresponding depsidone to olivetoric acid, physodic acid (var. *furfuracea*), which extends further south in Europe and into North Africa (Hale, 1968; Hawksworth and Chapman, 1971). This north–south trend is also apparent on a smaller geographical scale in the British Isles (Hawksworth and Chapman, 1971).

The colourless lichen products concerned in the formation of chemical races in lichens do not appear to have any inherent adaptive value, as the same compounds frequently occur in taxa characteristic of quite different ecological situations, often even within the same genus. It is consequently tempting to suggest that in those cases where habitat selection appears to be taking place this may be occurring in parallel with less readily detectable but genotypically determined adaptive physiological differences (see pp. 55–58). In *Haematomma ventosum* (with thamnolic acid but lacking zeorin) and *H. lapponicum* Räs. (lacking thamnolic acid but sometimes with zeorin), the former is essentially European while the latter is a circumpolar arctic–alpine species usually occurring at higher altitudes and latitudes than *H. ventosum* (Fig. 6). Thomson (1968) suggested that *H. ventosum* may have evolved from *H. lapponicum* and in view of their different ecological requirements it seems probable that physiological adaptations have occurred together with the changes in chemical components.

In some other cases in which chemotypes appear to have become adapted to slightly different climatic conditions slight morphological differences correlate with the differences in chemical components. Two examples of species pairs in this category in the British flora are *Parmelia borreri* (Sm.) Turn. (gyrophoric acid present, commonest in southern Britain, see Hawksworth, 1972b) and *P. subrudecta* Nyl. (lecanoric acid present, more widespread in Britain extending northwards into Ross-shire and Kincardine), and *P. omphalodes* (L.) Ach. (salazinic acid present, widespread) and *P. discordans* Nyl. (protocetraric acid present, western). Different ecological requirements may be related to both morphological and chemical trends in some instances and some examples in which this occurs have already been discussed above.

Some mention must also be made here of the misleading paper by Murray (1971) which claimed to have shown that mechanical shock could affect the chemical components of *Cladonia impexa* Harm. (syn. *C. pacifica* Ahti). Murray only attempted to name the chemicals concerned by microcrystal tests and her photographs of crystals are almost all incorrectly determined, most being merely different crystal forms of the single compound usnic acid.

PHYSIOLOGY

Progress in the study of lichen physiology has lagged considerably behind that of other major plant groups, primarily because of technical difficulties. The last decade has, however, seen a resurgence of interest in this field which has led to a much greater understanding both of the physiology of intact thalli and their isolated components. An excellent review of most aspects of lichen physiology has recently been prepared by Farrar (1973).

It has been recognized for many years that rates of photosynthesis and respiration in lichens are related to differences in the availability of water and the degree of illumination, but while several papers have now provided information on the physiological characteristics of species characteristic of extreme and more mesic habitats few studies have attempted to consider the possibility of physiologically defined ecotypes. In *Cladonia rangiferina* (L.) Wigg., however, Adams (1971) found that material from the Wisconsin pine barrens had higher optimum temperatures for net photosynthesis than material from more alpine sites, suggesting that ecophysiological differences existed between the populations.

Harris (1971) studied the effects of light intensity and moisture availability on three foliose lichens (*Hypogymnia physodes* (L.) Nyl., *Parmelia caperata* (L.) Ach. and *P. perlata* (Huds.) Ach.) from different parts of oak trees in a wood in Devon under laboratory conditions. He discovered, for example, that tree-top and tree-base collections of *P. caperata* had different optimal water contents of their thalli for maximum carbon assimilation (50–55% saturation for tree-top types and 70–80% for tree-base types). These pioneer studies demonstrated for the first time that physiological adaptations to micro-environmental conditions could occur within an individual species at a single site. Kershaw and Harris (1971) carried out comparable studies on material of *P. caperata* collected from

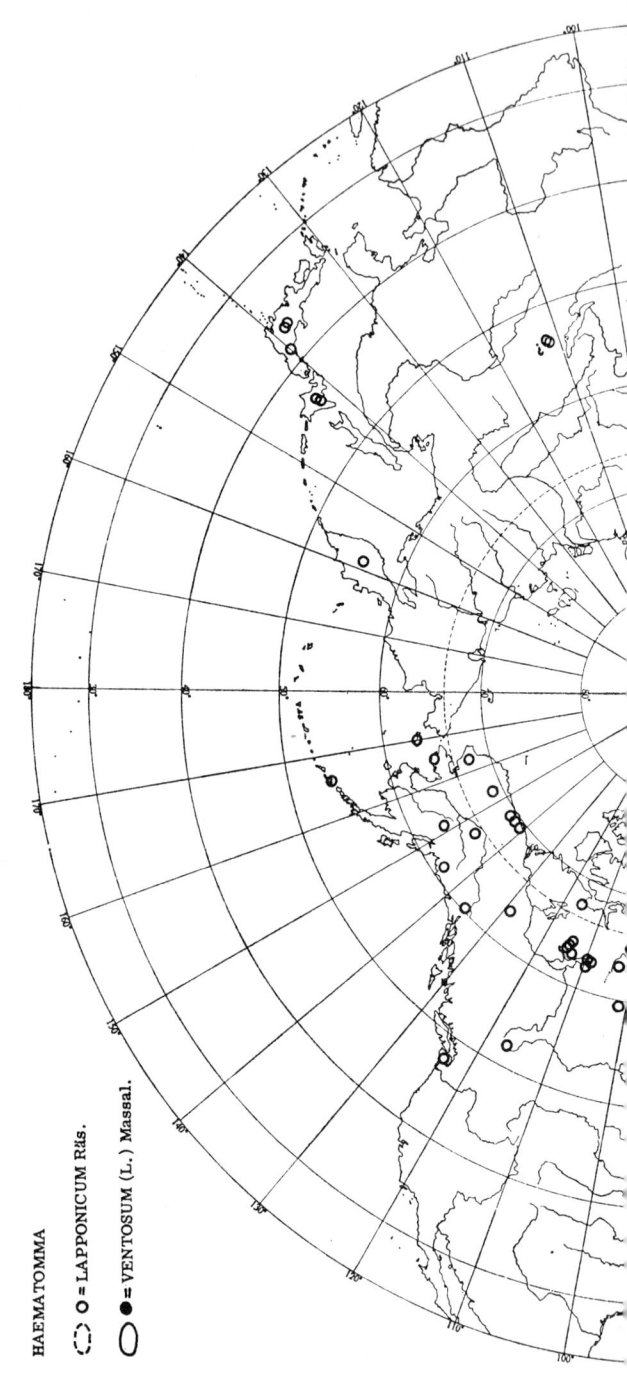

2. Ecological Factors and Species Delimitation in Lichens

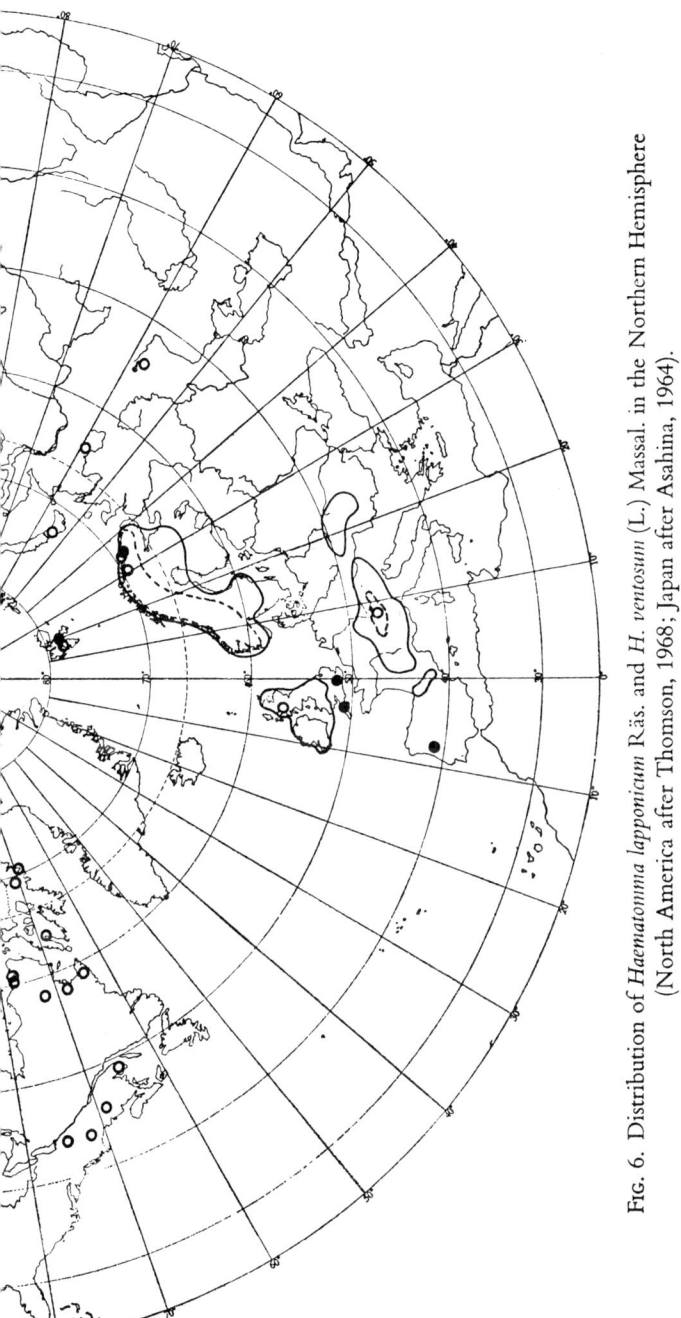

FIG. 6. Distribution of *Haematomma lapponicum* Räs. and *H. ventosum* (L.) Massal. in the Northern Hemisphere (North America after Thomson, 1968; Japan after Asahina, 1964).

areas in southern England with different mean annual rainfalls and found striking differences in the positions of their optimal moisture contents for maximum net carbon assimilation (Fig. 7). These two studies indicate either that these species consist of a number of distinct genotypes characterized by physiological adaptations to very precise ecological conditions, or that individuals are able to assume such adaptations by virtue of their phenotypic plasticity and retain these under laboratory conditions in the short term. In order to determine whether genotypic or phenotypic factors are involved long-term studies on

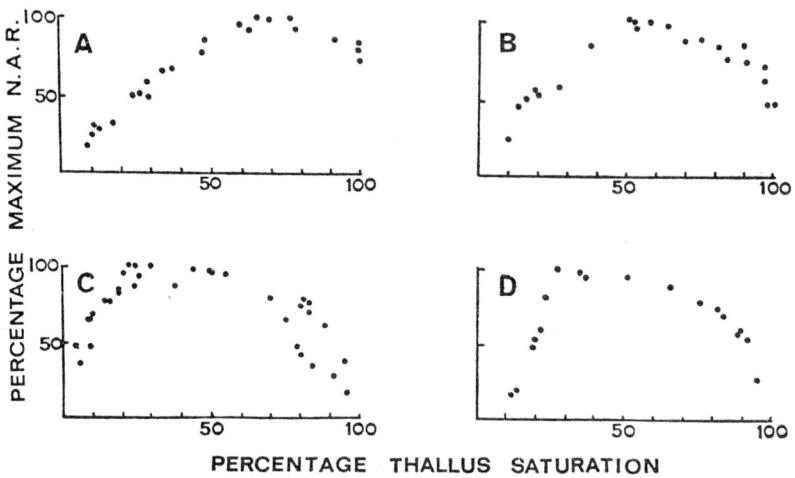

FIG. 7. The interaction between net assimilation rate and thallus water content for specimens of *Parmelia caperata* (L.) Ach. collected in (A) Devon, (B) Wiltshire, (C) Surrey and (D) Norfolk. Reproduced with permission from Kershaw and Harris (1971, p. 36).

different physiological races maintained in non-optimal laboratory conditions or transplanted into different environments in the field are required. If their optima were found to change to fit their new environments this would show the variation to be merely phenotypic but if these remained unaffected there would be further grounds for assuming that genotypic factors were involved.

Another instance in which physiologically adapted ecotypes may be occurring was reported by Gilbert (1971) who discovered that in an area affected by sulphur dioxide air pollution in Northumberland a local population of *Parmelia saxatilis* (L.) Ach. was able to maintain itself in an area from which this species was generally absent or undergoing a decline in response to increasing sulphur dioxide levels. Gilbert suggested that this population might represent an ecotype with enhanced resistance to sulphur dioxide but he did not carry out any physiological studies to substantiate this view. The effects of sulphur dioxide on respiration in this species have been studied by Baddeley et al. (1971), however, and these authors found that thalli of this species from a polluted area were slightly more tolerant of increasing sulphur dioxide levels than specimens from a less polluted region.

The growth-rates of some saxicolous lichens have received considerable attention in the course of the application of "lichenometric" dating techniques to glacial moraines. From these studies it is evident that the growth rate of the same species may vary very considerably in different climatic areas. In *Rhizocarpon geographicum* (L.) DC., for example, growth rates may vary from 2 to 93 mm per century (Andrews and Webber, 1964, p. 82). It is not known how far variations of this type are due to phenotypic plasticity alone, and transplant experiments are required to clarify the situation. It is interesting to note, however, that Lunan (1970) found that this species showed very similar growth rates on both slate and sandstone in the English Lake District.

It is well known that some species of lichens occur on different substrates in parts of their geographical ranges or are occasionally encountered on substrates or in ecological situations from which they are normally absent. Examples of substrate-switches in lichens are adequately reviewed by Brodo (1973) and are therefore not treated in detail here, and it is conceivable that at least some of these involve physiological adaptations which may be genotypically determined. In addition to adaptations to different types of substrate, adaptations to differences in the chemical nature of the substrate (e.g. acidity, heavy metal content) might also be expected to occur, and some of these might lead to the formation of physiological ecotypes. The effects of heavy metals on lichens are reviewed in detail by James (1973).

Cherkasskaya (1965) studied the "resinoids" in samples of *Evernia prunastri* collected from different species of trees and reported some significant differences in the amounts present. His investigation does not appear to have utilized a large number of specimens and consequently the possibility that his results indicate natural variation within the species itself cannot be discounted.

DISCUSSION AND CONCLUSIONS

In this paper I have attempted to review and discuss the taxonomic significance of some characters which may be either related to or affected by various ecological parameters. Individual species of lichens differ considerably in their ecological amplitudes (Table IV) but it is interesting to note that some of those which are most substrate- or host-specific are either allied to non-lichen-forming species or show tendencies to this state. It must be emphasized that

TABLE IV. Examples of variations in substrate and host requirements seen in different lichen species

Species	Substrate	Species	Substrate
Lepraria incana (L.) Ach.	siliceous, calcareous, exposed, sheltered and maritime rocks; peaty and calcareous soils; etc.	*Fulgensia fulgens* (Sw.) Elenk.	well drained sunny calcareous soils and dunes
Lecanora dispersa (Pers.) Sommerf.	siliceous and calcareous rocks, wood, bark, *Armeria*, *Limonium*, bone, leather, iron, etc.	*Icmadophila ericetorum* (L.) Zahlbr.	moist peaty soils and rotting bark
Parmelia sulcata Tayl.	siliceous rocks, wood, bark and on the ground	*Lecanora chlarotera* Nyl.	bark and wood in well-lit situations
Mycoblastus sanguinarius (L.) Norm.	siliceous coarse-grained rocks and acidic bark	*Schismatomma decolorans* (Turn. & Borr. ex Sm.) Clauz. & Vězda	dry barked areas and crevices of deciduous trees
Pertusaria corallina (L.) Arnold	siliceous rocks	*Lecanora populicola* (DC.) Duby	bark of *Populus* spp.
Lecanora calcarea (L.) Sommerf.	calcareous rocks and cement	*Lecidea turgidula* Nyl.	bark of coniferous trees
Rhizocarpon sphaericum (Schaer.) Migula	serpentine rocks	*Strigula elegans* (Fée) Müll. Arg.	leaves of trees of 99 genera from 51 families
Caloplaca thallincola (Wedd.) DR.	maritime siliceous rocks	*Lecidea insularis* Nyl.	thallus of *Lecanora rupicola*
Sclerophyton circumscriptum (Tayl.) Zahlbr.	dry recesses in siliceous maritime rocks	*Leptorhaphis epidermidis* (Ach.) Th. Fr.[a]	boles of *Betula* spp.
Stereocaulon nanodes Tuck.	siliceous rocks rich in lead or zinc	*Stenocybe pullatula* (Ach.) Stein[a]	bark of *Alnus glutinosa*
Toninia coeruleonigricans (Lightf.) Th. Fr.	calcareous soils, crevices in calcareous rocks; cement and mortar	*Calicium curtsii* Tuck.[a]	bark of *Rhus typhina*
		Calicium corynellum Ach.[a]	thallus of *Haematomma ventosum*
		Arthonia glaucomaria (Nyl.) Nyl.[a]	ascocarps of *Lecanora rupicola*

[a] Non-lichenized or partially lichenized species

lichens are not a distinct taxonomic entity, although on hypothetical grounds they may be particularly ancient (Ahmadjian, 1970b; Cain, 1972), but merely fungi drawn from diverse taxonomic groups united in their common method of nutrition (symbiosis). Santesson (1967) has pointed out, for example, that the genus *Buellia* de Not. contains not only independent lichens but also lichenicolous lichens, lichenicolous fungi (parasymbionts), lichen parasites and saprophytes. Similar trends are also seen in some other genera (e.g. *Acarospora* Massal., *Arthonia* Ach., *Arthopyrenia* Massal. em. Müll. Arg., *Bacidia* de Not. em. Zahlbr., *Calicium* Pers. em. de Not., *Catillaria* (Ach.) Th. Fr., *Lecidea* Ach., *Microthelia* Körb., *Rhizocarpon* Ram. em. Th. Fr., *Sphinctrina* Fr., *Thelocarpon* Nyl.) and it seems probable that some parasymbionts may have passed through a stage of lichenization in the course of their evolution and become host-specific. An interesting transition between lichens and lichenicolous lichens is seen in *Acarospora* subgen. *Xanthothallia* where some species start growth as lichenicolous lichens but later develop independent thalli (Poelt and Steiner, 1972). The tendency towards non-lichenization provides an interesting example of adaptation of the fungal components of lichens to very precise sets of ecological conditions in which nutrients have to be derived from a source other than a symbiotic alga.

The major factors affecting host-specificity in corticolous lichens have been reviewed by Barkman (1958) and, as emphasized by Fabiszewski (1968) and Brodo (1973), it appears to be the physical and chemical characteristics of the bark which seem to be the most important limiting factors. Adequate study is necessary before concluding that species are really specific to particular tree species as some species once thought to be confined to a single tree have since been discovered on trees of quite different genera. *Stenocybe septata* (Leight.) Massal., once thought to occur only on *Ilex*, is now also known from *Alnus*, *Corylus*, *Betula* and *Sorbus* (Coppins, 1972); *Pyrenula nitidella* (Flörke) Müll. Arg., formerly considered to be confined to *Fagus*, also occurs on *Acer*, *Castanea*, *Corylus* and *Quercus* in Britain; and *Dermatina quercus* (Massal.) Zahlbr. occurs on *Corylus* as well as on *Quercus*. Saxicolous species are affected by the texture and chemical nature of the rocks on which they occur but few studies have attempted to relate the occurrence of particular species to individual rock types. Most lichen ecologists have been content to speak of "siliceous" and "calcareous" rock species and communities but more critical work in this field is clearly needed. Hertel (1967) has demonstrated that some of the calcicolous species of *Lecidea* show marked preferences for different types of calcareous rocks. The preference of some species for substrates rich in heavy metals is discussed by James (1973).

One characteristic of the ecology of lichens is that species drawn from quite different taxonomic groups are frequently found growing in close association in particular ecological and environmental situations. Within a single, climatically more or less uniform, geographical region each substrate tends to assume a characteristic and often surprisingly uniform flora under comparable environmental conditions. These communities are often readily recognizable in the field and may be described as phytosociological taxa. The phytosociology of lichen communities and their taxonomic implications are discussed by Barkman elsewhere in this symposium (pp. 141–150) and so are not treated further here.

In some of the examples discussed above it seems clear that closely related taxa adapted to particular ecological situations are genotypically distinct (i.e. that they are "ecotypes" rather than "ecads") because the two types occasionally exist side by side in the same habitat. In many cases, however, it is uncertain whether or not genotypic factors are involved and in such situations it seems preferable to use the terms "morphotype" and "chemotype" as recommended by Santesson (1968), since these terms do not have any genetic connotations. When, after extensive studies of field populations in relation to ecological parameters, it is clear that morphotypes and chemotypes which seem likely to be genotypically different occur, they should be treated in the most appropriate taxonomic ranks (see Davis and Heywood, 1963, pp. 88–103). The detection of morphotypes and chemotypes adapted to particular ecological situations is not always an easy process and it seems likely that many such adaptations in lichens have yet to be described. Lichenologists must be constantly aware of the errors that may be incurred through incomplete sampling of ecoclines (see Bradshaw, 1962) and the extent to which characters may vary within single populations growing under apparently uniform ecological conditions. The range of variation in a few characters exhibited by a single population of *Alectoria fuscescens* on birch trunks in a small wood in Scotland is illustrated in Table V. Kappen and Schulze (1972) describe four distinctive but intergrading morphotypes in *Ramalina maciformis* (Del.) Nyl. which do not appear to be correlated with any discernible differences in their habitats and without taxonomic significance.

Stebbins (1950, pp. 197–198) considered that most geographical races may have arisen primarily as ecotypes and the detection of these in lichens might therefore be expected to add considerably to our understanding of speciation in this group. Phenotypic plasticity might, however, be expected to be particularly important in the lichens because their relatively slow rates of establishment and growth might enable them to produce morphotypes or undergo physiological changes so as to be in equilibrium with very precise environmental

TABLE V. Variation in some morphological and chemical characters in fifty specimens from a single population[a] of *Alectoria fuscescens* Gyeln.

Diameter of basal branches (mm)	0·10–0·35–0·70	Colour[b] of (a) apical branches	2·0-3·3-4·0
Largest angle between main dichotomies (degrees)	45–72·1–110	(b) basal branches	1·0-2·3-4·0
Maximum distance between dichotomies (mm)		PD[c] +red (a) medulla	30%
(a) basal branches	2·0–8·26–30·3	(b) soralia	82%
(b) apical branches	1·0–3·2–10·3	Fumarprotocetraric acid present (by paper chromatography)	100%
Maximum distance between soralia (mm)	1·5–3·3–7·0	Fragmentation regions present	82%
Total length of thalli (cm)	2·0–6·2–12·0	Soralial spinules present	52%
Crispate apices present	34%		

[a] Scotland, E. Inverness, Aviemore, S. end of Craigellachie N.N.R. (Nat. Grid Ref. 28/882118), alt. 700–725 ft., on young trees of *Betula pendula*, 3 August 1968, D. L. Hawksworth.

[b] Colour scored as units from 1 (greyish-white) to 4 (dark brown).

[c] Reaction with Steiner's stable *p*-phenylenediamine solution.

conditions. The ability to produce a wide range of phenotypic responses clearly is of considerable adaptive significance. In some instances the "Baldwin effect", in which advantageous phenotypic changes are succeeded by similar but genotypically determined changes, may be expected to have operated. Many lichens appear to be functionally apomicts and, as noted by Brodo (1973), the appearance of a genetically determined character may therefore be expected to become fixed in populations even if it is only locally advantageous. Many species of *Usnea* P. Browne ex Adans., for example, behave as apomicts and rarely form ascocarps at all, and in this genus large numbers of taxa which appear to be genetically distinct have relatively small geographical ranges.

I have already pointed out that techniques for the growth of intact lichen thalli under controlled environmental conditions are now becoming available (p. 32). I agree with Haynes and Morgan-Huws (1970) and Brodo (1973) that the time is now ripe for a vast expansion into the field of experimental ecology in the lichens which will not only help resolve some taxonomic problems but lead to a sounder understanding of the factors controlling ecological amplitudes and speciation in the group. It is to be hoped that modern lichen taxonomists will not be as slow to employ experimental work in the field and laboratory as many of their predecessors were to accept that chemical components constituted valid taxonomic criteria.

I hope that this paper will serve not only as a review of a difficult and complex aspect of lichenology but also to remind taxonomists of the continuing need to study populations in the field and to be conscious of the influence of ecological factors on the characters they use in the delimitation of taxa. Grove (1937, p. 368) observed that "the old weather-beaten field-naturalists . . . were in many respects nearer to nature, and to the truth, than a great many of their pallid indoor successors of today".

ACKNOWLEDGEMENTS

I am very grateful to Dr I. M. Brodo for making some of his unpublished work available to me; to the Nature Conservancy and Dr T. Ahti for the base maps used in Figs 2 and 6, respectively; to Dr F. Rose for information on the distribution of *Parmelia revoluta* in Britain; to Dr P. B. Topham for allowing me to make use of some of her collections of *Cladonia uncialis*; and to Miss E. Jelínková for sending me some of her specimens of *Lecanora conizaeoides*.

I am especially indebted to Mr P. W. James for stimulating discussions in the course of the preparation of this paper, allowing me to utilize some of his unpublished work, and his most valuable comments on a preliminary draft of this paper.

REFERENCES

DES ABBAYES, H. (1951). Traité de lichénologie. *Encycl. Biol.* **41,** i–x, 1–217.
ADAMS, M. S. (1971). Temperature response of carbon dioxide exchange of *Cladonia rangiferina* from the Wisconsin pine barrens, and comparison with an alpine population. *Am. Midl. Nat.* **86,** 224–227.
AHMADJIAN, V. (1967). "The Lichen Symbiosis." Blaisdell, Waltham, Mass., 152 pp.
AHMADJIAN, V. (1970a). Adaptations of Antarctic terrestrial plants. *In* "Antarctic Ecology" (M. W. Holdgate, ed.), Vol. 2, pp. 801–811. Academic Press, London and New York.
AHMADJIAN, V. (1970b). The lichen symbiosis: its origin and evolution. *In* "Evolutionary Biology" (T. Dobzhansky, M. K. Hecht and W. C. Steere, eds.), Vol. 4, pp. 163–184. Meredith Corporation, New York.
AHMADJIAN, V. and HEIKKILÄ, H. (1970). The culture and synthesis of *Endocarpon pusillum* and *Staurothele clopima*. *Lichenologist* **4,** 259–267.
AHTI, T. (1961). Taxonomic studies on Reindeer lichens (*Cladonia*, subgenus *Cladina*). *Suomal. eläin-ja kasvit. Seur. van. Julk.* **32**(1), 1–160.
AHTI, T. (1966). Correlation of chemical and morphological characters in *Cladonia chlorophaea* and allied lichens. *Ann. Bot. Fenn.* **3,** 380–390.
ALMBORN, O. (1965). The species concept in lichen taxonomy. *Bot. Notiser* **118,** 454–457.
ANDREWS, J. T. and WEBBER, P. J. (1964). A lichenometrical study of the northwestern margin of the Barnes Ice Cap: a geomorphological technique. *Geogrl Bull.* **22,** 80–104.
ASAHINA, Y. (1964). Lichenologische Notizen (§ 193). *J. Jap. Bot.* **39,** 165–171.
BADDELEY, M. S., FERRY, B. W. and FINEGAN, E. J. (1971). A new method of measuring lichen respiration: response of selected species to temperature, pH and sulphur dioxide. *Lichenologist* **5,** 18–25.
BARKMAN, J. J. (1958). "Phytosociology and Ecology of Cryptogamic Epiphytes." Van Gorcum, Assen, 628 pp.
BEN-SHAUL, Y., PARAN, N. and GALUN, M. (1969). The ultrastructure of the association between phycobiont and mycobiont in three ecotypes of the lichen *Caloplaca aurantia* var. *aurantia*. *J. Microscopie* **8,** 415–422.
BRADSHAW, A. D. (1962). The taxonomic problems of local geographical variation in plant species. *In* "Taxonomy and Geography" (D. Nichols, ed.), pp. 7–16. Systematics Association, London.
BRODO, I. M. (1973). Substrate ecology. *In* "The Lichens" (V. Ahmadjian and M. E. Hale, eds). Academic Press, London and New York, in press.
CAIN, R. F. (1972). Evolution of the fungi. *Mycologia* **64,** 1–14.
CHERKASSKAYA, V. S. (1965). Vliyanie substrata na biokhimichiskie osobennosti lishaĭnika *Evernia prunastin* (L.) Ach. *Bot. Zhurn. (Moscow)* **50,** 979–981.
COPPINS, B. J. (1972). Field meeting at Richmond, Yorkshire. *Lichenologist* **5,** 326–336.
CULBERSON, C. F. (1969). "Chemical and botanical guide to lichen products." University of North Carolina Press, Chapel Hill, 628 pp.
CULBERSON, C. F. (1970). Supplement to "Chemical and Botanical Guide to Lichen Products." *Bryologist* **73,** 177–377.
CULBERSON, W. L. (1967). Analysis of chemical and morphological variation in the *Ramalina siliquosa* species complex. *Brittonia* **19,** 333–352.
CULBERSON, W. L. (1969a). The behaviour of the *Ramalina siliquosa* group in Portugal. *Öst. bot. Z.* **116,** 85–94.

CULBERSON, W. L. (1969b). The chemistry and systematics of some species of the *Cladonia cariosa* group in North America. *Bryologist* **72**, 377–386.
CULBERSON, W. L. (1970). Chemosystematics and ecology of lichen-forming fungi. *Ann. Rev. Ecol. Systematics* **1**, 153–170.
CULBERSON, W. L. and CULBERSON, C. F. (1967). Habitat selection by chemically differentiated races of lichens. *Science, N.Y.* **158**, 1195–1197.
CULBERSON, W. L. and CULBERSON, C. F. (1970). A phylogenetic view of chemical evolution in the lichens. *Bryologist* **73**, 1–31.
DAVIS, P. H. and HEYWOOD, V. H. (1963). "Principles of Angiosperm Taxonomy." Oliver and Boyd, Edinburgh, 556 pp.
DEGELIUS, G. (1935). Das ozeanische Element der Strauch- und Laubflechten-flora von Skandinavien. *Acta phytogeogr. suec.* **7**, i–xii, 1–411.
DE SLOOVER, J. and LEBLANC, F. (1970). Pollutions atmosphériques et fertilité chez les mousses et chez les lichens épiphytiques. *Bull. Acad. Soc. lorr. Sci.* **9**, 82–90.
DIBBEN, M. J. (1971). Whole-lichen culture in a phytotron. *Lichenologist* **5**, 1–10.
DU RIETZ, G. E. (1924). Die Soredien und Isidien der Flechten. *Svensk bot. Tidskr.* **18**, 371–396.
FABISZEWSKI, J. (1968). Porosty Snieznika klodzkiego i gor Bialskich. *Monogr. Bot. (Warsaw)* **26**, 1–116.
FARRAR, J. F. (1973). Lichen physiology: progress and pitfalls. *In* "Air Pollution and Lichens" (B. W. Ferry, M. S. Baddeley and D. L. Hawksworth, eds), pp. 238–282. University of London, Athlone Press, London.
GALUN, M. (1963). Autecological and synecological observations on lichens of the Negev, Israel. *Israel J. Bot.* **12**, 179–187.
GALUN, M., MARTON, K. and BEHR, L. (1972). A method for the culture of lichen thalli under controlled conditions. *Arch. Mikrobiol.* **83**, 189–192.
GALUN, M., PARAN, N. and BEN-SHAUL, Y. (1970). An ultrastructural study of the fungus alga association in *Lecanora radiosa* growing under different environmental conditions. *J. Microscopie* **9**, 801–806.
GILBERT, O. L. (1971). Studies along the edge of a lichen desert. *Lichenologist* **5**, 11–17.
GOOD, R. (1964). "The Geography of the Flowering Plants" (3rd Edition). Longmans, London, 518 pp.
GRAHAM, W. L. (1969). The occurrence of the lichen complex, *Cetraria ciliaris*, in the straits region of Michigan. *Mich. Bot.* **8**, 67–71.
GROVE, W. B. (1937). "British stem- and leaf-fungi (Coelomycetes)," Vol. 2. Cambridge University Press, Cambridge, 407 pp.
GRUMMANN, V. J. (1941). Morphologische, anatomische und entwicklungsgeschichtliche Studien über Bildungsabweichungen bei Flechten. *Reprium nov. Spec. Regni veg., Beih.* **122**, 1–128.
HALE, M. E. (1954). First report on lichen growth rate and succession at Aton Forest, Connecticut. *Bryologist* **57**, 244–247.
HALE, M. E. (1963). Populations of chemical strains in the lichen *Cetraria ciliaris*. *Brittonia* **15**, 126–133.
HALE, M. E. (1968). A synopsis of the lichen genus *Pseudevernia*. *Bryologist* **71**, 1–11.
HALE, M. E. (1972). *Parmelia pustulifera*, a new lichen from southeastern United States. *Brittonia* **24**, 22–27.

HARRIS, G. P. (1971). The ecology of corticolous lichens II. The relationship between physiology and the environment. *J. Ecol.* **59**, 441–452.

HAWKSWORTH, D. L. (1968). A note on the chemical strains of the lichen *Ramalina subfarinacea*. *Bot. Notiser* **121**, 317–320.

HAWKSWORTH, D. L. (1969). The scanning electron microscope. An aid to the study of cortical hyphal orientation in the lichen genera *Alectoria* and *Cornicularia*. *J. Microscopie* **8**, 753–760.

HAWKSWORTH, D. L. (1970). The chemical constituents of *Haematomma ventosum* (L.) Massal. in the British Isles. *Lichenologist* **4**, 248–255.

HAWKSWORTH, D. L. (1971). Types of chemical variation in lichens and their taxonomic treatment. *Abstr. First International Mycological Congress*: 41. First International Mycological Congress, Exeter.

HAWKSWORTH, D. L. (1972a). Regional studies in *Alectoria* (Lichenes) II. The British species. *Lichenologist* **5**, 181–261.

HAWKSWORTH, D. L. (1972b). The natural history of Slapton Ley Nature Reserve IV. Lichens. *Fld Stud.* **3**, 535–578.

HAWKSWORTH, D. L. (1973). Mapping studies. *In* Air Pollution and Lichens" (B. W. Ferry, M. S. Baddeley and D. L. Hawksworth, eds), 38–76. University of London, Athlone Press, London.

HAWKSWORTH, D. L. and CHAPMAN, D. S. (1971). *Pseudevernia furfuracea* (L.) Zopf and its chemical races in the British Isles. *Lichenologist* **5**, 51–58.

HAWKSWORTH, D. L., ROSE, F. and COPPINS, B. J. (1973). Changes in the lichen flora of England and Wales attributable to pollution of the air by sulphur dioxide. *In* "Air Pollution and Lichens" (B. W. Ferry, M. S. Baddeley and D. L. Hawksworth, eds), 330–367. University of London, Athlone Press, London.

HAYNES, F. N. and MORGAN-HUWS, D. I. (1970). The importance of field studies in determining the factors influencing the occurrence and growth of lichens. *Lichenologist* **4**, 362–368.

HERTEL, H. (1967). Revision einiger calciphiler Formenkreis der Flechtengattung *Lecidea*. *Nova Hedwigia, Beih.* **24**, 1–155.

HILL, D. J. and WOOLHOUSE, H. W. (1966). Aspects of the autecology of *Xanthoria parietina* agg. *Lichenologist* **3**, 207–214.

HUNECK, S. (1971). Chemie und Biosynthese der Flechtenstoffe. *Fortschr. Chem. org. Nat.* **29**, 210–306.

JAMES, P. W. (1970). The lichen flora of shaded acid rock crevices and overhangs in Britain. *Lichenologist* **4**, 309–322.

JAMES, P. W. (1973). The effect of air pollutants other than hydrogen fluoride and sulphur dioxide on lichens. *In* "Air Pollution and Lichens" (B. W. Ferry, M. S. Baddeley and D. L. Hawksworth, eds), 143–175. University of London, Athlone Press, London.

KAPPEN, K. and SCHULZE, E-D. (1972). *Ramalina maciformis* (Del.) Nyl. fertile in the Western Negev, Israel. *Lichenologist* **5**, 323–325.

KÄRENLAMPI, L. (1964). Preliminary notes on the variability of *Cladonia uncialis* (L.) Wigg. in Eastern Fennoscandia. *Ann. Bot. Fenn.* **1**, 220–223.

KÄRENLAMPI, L. and PELKONEN, M. (1971). Studies on the morphological variation of the lichen *Cladonia uncialis*. *Rept Kevo Subarctic Res. Stn* **7**, 47–56.

KERSHAW, K. A. and HARRIS, G. P. (1971). Simulation studies and ecology: use of the model. *In* "Statistical Ecology" (G. P. Patil, E. C. Pielou and W. E. Waters, eds), Vol. 3, pp. 23–42. Pennsylvania State University Press, University Park, Pennsylvania.

KERSHAW, K. A. and MILLBANK, J. W. (1969). A controlled environment lichen growth chamber. *Lichenologist* **4**, 83–87.

KERSHAW, K. A. and MILLBANK, J. W. (1970). Isidia as vegetative propagules in *Peltigera aphthosa* var. *variolosa* (Massal.) Thoms. *Lichenologist* **4**, 214–217.

KRISTINSSON, H. (1969). Chemical and morphological variation in the *Cetraria islandica* complex in Iceland. *Bryologist* **72**, 344–357.

KUROKAWA, S. (1962). A monograph of the genus *Anaptychia*. *Nova Hedwigia, Beih.* **6**, 1–115.

LAUNDON, J. R. (1958). The lichen vegetation of Bookham Common. *London Nat.* **37**, 66–79.

LAUNDON, J. R. (1965). Lichens new to the British flora: 3. *Lichenologist* **3**, 65–71.

LAUNDON, J. R. (1967). A study of the lichen flora of London. *Lichenologist* **3**, 277–327.

LAUNDON, J. R. (1970). Lichens new to the British flora: 4. *Lichenologist* **4**, 297–308.

LAUNDON, J. R. (1973). Urban lichen studies. *In* "Air Pollution and Lichens" (B. W. Ferry, M. S. Baddeley and D. L. Hawksworth, eds), 109–123. University of London, Athlone Press, London.

LOOMAN, J. (1964). Ecology of lichen and bryophyte communities in Saskatchewan. *Ecology* **45**, 481–491.

LUNAN, D. A. (1970). Lichenometric analysis in the Lake District. *Brycgstowe* [*J. Univ. Bristol Geogrl Soc.*] **5**, 56–57.

MURRAY, S. A. (1971). Modification of lichen substances and morphology induced by mechanical shock in *Cladonia pacifica*. *Experientia* **27**, 11–13.

PAULSON, R. and HASTINGS, S. (1914). A wandering lichen. *Knowledge* **37**, 319–323.

PEVELING, E. (1970). Die Darstellung der Oberflächenstrukturen von Flechten mit dem Raster-Elektronmikroskop. *Votr. Bot. Ges.* [*Dtsch. bot. Ges.*], N.F. **4**, 89–101.

PIŠÚT, I. and JELÍNKOVÁ, E. (1971). Über die Artberechtigung der Flechte *Lecanora conizaeoides* Nyl. ex Cromb. *Preslia* **43**, 254–257.

POELT, J. (1970). Das Konzept der Artenpaare bei den Flechten. *Votr. Bot. Ges.* [*Dtsch. bot. Ges.*], N.F. **4**, 187–198.

POELT, J. (1972). Die taxonomische Behandlung von Artenpaaren bei den Flechten. *Bot. Notiser* **125**, 77–81.

POELT, J. and STEINER, M. (1972). Über einige parasitische gelbe Arten der Flechtengattung *Acarospora* (Lecanorales, Acarosporaceae). *Annl. Naturhist. Mus. Wien* **75**, 163–172.

RICHARDSON, D. H. S. (1967). The transplantation of lichen thalli to solve some taxonomic problems in *Xanthoria parietina* (L.) Th. Fr. *Lichenologist* **3**, 386–391.

RICHARDSON, D. H. S. (1971). Lichens. *In* "Methods in Microbiology" (C. Booth, ed.), Vol. 3, pp. 267–293. Academic Press, London and New York.

RUNDEL, P. W. (1969). Clinal variation in the production of usnic acid in *Cladonia subtenuis* along light gradients. *Bryologist* **72**, 40–44.

RUNDEL, P. W. (1972). Notes on the ecology and chemistry of *Ramalina montagnei* and *R. denticulata* in the U.S. Virgin Islands. *Bryologist* **75**, 69–72.

RUNEMARK, H. (1956). Studies in *Rhizocarpon* I. Taxonomy of the yellow species in Europe. *Opera Bot.* **2(1)**, 1–152.
SANTESSON, R. (1967). On taxonomical and biological relations between lichens and non-lichenized fungi. *Bot. Notiser* **120**, 497–498.
SANTESSON, R. (1968). Lavar. Some aspects on lichen taxonomy. *Svensk Natur.* **1968**, 176–184.
SCHADE, A. (1966). Über die Artberechtigung der *Cladonia subrangiformis* Sandst. sowie das Auftreten von Calciumoxalat-Exkreten bei ihr und einigen anderen Flechten. *Nova Hedwigia* **11**, 285–308.
SCHADE, A. (1970). Über Herkunft und Vorkommen der Calciumoxalat-Exkrete in kortizikolen Parmeliaceen. *Nova Hedwigia* **19**, 159–186.
SERNANDER, R. (1907). Om några former för artoch varietetsbildundning hos lofvarna. *Svensk bot. Tidskr.* **1**, 135–186.
SHIBATA, S. and CHIANG, H. C. (1965). The structure of cryptochlorophaeic acid and merochlorophaeic acid. *Phytochemistry* **4**, 133–139.
SIPLE, P. A. (1938). The second Byrd Antarctic Expedition—Botany I. Ecology and geographical distribution. *Ann. Mo. bot. Gdn* **25**, 467–514.
STEBBINS, G. L. (1950). "Variation and Evolution in Plants." Columbia University Press, New York and London, 643 pp.
SWINSCOW, T. D. V. (1962). Pyrenocarpous lichens: 3. The genus *Porina* in the British Isles. *Lichenologist* **2**, 6–56.
SWINSCOW, T. D. V. (1965). Pyrenocarpous lichens: 8. The marine species of *Arthopyrenia* in the British Isles. *Lichenologist* **3**, 55–64.
SYERS, J. K., BERNIE, A. C. and MITCHELL, B. D. (1967). The calcium oxalate content of some lichens growing on limestone. *Lichenologist* **3**, 409–414.
THOMSON, J. W. (1948). Experiments upon the regeneration of certain species of *Peltigera*; and their relationship to the taxonomy of this genus. *Bull. Torrey bot. Club* **75**, 486–491.
THOMSON, J. W. (1968). *Haematomma lapponicum* Räs. in North America. *J. jap. Bot.* **43**, 305–310.
THOMSON, J. W. (1972). Distribution patterns of American arctic lichens. *Can. J. Bot.* **50**, 1135–1156.
TRASS, H. (1969). Floristical and phytogeographical study of the Estonian lichen-flora. *In* "Plant Taxonomy, Geography and Ecology in the Estonian S.S.R." (L. Laasimer, ed.), pp. 38–48. Valgus, Tallinn.
WEBER, W. A. (1962). Environmental modification and the taxonomy of the crustose lichens. *Svensk. bot. Tidskr.* **56**, 293–333.
WEBER, W. A. (1967). Environmental modification in crustose lichens II. Fruticose growth forms in *Aspicilia*. *Aquilo, ser. Bot.* **6**, 43–51.
WEBER, W. A. (1968). A taxonomic revision of *Acarospora*, subgenus *Xanthothallia*. *Lichenologist* **4**, 16–31.
WETHERBEE, R. (1969). Population studies in the chemical species of the *Cladonia chlorophaea* group. *Mich. Bot.* **8**, 170–174.
WETMORE, C. M. (1970). The lichen family Heppiaceae in North America. *Ann. Mo. bot. Gdn* **57**, 158–209.
ZAHLBRUCKNER, A. (1931). "Catalogus Lichenum Universalis," Vol. 7. Borntraeger, Leipzig. 784 pp.

3 | Adaptive Evolution in a Tropical-Alpine Environment

O. HEDBERG

Institute of Systematic Botany, University of Uppsala, Sweden

Abstract: Adaptive evolution can profitably be studied in extreme environments exerting strong selection pressures. Few environments can rival the harshness of the alpine belt in tropical mountains, as studied by the author in East Africa. Its climate combines intense insolation and high surface temperatures in daytime with equally intense outward radiation at night accompanied by frost and solifluction all the year round, and its flora is poor in species. Field studies of this afroalpine flora revealed diverse conspicuous morphological adaptations exploiting different ways of maintaining the water balance in a tropical-alpine environment. The plants possessing them can be grouped into a number of life-forms all of which also occur in other environments with similar climates. Adaptive evolution seems to have been accelerated here by the fact that most species occur in small and partly isolated populations of fluctuating size, and by strong selection pressures. Studies of this evolutionary process may facilitate our understanding of the origin and differentiation of the angiosperms.

INTRODUCTION

For a student of adaptive evolution it is natural to seek out an extreme environment with strong selection pressures where conspicuous adaptations are likely to be found, such as arctic or alpine habitats (cf. e.g. Bliss, 1962; Savile, 1972). A classical example of such conspicuous adaptation is offered by the genus *Soldanella* L. in central Europe; it has become adapted to long-lasting snow-cover by being able to grow at very low temperatures and by completing the development of its new shoots to such an extent during the preceding autumn so that it can start flowering even before the snow-cover has gone. Its early development is also facilitated, of course, by the "greenhouse effect" of the macrocrystalline snow (cf. Savile, 1972, p. 23). Another way of coping with long-lasting snow-cover is demonstrated by the Arctic annual species *Koenigia islandica* L., which through its extreme reduction is able to complete its whole life cycle in 4–6 weeks (Löve and Sarkar, 1957). An opposite ecological extreme is represented by *Carex glacialis* Mack. and other inhabitants of wind-swept hilltops in the alpine belt, which get no snow protection in winter but must withstand the full force of the winter gales and the abrasive action of drifting snow. In spite of these vicissitudes, primarily caused by the long and hard winter, most alpine and arctic habitats normally offer an uninterrupted growing season during the summer, however, varying in extent from a few weeks to several months.

THE AFROALPINE ENVIRONMENT

Climatic vicissitudes of a different dimension obtain at high levels on tropical mountains. Under natural conditions the vegetation of such mountains displays a regular zonation, as described for East Africa by Hedberg (1951). In their alpine belt seasonal temperature variations are almost negligible whereas diurnal variations are very conspicuous—in other words there is "summer every day and winter every night" (Hedberg, 1964a, p. 18). During much of the day the insolation tends to be very intense, whereas at night the outward radiation brings about a rapid temperature decrease with accompanying night frost, causing pronounced (micro-)solifluction phenomena on open soils. The afroalpine climate is cold enough for glaciers to exist on three of the highest East African mountains (Ruwenzori, Mt Kenya and Kilimanjaro; cf. Hedberg, 1951; see Plate I). The climatic and edaphical peculiarities of the afroalpine belt were described in Hedberg (1964a, b); concerning Mt Kenya see also Coe (1967). The violent changes between intense insolation in daytime and recurrent nightly frosts and solifluction evidently pose very difficult problems for plants and animals alike (for animals see also Salt (1954)). It is noteworthy that in spite of the richness of the flora at lower levels in the surroundings the number of vascular plant species occurring in the alpine belt of each East African mountain only amounts to between 70 and 150, and that most of these belong to alien flora elements (Hedberg, 1957, 1961, 1965).

MORPHOLOGICAL ADAPTATIONS IN AFROALPINE PLANTS

Looking at the afroalpine flora in the field one soon discovers a number of morphological features that are unknown or rare at lower levels on the mountains and in their surroundings (Plate I). The most conspicuous of these are

---

Plate I

Kenya, Mt Kenya, Teleki Valley. View of the central peaks from about 4150 m altitude. In the foreground trees of *Senecio keniodendron* R. E. Fr. & Th. Fr. jr.; a little further in *Festuca pilgeri* grassland, with scattered inflorescences of *Lobelia telekii* Schweinf. Four glaciers are visible on the peak. Photograph O. Hedberg, July 1948.

3. Adaptive Evolution in a Tropical-Alpine Environment

giant leaf rosettes, dense grass tussocks, acaulescent growth, dense cushion growth, and sclerophylly, and I shall offer a few examples and observations for each of these.

1. Giant Leaf Rosettes

The most famous plants of the high East African mountains are doubtless the Giant Senecios and Giant Lobelias. These "megaphytic" plants have sparsely branched or unbranched woody or herbaceous stems each terminated by a permanently growing huge leaf rosette. A worthy representative is *Senecio keniodendron* R. E. Fr. & Th. Fr. jr. (Plate I, foreground), the rosettes of which are about 1 m in diameter. A cross-section of such a rosette (Plate IIA) demonstrates that the shoot apex is surrounded by a layer of young leaves and leaf primordia more than 10 cm thick, which apparently provide good shelter against environmental changes. This protection is further enhanced by the fact that at night most leaves fold up around the centre so that the rosette forms a giant "night bud" which unfolds again at sunrise. Early morning temperature readings gave a value of $+2°C$ at the surface of the dense central cone while the air temperature among the outermost leaves was $-4°C$. Similar giant leaf rosettes folding up at night and unfolding in the morning occur in the other alpine species of the subgenus *Dendrosenecio* Hauman ex Hedb. Equally impressive are the rosettes in *Lobelia telekii* Schweinf. which in the daytime appear quite dense (Plate IIIB), whereas at night they fold up entirely and become cabbage-like (Plate IIIA). Temperature readings demonstrate that these rosettes

---

PLATE II

A. Young tree of *Senecio keniodendron* R. E. Fr. & Th. Fr. jr. in longitudinal section. The tree is about 1·7 m tall and carries a huge terminal leaf rosette, in the centre of which appears a young inflorescence surrounded by a mass of densely folded young leaves. The stem contains a thick pith (greyish in the photo) surrounded by a thin layer of secondary wood (white), and a thin cortex (grey), outside which there is a dense insulating mantle of marcescent dry leaves. In this plant the water-conducting tissues of the stem as well as the shoot apex and the young inflorescence are efficiently protected against the recurrent night frosts. Kenya, Mt Kenya, Teleki Valley, 4200 m. Photograph O. Hedberg, 1948. (From Hedberg, 1964a.)

B. Grove of *Senecio erici-rosenii* R. E. Fr. & Th. Fr. jr. at about 3400 m altitude in a ravine on the W. side of Muhavura, Kigezi Distr., Uganda. In this and other species of the ericaceous belt the stems are thin and often comparatively richly branched, and the leaf rosettes are comparatively lax with caducous leaves. Photograph O. Hedberg, 1948. (From Hedberg, 1964a.)

also have an excellent capacity for temperature insulation. An even more elaborate method to achieve this has been developed in *Lobelia keniensis* R. E. Fr. & Th. Fr. jr. and the other species of the *Lobelia deckenii* group. Their leaf rosettes are normally filled with water, secreted from the leaf bases and reaching to the top of the young leaves in the centre (Plate IIIc). In the evening the rosettes fold up and if the night is cold a thin layer of the water will freeze; but the shoot apex is situated at the bottom of the water reservoir and is efficiently protected against frost.

In most Giant Senecios and Giant Lobelias the stem is well protected against violent temperature changes, either by a cylinder of marcescent dry leaves as in *Senecio keniodendron* R. E. Fr. & Th. Fr. jr. (Plate IIA), or by a thick corky cortex, as in *Senecio barbatipes* Hedb., which means that water transport through it may proceed unimpeded by freezing. Temperature readings indicate that the stem is kept constantly at temperature well above zero even if at night the surrounding air is colder than $-5°C$.

2. Dense Grass Tussocks

Another very characteristic feature of the afroalpine vegetation is the appearance of the afroalpine grasses, most of which form large and dense tussocks of a dull brownish-green colour (cf. Plates I, IIA, IVA). A closer inspection reveals that most leaves and culms of such a tussock are dead and decaying. The living parts tend to be largely concentrated to the centre of the tussocks, where they are protected by a felt-like mass of decaying leaf- and culm-bases which

---

PLATE III

A. Leaf rosette of *Lobelia telekii* Schweinf. in "night position", with tightly folded leaves, photographed immediately after sunrise. The leaves appeared quite stiff when touched and were densely covered by hoar frost along the margins; the air temperature was still a few degrees centigrade below freezing. Width of rosette *ca.* 0·4 m. Kenya, Mt Kenya, Teleki Valley, 4200 m. Photograph O. Hedberg, 1948. (From Hedberg, 1964a.)
B. Leaf rosette of the same species in "day position". Most of the leaves are unfolded, but the youngest ones are firmly pressed together around the shoot apex. Width of rosette *ca.* 0·5 m. Kenya, Mt Kenya, Teleki Valley, 4200 m. Photograph O. Hedberg, 1948. (From Hedberg, 1964a.)
C. Leaf rosette of *Lobelia keniensis* R. E. Fr. & Th. Fr. jr. in "day position" with patent leaves. Between the central leaves appears the surface of the "water reservoir" described in the text. Diameter of rosette *ca.* 0·5 m. Kenya, Mt Kenya, Teleki Valley, 4200 m. Photograph O. Hedberg 1948. (From Hedberg, 1964a.)

provide excellent temperature insulation for the innovation shoots in the centre. Early morning temperature measurements gave a value of +2·5°C for the centre of a tussock while the temperature at its surface was still −5°C.

3. Acaulescent and Prostrate Growth

A third peculiarity of the afroalpine flora is that in a number of species of diverse groups the elongation of the flowering stems is impeded so that they become acaulescent. A good example is afforded by *Carduus chamaecephalus* (Vatke) Oliv. & Hiern, the capitula of which are crowded at the centre of the leaf-rosette. The same habit occurs in occasional species of other genera, such as *Oreophyton falcatum* (A. Rich.) O. E. Schulz (Cruciferae, Plate VA), *Haplosciadium abyssinicum* Hochst. (Umbelliferae), and *Wahlenbergia pusilla* Hochst. ex A. Rich. (Campanulaceae). These plants evidently exploit the congenial day temperature at the soil surface at the same time as the conduction of water to their inflorescences is facilitated by the brevity of the stem and by its protected situation in the centre of the leaf-rosette. The prostrate stems occurring in species of amongst others *Trifolium* (Plate VC) and *Veronica* are apparently analogous adaptations to exploit the favourable day temperatures at the soil-surface.

4. Dense Cushions

An ecological parallel to the acaulescent and prostrate growth forms mentioned above is provided by species like *Agrostis sclerophylla* C. E. Hubb., *Sagina afroalpina* Hedb. and *Myosotis keniensis* Th. Fr. jr. These form dense cushions consisting of richly branched interwoven stems and roots with the surface covered by densely aggregated leaves (Plate IVB). These cushion plants also exploit the favourable microclimate at the soil-surface during the day while their compact growth contributes to protect their water-conducting tissues against frost.

PLATE IV

A. Dense grass tussocks formed by *Festuca pilgeri* St-Y. subsp. *supina* (Pilg. ex St-Y.) Hedb. on open solifluction soil. Height of culms *ca.* 0·4 m. Tanzania, Kilimanjaro, in the Saddle between Kibo and Mawenzi, 4450 m. Photograph O. Hedberg, June 1948.
B. Large tussock of *Sagina afroalpina* Hedb. in longitudinal section, showing central soil core. On moist soil with locally intense solifluction along a streamlet. Kenya, Mt Kenya, Teleki Valley, 4200 m. Photograph O. Hedberg, 1948. (From Hedberg, 1964a.)

5. Sclerophyllous Growth

The only woody plants of the afroalpine belt except the Giant Senecios (and two of the Giant Lobelias) are a number of thin-stemmed, sclerophyllous shrubs. Species with this growth form are fairly numerous here (as well as in the ericaceous belt lower down) and occur in a number of genera, such as *Philippia* and *Blaeria* (Ericaceae), *Bartsia* (Scrophulariaceae), *Anthospermum* (Rubiaceae), *Helichrysum*, *Stoebe* and *Euryops* (Compositae, cf. Plate V<small>D</small>). The thin stems of these plants have no efficient protection against frost, but since their leaves are small and xeromorphic they may evidently stand nightly freezing with complete cessation of water conduction without suffering any damage.

AFROALPINE LIFE-FORMS OF PLANTS

Students of climatic adaptations in plants have often attempted to classify the species studied into physiognomic life-forms, each of which is believed to result from parallel climatic adaptations in different taxonomic groups of plants. The most widely known life-form system was devised by Raunkiaer (1907, 1908, 1934, etc.) and was based upon "the adaptation of the plant to survive the unfavourable season, having special regard to the protection of the surviving buds or shoot-apices" (Raunkiaer, 1934, p. 112). In a tropical-alpine climate where the unfavourable season is neither winter nor dry season but the night Raunkiaer's life-form system is not so easily applied. The giant night buds of *Senecio keniodendron* cannot unhesitatingly be compared with the winter buds of *Fagus* or *Anemone*. But the five morphological types I have just described do evidently represent five distinct life-forms well adapted to tropical-alpine conditions (Fig. 1), namely giant rosette plants, tussock grasses, acaulescent

---

PLATE V

A. *Oreophyton falcatum* (A. Rich.) O. E. Schultz. 75% natural size. Uganda, Elgon, on the W. part of the crater rim, in rock crevice. Photo Å. Holm, May 1948. (From Hedberg, 1957.)

B. *Dianthoseris schimperi* Sch. Bip. ex A. Rich. Specimen grown in a greenhouse, maintaining under these conditions its acaulescent habit. About natural size. Material from Teleki Valley on Mt Kenya. Photograph O. Hedberg, 1959. (From Hedberg, 1964a.)

C. A fruiting specimen of *Trifolium elgonense* Gillett, with prostrate stem and sessile one-flowered inflorescences. Uganda, Elgon, on thin soil near the track just W. of the western entrance to the crater, *ca.* 3600 m. Photograph O. Hedberg, 1967.

D. *Euryops dacrydioides* Oliv., top of flowering shrub. Note the scale-like xeromorphic leaves. Tanganyika, Kilimanjaro, S. slope of the Saddle above Peter's Hut, 4200 m. Photograph O. Hedberg, 1948. (From Hedberg, 1964a.)

rosette plants, cushion plants, and sclerophyllous shrubs (cf. Hedberg, 1964a, b). From a physiological point of view these adaptations apparently represent different ways of maintaining the water balance in a climate with recurrent night frosts all the year round ("Frostwechselklima" in the sense of Troll, 1943; cf. Hedberg, 1964a).

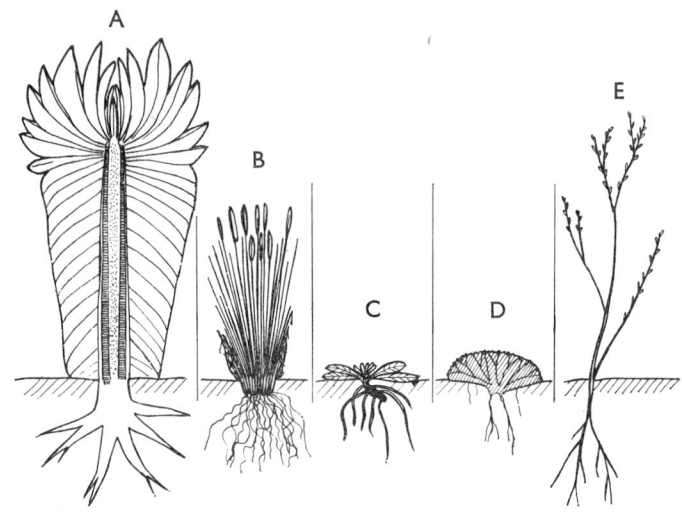

FIG. 1. Diagrammatic drawings of the five most important phanerogamic life-forms of the afroalpine belt. A. Giant rosette plant; B. Tussock grass; C. Acaulescent rosette plant; D. Cushion plant; E. Sclerophyllous shrub. (From Hedberg, 1964b.)

Admittedly the five life-forms I have described may be equally suited to "Frostwechselklimas" in other parts of the world, and some of them may be matched also in other types of climate, as the tussock grasses. A few of the relevant species may therefore have been pre-adapted to this type of climate before they became members of the afroalpine flora. But in view of the high degree of endemism in this flora and its marked geographic isolation from other areas with "temperate" climate any large scale exchange with other parts of the world seems rather unlikely (cf. Hedberg, 1961, 1969, 1970). Furthermore, several of the species concerned belong to groups endemic to the East African mountains, or at least have their closest relatives at lower levels on the same mountains, and may hence be expected to have developed their tropical-alpine adaptations in this part of the world. A few examples will be given below for each of the life-forms concerned.

3. Adaptive Evolution in a Tropical-Alpine Environment

EXAMPLES OF PRESUMED AUTOCHTHONOUS ORIGIN OF AFROALPINE LIFE-FORMS

1. Giant Rosette Plants

The most celebrated example of autochthonous origin is furnished by the giant groundsels, the subgenus *Dendrosenecio* being endemic to East African high mountains (Fig. 2). This group is particularly illuminating because it occurs in a

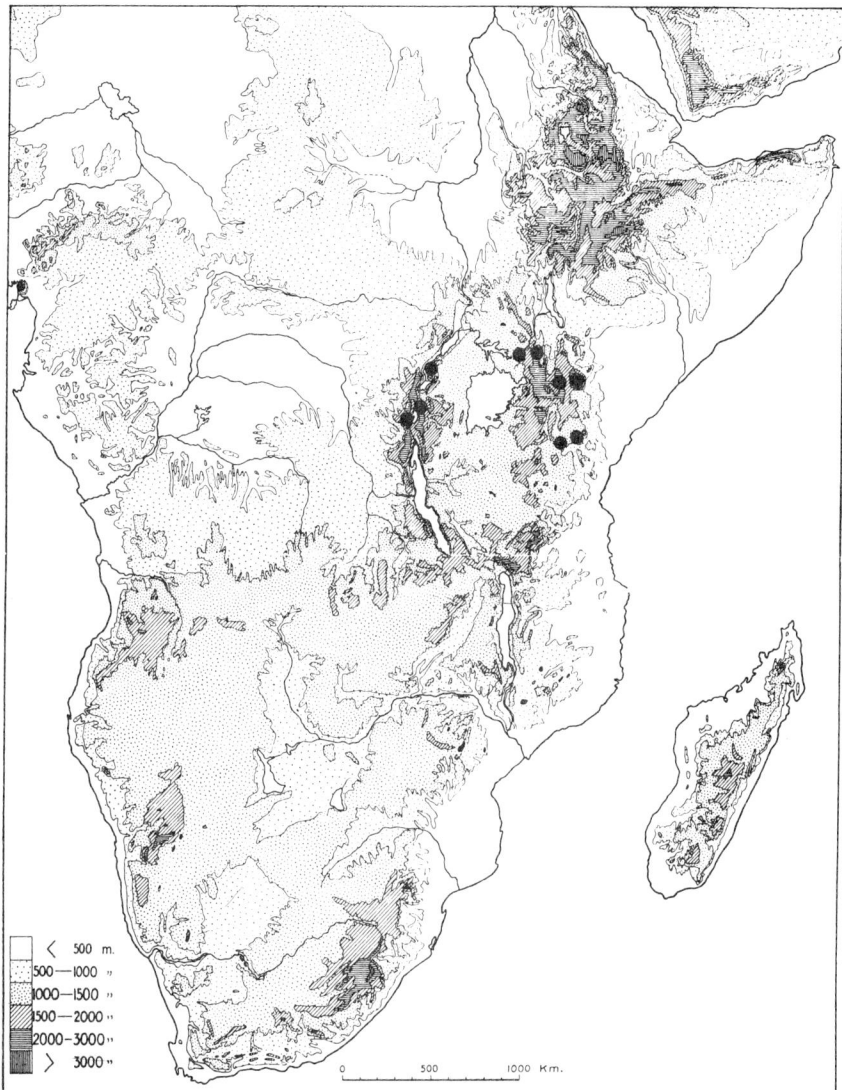

FIG. 2. Distribution map for *Senecio* L. subgenus *Dendrosenecio* Hauman ex Hedb. (From Hedberg, 1964b.)

number of geographically and altitudinally vicarious species, resulting from parallel adaptive evolution superimposed upon random differentiation of geographically isolated populations (cf. Hedberg, 1969).

Low-level taxa are thin-stemmed with lax and open leaf rosettes, thin leaves and relatively small capitula (Plate IIB), whereas the high-level species have thick and compact leaf rosettes folding up at night and formed by coriaceous leaves which in most species remain attached to the stem long after they have died (Plates I and IIA). Equally striking examples of progressive adaptation to tropical-alpine conditions are found among the Giant Lobelias, the low-level species of which have thin and sometimes branched stems with lax inflorescences whereas species from high levels have thick stems and very compact leaf rosettes of the sort described above (Plate IIIA, B), culminating in the water-filled rosettes and dense inflorescences of the *Lobelia deckenii* group (Plates IIIc and VIA).

2. Tussock Grasses

Other good examples of progressive adaptation to tropical-alpine conditions are found among the tussock grasses. Thus *Festuca abyssinica* Hochst. ex A. Rich. and *Poa leptoclada* Hochst. ex A. Rich. at low levels often occur in solitary culms or loosely tufted, whereas at high levels they form very dense tussocks. *Festuca pilgeri* St-Y. and *Poa kilimanjarica* (Hedb.) Mgf-Dbg., on the other hand, always seem to form dense tussocks.

3. Acaulescent Rosette Plants

A good example of an acaulescent rosette plant likely to have arisen in this environment is *Dianthoseris schimperi* Sch. Bip. ex A. Rich. (Compositae—Plate VB), the only species of a genus entirely restricted to the afroalpine belt. Its closest relationship appears to be with *Crepis* L., which is also represented on

---

PLATE VI

A. Flowering specimen of *Lobelia keniensis* R. E. Fr. & Th. Fr. jr. The flowers are barely visible between the bracts in the compact spike-like inflorescence. Height of specimen *ca.* 1·3 m. Kenya, Mt Kenya, W. slope in the lower part of the alpine belt, about 3700 m above sea level. Photograph O. Hedberg, 1948.

B. Flowering specimen of *Puya raimondii* Harms with compact spike-like inflorescence. This South American giant rosette plant forms a beautiful ecological parallel to the Giant Lobelias of East Africa. Note also the similarity of the grass tussocks in this photograph to those in Plate IV A. N.-Central Peru, valley of Rio Ingemo (Calleyon de Huaylas) 4200 m, June 1969. Photograph G. Pontecorvo.

these mountains. Two other monotypic genera restricted to the upper parts of the high East African mountains which have adopted this life form are *Oreophyton* O. E. Schulz (Cruciferae, Plate VA) and *Haplosciadium* Hochst. (Umbelliferae). Occasional acaulescent rosette plants further occur in the genera *Wahlenbergia* Schrad. (Campanulaceae) and *Carduus* L. (Compositae). There are also good examples of progressive adaptations involving prostrate growth. Thus *Veronica gunae* Schweinf. ex Engl. is entirely prostrate, whereas its close relative *V. glandulosa* Hochst. ex Bth. is erect or climbing. Of particular interest in this connection is the genus *Trifolium* L., two afroalpine species of which (*T. acaule* Steud. ex A. Rich. and *T. elgonense* Gillett) combine prostrate growth with sessile inflorescences (Plate Vc).

4. Cushion Plants

The *Trifolium* species mentioned above form a transition to the cushion plants, which are characterized by profusely branched stems with densely compressed branches, each of which has very brief internodes and is terminated by a leaf rosette. A prerequisite for good cushion growth is a particular kind of profuse branching with branches of comparable dimensions (Rauh, 1939, p. 483), and the capacity for this type of branching is apparently rare in the afroalpine flora. Of the three species of cushion plants quoted above *Sagina afroalpina*, at least, appears to have developed its adaptations autochthonously, and the same holds good for the less "perfect" cushion plant *Swertia subnivalis* Th. Fr. jr. Their closest relatives appear to be *Sagina abyssinica* Hochst. ex A. Rich. and *Swertia volkensii* Gilg, respectively.

5. Sclerophyllous Shrubs

The least exclusive of the afroalpine life-forms is the sclerophyllous shrub, since morphologically almost identical types occur in several other types of climate. For most afroalpines belonging here it is therefore difficult to establish whether this habit has developed on the tropical African mountains or elsewhere. Perhaps the best example of autochthonous origin is provided by the shrubby *Alchemilla* species like *A. elgonensis* Mildbr., *A. argyrophylla* Oliv., and *A. johnstonii* Oliv.

ZOOLOGICAL ADAPTATIONS

It must also be emphasized that these botanical adaptations to the afroalpine environment have their zoological counterparts. Numerous invertebrates have become adapted to a "cryptozoic" life under stones and among leaves, since in the open they are either frozen stiff at night or heated to death by intense sunshine during the early part of the day (Salt, 1954). Thus a number of spider taxa appear to be largely confined to the ameliorated microclimatic niche provided by the dry leaf-cylinders of the Giant Senecios, and they have developed a number of vicarious taxa paralleling those of *Dendrosenecio* (Holm, 1962). The sunbird *Nectarinia johnstonii* Shelley seems to feed largely on the flowers of the Giant Lobelias and often builds its nests in the temperature-insulating tussocks of *Festuca pilgeri*, etc., whereas the rock hyrax has extensive underground dwellings (cf. Coe, 1967). Further examples of zoological adaptations are given by Salt (1954).

COMPARISON WITH TROPICAL-ALPINE LIFE-FORMS ELSEWHERE

Strong additional support for the contention that the five life-forms described above result from ecological adaptation to the afroalpine climate is furnished by the fact that exactly the same life-forms occur in the climatically equivalent paramo of the South American Andes, although represented there by other genera and families of plants. A beautiful parallel to the Giant Senecios is furnished by the *Espletia* species of the paramo, in the same way as the Giant Lobelias are paralleled by the Bromeliad *Puya raimondii* Harms (Plate VI). Big and dense grass tussocks exactly mimicking those of the afroalpine belt are characteristic of the paramo, although they are formed by other genera of grasses (cf. Plate VIB).

Moreover acaulescent rosette plants, cushion plants and sclerophyllous shrubs are very conspicuous in the South American paramo (cf. Cuatrecasas, 1934, 1958). In spite of the almost total floristic dissimilarity between afroalpine vegetation and the paramo, their dominant life-forms are so closely similar that they must be ascribed to parallel adaptive evolution in response to tropical-alpine conditions (cf. Hedberg, 1964a).

RATE AND MODE OF ADAPTIVE EVOLUTION IN THE AFROALPINE ENVIRONMENT

The differentiation of vicarious taxa in the afroalpine flora has earlier been demonstrated to have proceeded comparatively fast, and is assumed to have resulted largely from natural selection in connection with genetic drift acting upon geographically isolated and originally often very small samples of the gene-pools concerned (Hedberg, 1969). In combination with the strong selection pressure exerted by the rigorous tropical-alpine climate this will also have promoted rapid adaptive evolution, which seems to have proceeded, in many cases, parallel to the differentiation of geographically vicarious populations, as in the Giant Senecios already mentioned. One case where differential adaptive evolution has occurred in combination with geographical isolation seems to be the species pair *Lobelia wollastonii* E. G. Bak.—*L. telekii* Schweinf. (Hedberg, 1964a, p. 144).

The large changes in population size likely to have been caused by volcanic eruptions, and Pleistocene changes of climate, may also have contributed to promote rapid evolution (cf. Hedberg, 1970), although the extent of changes in vegetation zonation on the mountains during the Pleistocene may at times have been over-estimated (Hedberg, 1972). The volcanic eruptions may also have catalysed evolution by increased mutation frequency caused by heat shocks (Petterson, 1961). The rate of evolution differs appreciably in different taxonomic groups, and is largely connected with differences in genetic system (cf. Hedberg, 1957, p. 376; 1969, p. 143).

ADAPTIVE EVOLUTION IN TROPICAL MOUNTAINS AS A KEY TO THE UNDERSTANDING OF ANGIOSPERM EVOLUTION

In conclusion I wish to point out that studies of adaptive evolution on tropical mountains may be of interest not only for the understanding of morphological peculiarities and evolution in mountain plants but also in a wider context. Most modern students of paleobotany and angiosperm evolution seem to agree that "the angiosperms originated and for a rather prolonged period evolved under montane tropical conditions" (Takhtajan, 1969, p. 164). Their evolution and early differentiation are believed to have been facilitated by the multitude of ecological niches coexisting in a mountain habitat, and by their distribution in small and partly isolated populations, leading to genetic drift and adaptive radiation. "Even in the very earliest stages of their ecological evolutions the angiosperms must have been adapted to the various altitudinal belts, with their various climatic and edaphical conditions" (*op. cit.*, p. 168). The angiospermous flora invading temperate latitudes during the Cretaceous expansion would therefore have been pre-adapted at temperate altitudes on tropical and subtropical mountains. The primary centre of origin for the temperate flora of the northern hemisphere is considered by Takhtajan to be S. E. Asia, around the eastern end of the Himalayas, where a huge and complicated system of mountains has contributed excellent conditions for both differentiation and migration of mountain plants. For obvious reasons it is impossible to trace the early differentiation of the angiosperms in any detail: in the almost complete absence of relevant fossils we have to rely in the main on morphological and phytogeographical comparisons. Some appreciation of the mode of origin and differentiation of the earliest angiosperms may nevertheless be gained by detailed studies of those groups of recent plants where species apparently descending from a montane stock have managed to acquire a wide distribution at lower altitudes. Even if the gene-pools of this material differ very much from those of the angiosperm progenitors, and the competition in the lowland presumably is much harder than at the time of the Cretaceous expansion, the mechanisms of evolution and migration must be the same.

A good example of comparatively late invasion of lowland areas by a plant species that arose in subtropical high mountains is furnished by the genus *Koenigia* L., which I am currently revising. Five species out of the six belonging to this genus are confined to high mountain areas around the Eastern Himalayas, where they are adapted to different altitudes. Only the sixth one, *Koenigia islandica* L., has, by virtue of its extreme reduction and adaptation to a short vegetation period, also been able to achieve a wide distribution in the Arctic and on high mountains at high latitudes; it has even penetrated to southern South America (Fig. 3).

The pronounced geographical isolation of the high East African mountains makes it much less likely to find an analogous example here, and I cannot confidently state any case where a group primarily adapted to afroalpine conditions has later spread to temperate areas at other latitudes. A possible one might, of course, be part of the Ericaceae—the genera *Philippia* Klotzsch and *Blaeria* L. seem to have their centre of diversity in the East African mountains, and the only tree-forming species of *Erica* L. are widely distributed here with *Erica arborea* L. extending to the Mediterranean area (see distribution map in Hedberg et al., 1961, Fig. 21). The proliferation of small shrubby Ericas in South Africa (cf. Baker and Oliver, 1967) may represent a secondary centre of diversity. But this case obviously needs much more study.

Takhtajan and other students of angiosperm evolution have concentrated most of their interest on the montane forests as the most likely habitat for angiosperm evolution. Without denying the importance of these forests I would like also to focus some attention on higher altitudes. Where would conditions for evolution of herbaceous plants have been more favourable than above the forest limit? Such tropical-alpine adaptations as I have described here might of course be expected to represent blind alleys without further evolutionary possibilities but in some cases at least they may have pre-adapted the plants possessing them to life in other climates at different latitudes. The climatic re-adaptations required may not always present too formidable obstacles—thus the manifestation of winter hardiness versus night hardiness in hybrids between strains of *Arabis alpina* L. from Mt Kenya and from northern Sweden segregated in a very simple manner (Hedberg, 1962). I therefore submit that further detailed studies of adaptive evolution in tropical and subtropical mountains may contribute substantially towards our understanding of "that abominable mystery, the origin of flowering plants".

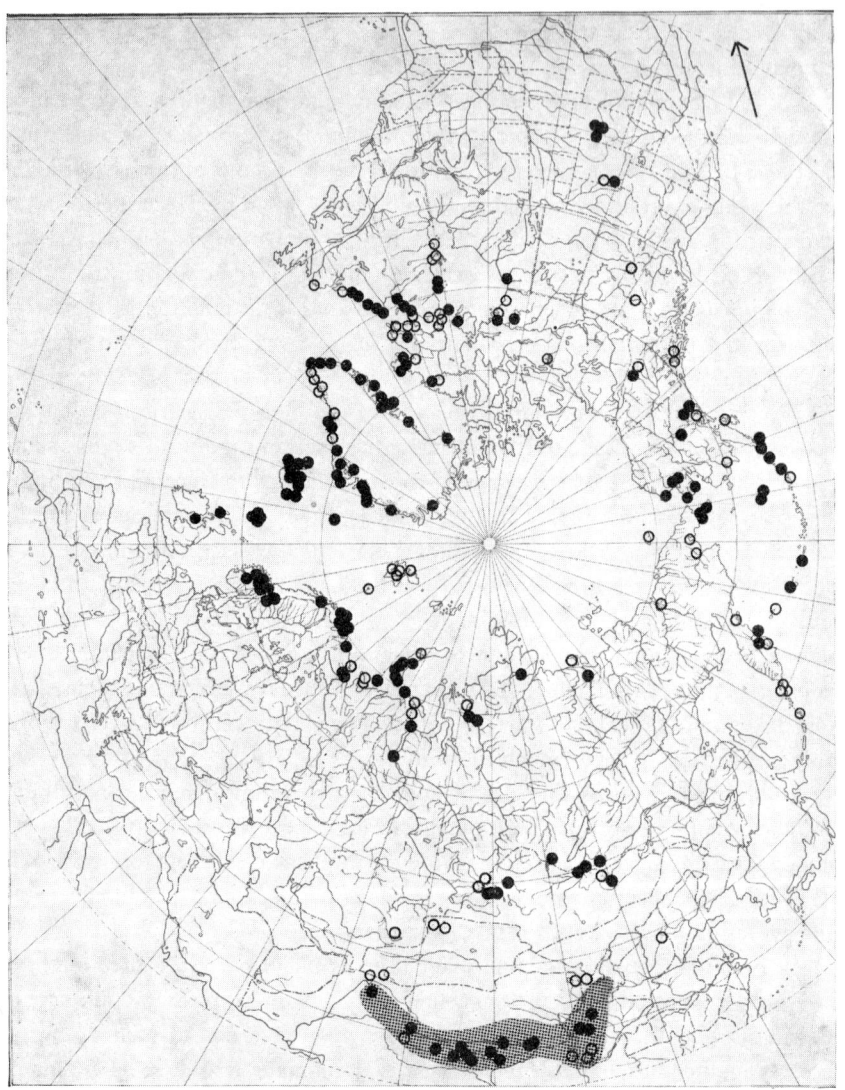

FIG. 3. Distribution map for the genus *Koenigia* L., based on herbarium material seen by the author. The dots refer to verified localities for *Koenigia islandica* L., the open rings to imprecise localities for the same species, and the hatching in Southern Asia denotes the area of the remaining five species of the genus. *K. islandica* also has an isolated enclave in Tierra del Fuego (indicated by an arrow).

REFERENCES

Baker, H. A. and Oliver, E. G. H. (1967). "Ericas in Southern Africa." Cape Town.
Bliss, L. C. (1962). Adaptations of arctic and alpine plants to environmental conditions. *Arctic* **15**, 117–144.
Coe, M. J. (1967). The ecology of the alpine zone of Mt Kenya. *Monogr. Biol.* **17** (136 pp). Junk, The Hague.
Cuatrecasas, J. (1934). Observaciones geobotánicas en Colombia. *Trab. Mus. Nacional Ciencias Naturales Madrid, Ser. Bot.* **27**, 1–144.
Cuatrecasas, J. (1958). Aspectos de la vegetación natural de Colombia. *Rev. Acad. Colombiana Cienc. Exactas* **10**(40), 221–268, pls. I–XXXV.
Hedberg, O. (1951). Vegetation belts of the East African mountains. *Sv. Bot. Tidskr.* **45**, 140–202. Uppsala.
Hedberg, O. (1957). Afroalpine vascular plants. A taxonomic revision. *Symb. bot. upsal.* **15**(1), 1–411.
Hedberg, O. (1961). The phytogeographical position of the afroalpine flora. *Rec. Adv. Bot.* **1**, 914–919. Toronto.
Hedberg, O. (1962). Intercontinental crosses in Arabis alpina L. *Caryologia* **15**, 253–260.
Hedberg, O. (1964a). Features of afroalpine plant ecology. *Acta phytogeogr. suec.* **49**, 144 pp. Uppsala.
Hedberg, O. (1964b). Etudes écologiques de la Flore afroalpine. *Bull. Soc. Bot. Belg.* **97**, 5–18.
Hedberg, O. (1965). Afroalpine flora elements. *Webbia* **19**, 519–529.
Hedberg, O. (1969). Evolution and speciation in a tropical high mountain flora. *J. Linn. Soc. (Biol.)* **1**, 135–148.
Hedberg, O. (1970). Evolution of the afroalpine flora. *Biotropica* **2**, 16–23.
Hedberg, O. (1972). Pollen-analytical evidence on the Quaternary history of the flora of Tropical Africa. *Birbal Sahni Inst. Proc.* (in press).
Hedberg, O. et al. (1961). Monograph of the genus *Canarina* L. (Campanulaceae). *Svensk Bot. Tidskr.* **55**, 17–62.
Holm, Å. (1962). The spider fauna of the East African mountains. *Zool. Bidr. Uppsala* **35**, 19–204.
Löve, A. and Sarkar, P. (1957). Heat tolerance of *Koenigia islandica*. *Bot. Notiser* **110**, 478–481.
Petterson, B. (1961). Mutagenic effects of radiant heat shocks on phanerogamous plants. *Nature, Lond.* **191** (4794), 1167–1169.
Rauh, W. (1939). Über polsterförmigen Wuchs. In "Nova Acta Leopoldina," N.F. **7**(49), 267–508, Taf. 23–44. Halle (Saale).
Raunkiaer, C. (1907). "Planterigets livsformer og deres betydning for geografien." Kjobenhavn and Kristiania.
Raunkiaer, C. (1908). Livsformernes Statistik som Grundlag for biologisk Plantegeografi. *Bot. Tidsskr.* **29**, 42–83.
Raunkiaer, C. (1934). "The Life Forms of Plants and Statistical Plant Geography." Oxford.
Salt, G. (1954). A contribution to the ecology of upper Kilimanjaro. *J. Ecol.* **42**, 375–423.

SAVILE, D. B. O. (1972). "Arctic Adaptations in Plants." Canada Dept. Agric. Res. Branch Monogr. **6** (81 pp). Ottawa.

TAKHTAJAN, A. (1969). "Flowering Plants—Origin and Dispersal." Oliver and Boyd, Edinburgh.

TROLL, C. (1943). Die Frostwechselhäufigkeit in den Luft- und Bodenklimaten der Erde. *Meteorol. Z.* **60,** 161–171.

4 | Seed Germination in Relation to Ecological and Geographical Distribution

P. A. THOMPSON

Jodrell Laboratory, Royal Botanic Gardens, Wakehurst Place, Sussex, England

Abstract: Germination represents a critical event in a plant's life cycle and its timing largely predetermines the chances of survival of a seedling to maturity. Amongst mechanisms responsible for coordinating germination with physical parameters of the environment two of the most important are the temperature response characteristics and the proportion of seeds with dormancy restrictions on growth of one kind or another.

Characteristics of the temperature response may be closely correlated with the geographical distribution of different species. Comparisons between populations may reveal considerable variation from one part of a range to another, or only minor differences over extensive geographical areas. It would appear that adaptive evolution of variations in temperature responses are constrained by the overall genotype of a species, which may restrict variation within narrow limits. A few species, notable for their adaptive capacity in other directions, expressed by morphological diversity or habitat exploitation, also display major variations between the germination responses to temperature of different populations.

Responses to temperature are modulated by the presence of a proportion of dormant individuals which require pretreatments before they will germinate. This proportion may be highly variable from one population to another, from plant to plant within a population, or even from capsule to capsule within a plant. It is arguable that variations in dormancy allied with relatively small changes in germination responses to temperature constitute the major components of the adaptive mechanisms responsible for the delimitation of range and habitat preferences typical of the ecological requirements of many species.

Germination is an event which marks the transition from the relatively safe state of the dormant embryo protected within the seed coat to the metabolically highly active and vulnerable form of the young seedling. The embryo within many seeds is capable of surviving for long periods under conditions which include extremes of heat, cold or drought. The young seedling, however, is typically dependent for survival on narrowly defined limits of moisture tension and temperature, and adverse conditions lasting only a few hours may cause death. Thus the season of the year in which the seed germinates acts as a predisposing factor influencing, and sometimes determining, the ability of the seedling to survive to maturity. The mechanisms regulating germination are therefore of major importance amongst the many processes which constitute the plant's adaptation to its environment; the survival of a species in a particular locality depends on its possession of genetically controlled responses within the seed which react to external stimuli in such a way that, when germination occurs, it does so at a season which results in the survival to maturity of a proportion of the seedlings.

The most favourable times of the year for germination vary from one geographical locality to another depending on the climatic parameters of the seasons in different parts of the world. They will vary also according to the nature of the life cycle of different species, and adaptive features of their anatomy, morphology and physiology. Thus within a particular locality Went and his colleagues (Went, 1949; Went and Westergaard, 1949; Went et al., 1952) have shown how the annual flora of Californian desert areas is divided into winter and summer components separated by differences in the temperatures at which they germinate and the photoperiodic adaptations and temperature tolerances of the developing plants. Mayer and Poljakoff-Mayber (1963) in a review of the ecology of germination cite examples of responses in a variety of species which appear to be directly correlated with features of their natural habitats. They also stress the complexity of interactions between the endogenous mechanisms which control germination and the exogenous stimuli produced by the seed's environment, and comment on the difficulty of relating observed responses in laboratory experiments to the situation of the seed under natural conditions.

Many factors regulate the timing of germination but amongst the most important are, firstly, the characteristics of the ambient daily temperature cycles and, secondly, the endogenous processes which regulate the nature and level of dormancy within the seed. Reviews of the interactions of these two with each other and with other factors controlling germination including moisture tensions, light and mineral nutrition have been published by Vegis (1965), Lang (1965) and Stokes (1965). These and other reviews, and many publications, document the ways in which germination responses have been defined in laboratory experiments. It is reasonable to suppose that the same factors act to regulate germination under natural conditions, but the relative importance and course of action of germination regulators may be quite different under field conditions from those found in the laboratory. Examples of such differences have been reported by Wesson and Wareing (1969) for the light responses of a number of British wild plants, and by Salisbury (1965) and

Morley (1958) for the progression of the post-harvest after-ripening changes which affect dormancy. Thus although it may now be possible to list and categorize many of the responses by which germination is controlled both in the laboratory and in the field it must be recognized that there is still little information on the ways in which these controls interact with the changing physical features of the seed's natural environment to regulate the timing of germination.

In certain regions such as deserts or geographical areas with definable and consistent seasons, which include extreme conditions favouring or damning seedling survival, correlations between germination responses in the laboratory and natural conditions may appear to be fairly evident. For example, a number of reports including those of Went (1957), Koller (1955) and Koller and Negbi (1959) present evidence for the presence of water-soluble inhibitors in the seed coats of desert plants, which require minimum levels of precipitation for dissolution and which, it can be argued, would act under natural conditions to prevent germination in response to periods of rain insufficient to support the development of seedlings to maturity. Datta *et al.* (1970) have described variations in the germination behaviour of the caryopses of *Aegilops ovata* L. depending on their position on the inflorescence and the presence or absence of inhibitors, and have correlated these variations, and the range of temperatures favouring germination, with the natural occurrence of the species as a weedy annual in Israel. In areas with less well defined climatic features it may be more difficult to identify those aspects of the germination response which act to synchronize germination with particular seasons of the year, and which therefore play a considerable part in the sum of the processes determining the ability of a population to survive in a particular locality.

Within a relatively restricted area such as Europe it is possible to define a number of phyto-geographical zones and to distinguish one zone from another by criteria based on climatic features and the compositions of their floras. Similarly, on a more local scale ecological habitats may be characterized by the presence or absence of particular species or types of plant community. Yet within Europe physical barriers to plant distribution are relatively minor, and between ecological habitats they are often practically negligible, so that in most cases it is arguable that the occurrence or absence of a population of a species in a particular locality depends very little on opportunities provided by the chances of dispersal over long distances, but is a measure of its physiological, anatomical and morphological competence to survive in that area. This implies the existence of measurable differences between the floras of differing phyto-geographical zones or ecological habitats and such differences have been identified for a number of physiological processes such as photoperiodic responses, rates of photosynthesis and net assimilation rates, and their genecological significances discussed in reviews by Gregor (1946), Turrill (1946), Vegis (1963), Heslop-Harrison (1964) and Hiesey and Milner (1965).

Genecological studies on germination responses may be made by selecting a particular group of taxa distributed over a geographical area containing varied phytogeographical zones or ecological habitats, and examining the responses

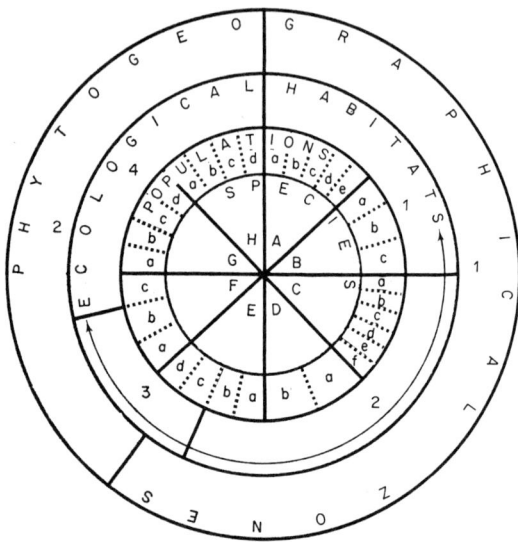

FIG. 1. Model illustrating possible correlations between populations of different species and their occurrence in different ecological habitats or geographical zones. Germination responses may be compared within or between species, and correlated with particular common or distinct features of the environments of the areas in which the species occur.

of populations of species, or groups of species, in relation to the climatic characteristics of the areas in which they occur. If this is done with sufficient populations of enough species, a sampling pattern can be built up within which correlations may be made in a number of different directions. This is shown diagrammatically, and in a very simplified form, in Fig. 1 for populations of eight species occurring in four definable ecological habitats in two phytogeographical zones. If comparisons were to be made between germination responses of these populations it might be expected that similarities would be identified between species A and B, and between species C and D, since all the populations of each pair occur in similar habitats and phytogeographical zones, but that differences due to different habitat preferences would occur which would distinguish populations of the pair AB from populations of the pair CD. Similarly it might be expected that populations of species G and H would be similar to each other but distinctively different from both AB and CD. In Fig. 1 species E and F represent variable species in which populations may occur in a

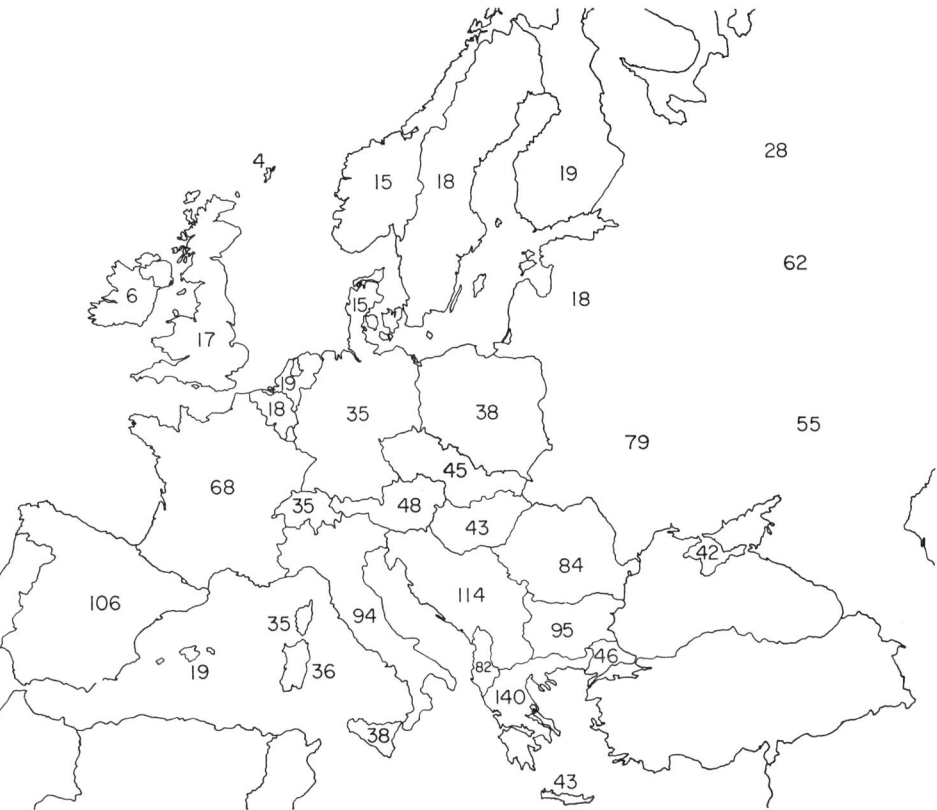

FIG. 2. Distribution of species of Silenoideae in Europe. Figures indicate the number of species, as recorded in *Flora Europaea*, which occur within the boundaries of different countries, islands or other districts of the continent.

variety of ecological habitats or phytogeographical zones, and which would lend themselves to correlations made either between populations within a species, or between populations of different species within similar habitats or zones.

Populations occurring within Europe of taxa in the subfamily Silenoideae of the family Caryophyllaceae were examined in germination tests designed to establish correlations between performance and geographical distribution. The overall distribution of this subfamily in Europe is illustrated in Fig. 2. This records the number of different species occurring within the boundaries of each country, or subdivisions of very large countries, following the distributions and delimitations of species in "*Flora Europaea*" (Tutin *et al.*, 1964) and ignoring climatic or phytogeographic divisions of the Continent. In spite of the limitations of such a method the pattern of distribution is a coherent one producing well defined differences between geographical areas. Thus the countries in which the greatest numbers of species occur are those of the Balkan Peninsula and Spain, and relatively large numbers are found throughout those countries and islands which border or are within the Mediterranean basin. Areas to the north of the Mediterranean contain far fewer different species and the numbers decline progressively from east to west. The parts of Europe in which the fewest species occur are the north-western oceanic borders of France, the Low Countries, Scandinavia and the British Isles. An analysis of the distribution of species in all parts of Europe in relation to their occurrence in Greece (the country containing the greatest number of species) is shown in Fig. 3. It is notable that, in those areas in which few species occur, a high proportion are widely distributed and are found also in Greece. However, in parts of the Mediterranean basin, and south-western U.S.S.R. where relatively large numbers of species occur, much smaller proportions are also found in the Greek flora. This is accounted for by a higher level of endemic species, or species with restricted ranges, in these areas.

Previous publications (Thompson, 1970a, b, c) have presented evidence to show that the germination responses of seeds which have been stored at room temperatures in the laboratory may be correlated with the areas of origin and geographical ranges of the species from which the seed was collected. Thus species of Caryophyllaceae which occur in areas with a Mediterranean climate germinate rapidly at median temperatures ($10-20°C$), have low minima ($<5°C$), and rather low maxima (*ca.* $25\pm5°C$) for germination. The germination responses of species restricted to parts of Europe with a markedly Continental climate, particularly those from steppeland areas, resemble Mediterranean species to the extent that they germinate rapidly at median temperatures but differ in their higher minima (*ca.* $5°C$), and their much higher maxima ($37-42°C$). Those species that occur in deciduous woodland zones of Europe, which comprise the majority of the few species occurring in western oceanic regions of the Continent, characteristically germinate less rapidly at median temperatures (3–5 days at $15-20°C$), and possess relatively high minima ($10-15°C$) and moderate maxima (*ca.* $35°C$). Species that are confined in their distribution to high altitudes or high latitudes form a more diverse group than the others described above, but tend to germinate relatively slowly, with high minima (sometimes exceeding 15°C) and maxima varying from one species to another from *ca.* 27°C to 35°C or above.

A few species in the subfamily Silenoideae, of which *Silene vulgaris* is the most extreme example, are widely distributed throughout Europe, and may be

FIG. 3. Pattern of distribution of species of Silenoideae in Europe in relation to their occurrence in Greece. Figures represent the percentages of species occurring in each country which also occur in the Greek flora.

found in a great variety of phytogeographical zones and ecological habitats. This variety is reflected in the germination responses of different populations of this species shown in Fig. 4. Populations from the Mediterranean, Continental Europe, Scandinavia and the British Isles (Fig. 4A) displayed widely different germination responses, which suggested a considerable degree of sub-specific

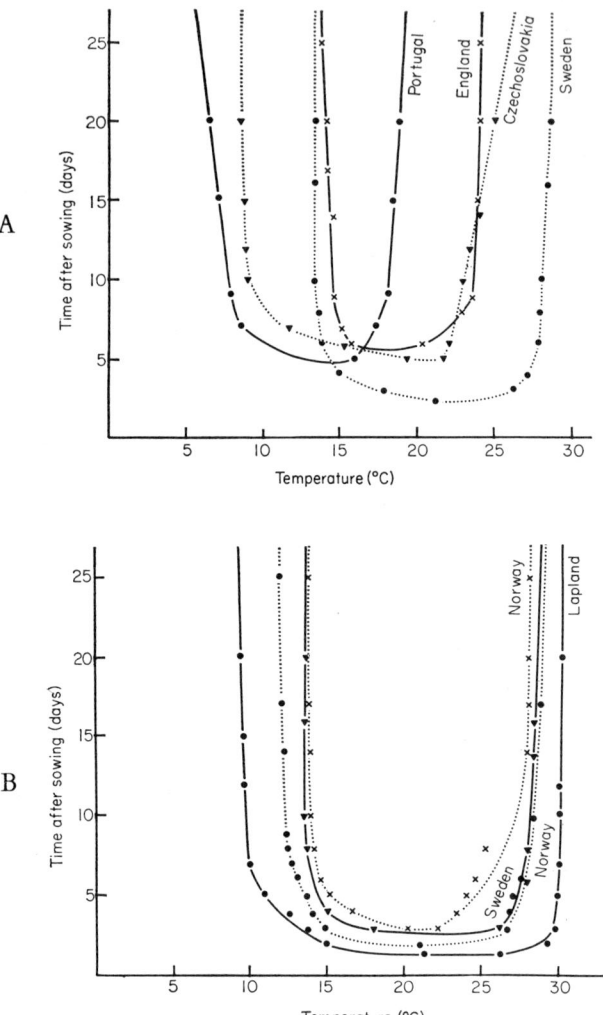

FIG. 4. Comparisons of the germination responses of stored seed of *Silene vulgaris* (Moench) Garcke, plotted as the maximum and minimum temperatures on successive days after sowing at which 50% of the final maximum germination rate was reached.

A. Populations collected from widely separate locations in Europe. (a) Coimbra, Portugal, ●——●; (b) Bratislava, Czechoslovakia, ▼······▼; (c) Dalecarlia, Sweden, ●······●; (d) Cliveden, England, ×——×.

B. Populations collected within Scandinavia. (a) Hordaland, Norway, ×······×; (b) Dalecarlia, Sweden, ▼——▼; (c) Aker, Norway, ●······●; (d) Kemi Lapland, Finland, ●——●.

adaptation to the different parts of Europe in which they were collected. Populations from Scandinavia (Fig. 4B) were markedly similar to each other, even though the geographical range over which they were collected was considerable. This suggests that quite small changes in their germination responses were sufficient to permit the maintenance of populations in a particular area of Europe in which phytogeographical and ecological variations are limited relative to the range of variation of the whole continent.

TABLE I. Comparisons of the germination responses of seeds of species distributed around the Mediterranean basin. Each batch of seed was tested at a range of temperatures immediately after it was harvested and retested after a period of dry storage at laboratory temperatures.

Species	State of seed	Temperature conditions of test: day/night °C					
		25/25	25/15	25/5	20/10	15/15	15/5
Silene pendula	fresh	15	70	—	—	89	93
	stored[a]	91	97	99	98	97	100
Silene perfoliata	fresh	0	11	73	94	93	94
	stored[a]	91	89	93	97	96	97
Silene colorata	fresh	0	52	0	80	77	87
	stored[a]	94	98	94	97	99	97
Silene nocturna	fresh	0	7	41	51	79	73
	stored[b]	69	88	93	91	100	95

[a] Stored for five months between tests.
[b] Stored for two months between tests.

It may appear surprising that correlations should occur between the germination responses of seed stored in the laboratory and climatic features of the geographical regions in which species are distributed, since storage conditions bear little resemblance to conditions experienced by these species in their natural habitats. However, the seeds of those species that grow around the Mediterranean and in the drier parts of continental Europe normally experience dry, warm or hot conditions for a period between maturation and germination, and there may be some analogy between the physiological effects caused by the natural experiences of such species and conditions of storage in a laboratory. Table I compares germination immediately after harvest with that of stored seed of four species which are widely distributed around the Mediterranean.

Germination of freshly harvested seed of each of these species was restricted at high temperatures, and a considerable proportion of seeds remained dormant in any test which included 25°C as part of the daily temperature cycle. After a period of storage germination rates were increased at all temperatures and even at 25°C only small numbers of seeds remained dormant.

Seeds of species distributed in more northern or western areas would be less likely to experience periods of hot, dry conditions for any length of time, and in some of these species a period of storage in the laboratory appears to have little

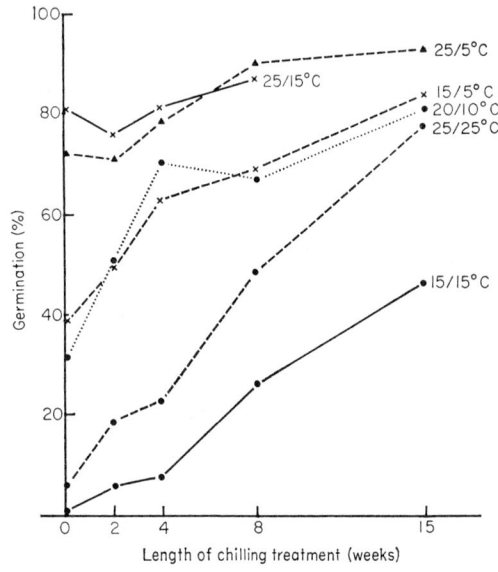

FIG. 5. Effects of chilling treatments of various durations at 2°C on subsequent germination at different temperatures of seed of *Silene alba* (Miller) E. H. L. Krause.

effect on the proportion of seeds remaining dormant after sowing. In these areas low temperature during the winter may be a major factor acting to reduce dormancy, with the result that a high proportion of seeds are capable of germinating when soil temperatures rise during the following spring. The effects of low temperature chilling treatments at 2°C on a population of *Silene alba* collected in south-eastern England are shown in Fig. 5. The germination of freshly harvested seed (zero on the "x" axis) was very variable depending on test conditions, but markedly improved when seed was exposed to fluctuating temperature cycles. Chilling treatments increased germination rates, but their effectiveness depended both on their duration and the temperatures of subsequent treatments. Even after a lengthy chilling treatment none of the

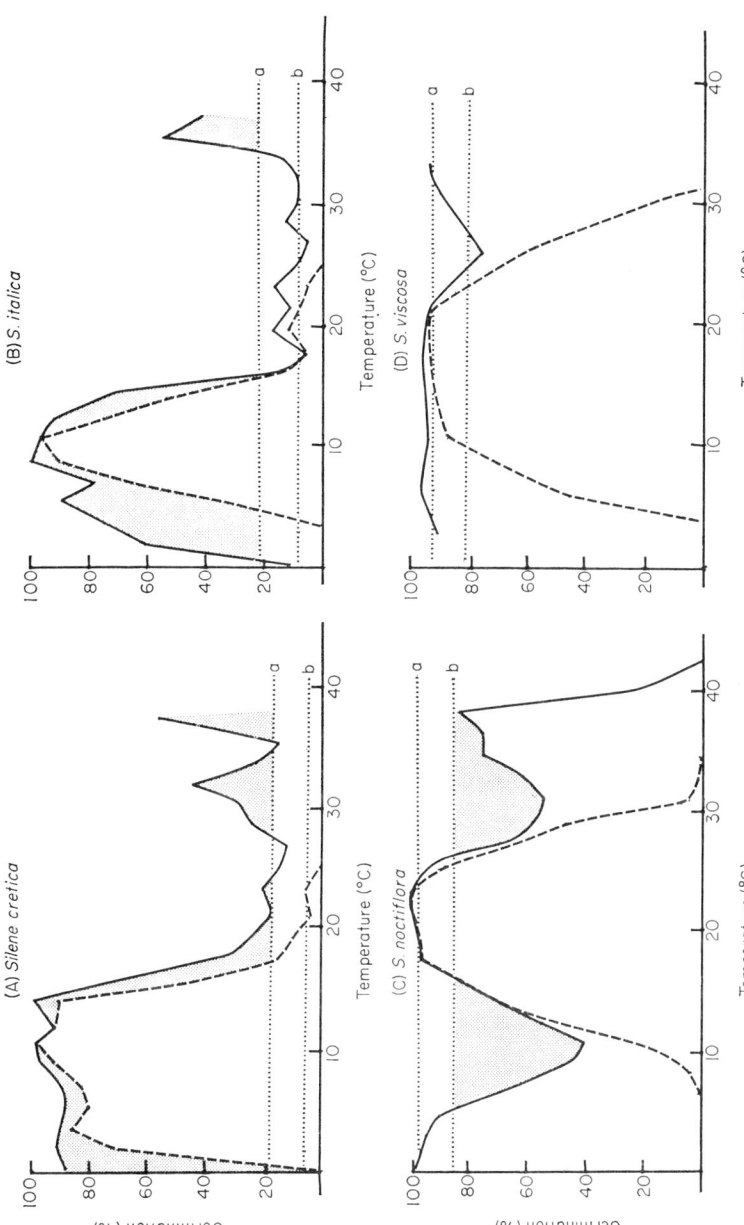

FIG. 6. Germination of four species tested on thermo-gradient bars immediately after harvest (- - -), and then transferred after 28 days to 21±1°C (——). The dotted lines "a" and "b" indicate, respectively, the upper and lower fiducial limits (P = 0·05) of the germination percentage at 21°C on the thermo-gradient bars. Germination responses exceeding or failing to attain these fiducial limits after transfer are interpreted as indications of promotive or inhibitory effects during treatment on the thermo-gradient bars (indicated by stippling).

subsequent conditions raised germination to 100%, and the least successful treatment, a constant temperature of 15°C, resulted in germination rates of less than 50%.

Germination tests made on freshly harvested seeds display greater differences in the number of seeds remaining dormant and the characteristics of their temperature responses than do tests with stored seed. The results of tests on four species with distinctively different geographical distributions, shown in Fig. 6, illustrate some of the variations that occur. Mediterranean species, e.g. *Silene cretica* and species from areas neighbouring the Mediterranean, e.g. *S. italica*, usually germinate at relatively low temperatures with temperature maxima of *ca.* 15°C, and minima below 5°C. Species with more northern distribution, e.g. *S. noctiflora* or *S. viscosa*, have higher optimal ranges for germination in which maxima may occur at around 30°C, and minima which usually exceed 5°C and may be as high as 15°C. Temperatures unfavourable for germination may induce physiological changes in the seed which modify subsequent germination responses. Thus when seeds used in the thermo-gradient bar experiments shown in Fig. 6 were subsequently transferred to room temperature (21°±1°C) responses were frequently different from those found for seed sown directly on the thermo-gradient bars at a temperature of 21°C. Germination of *S. cretica* and *S. italica* was considerably increased by some of the pretreatment temperatures, particularly those at extremes of the temperature scale. Germination of *S. noctiflora* was severely curtailed by some pretreatment temperatures, extending over relatively wide ranges on either side of those temperatures which favoured germination, and *S. viscosa* showed few and minor alterations in responses at 21°C as a result of pretreatment on the thermo-gradient bar.

Interpretation of results of this kind in terms of the conditions experienced by seeds in natural habitats depends on the possession of some information on the level of variability occurring between populations of different species occupying similar habitats, or with similar geographical distributions, and between and within populations of particular species. The germination responses to temperature of three species of annual Silenoideae collected on Andros in the eastern Mediterranean (see Fig. 7) display clearly marked affinities in the range of temperatures favouring germination, and the positions of optimal ranges. However, the proportion of seeds that remain dormant and fail to germinate is quite different from one species to the next. Comparisons in Fig. 8 of four populations from each of three species distributed in deciduous woodland zones of Europe, including western Europe and the British Isles, occupy markedly different temperature ranges from those found for the species from Andros, but

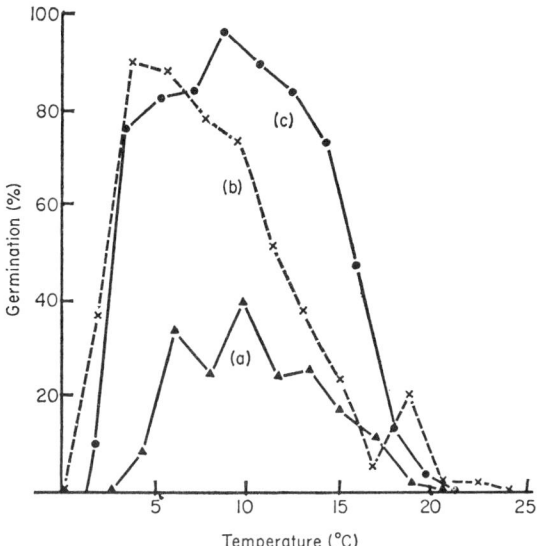

FIG. 7. Germination responses on thermo-gradient bars of newly harvested seed of three annual species of Silenoideae collected originally from Andros (S.W. Aegean Sea). A. *Petrorhagia velutina* (Guss.) P. W. Ball & Heywood; B. *Silene gallica* L.; C. *Silene cretica* L.

smaller overall differences between the three species (Fig. 9) occur in the range of temperatures at which they germinate. Optimal temperatures vary from one species to another and the level of dormancy varies very greatly from one population to another for each species.

Variations in dormancy confound comparisons between the temperature responses of different populations of a species, but their effect may be removed by calculating germination responses for each population as a percentage of its optimum performance. When this is done (see Fig. 10) it can be seen that the temperature responses of widely dispersed populations of *Silene dioica* are closely similar with a characteristic reduction in germination rates between 16°C and 21°C, and that considerable variations occur in the relative germination rates at different temperatures of the populations of *S. alba*. Since *S. alba* has a wider ecological and geographical range than *S. dioica* its greater variability is not surprising. Additional tests made with eight populations of *S. dioica* and fifteen populations of *S. alba*, using stored seed, also revealed much greater diversity from one population to another within *S. alba* than within *S. dioica* (see Fig. 11).

Silene alba and *S. dioica*, although closely related and frequently introgressing,

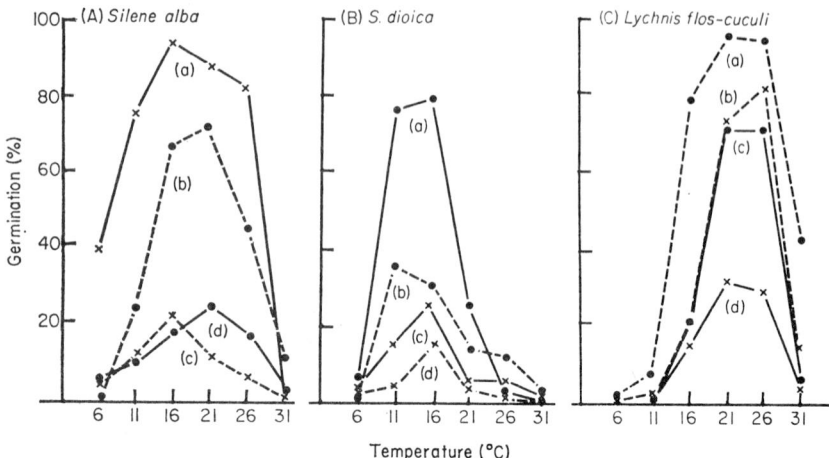

Fig. 8. Comparisons of the germination responses of newly harvested seeds of four populations of each of three species tested at various temperatures in incubators. A. *Silene alba* (Miller) E. H. L. Krause. (a) Hungary, ×——×; (b) Carpathian mountains, U.S.S.R., ●- - -●; (c) Austria, ×- - -×; (d) England, ●——●.
B. *Silene dioica* (L.) Clairv. (a) Sweden, ●——●; (b) Sweden, ●- - -●; (c) England, ×——×; (d) Poland, ×- - -×.
C. *Lychnis flos-cuculi* L. (a) Czechoslovakia, ●- - -●; (b) Poland, ×- - -×; (c) Norway, ●——●; (d) England, ×——×.

occupy habitats that are distinctively different, and possess important differences in their life cycles. The former, which is capable of flowering as an annual, occurs on dry banks, or grassland in sunlit areas, and very frequently as a ruderal in areas disturbed by man, or as a weed of certain cultivated crops. *Silene dioica*, which depends on a winter chilling treatment to initiate and develop flowers, is typically a deciduous woodland or hedgerow species in shaded positions, except in high latitudes or at high altitudes where it grows in the open. Figure 9 showed that freshly shed seed of populations of *S. alba* germinated better than seed of *S. dioica* at temperatures above 20°C, and Fig. 11 showed that after a period of storage seeds of *S. alba* also germinated better at higher temperatures than seed of *S. dioica*. Nevertheless, differences between the two species did not appear to be very great considering the well defined differences in their habitat preferences. The results of further tests to examine the magnitude of variation between populations of the two species, shown in Fig. 12A, revealed considerable overlap in the responses of different populations of the two species at temperatures below 20°C. Means calculated from all the populations for each species (Fig. 12B) occupied different positions and analysis

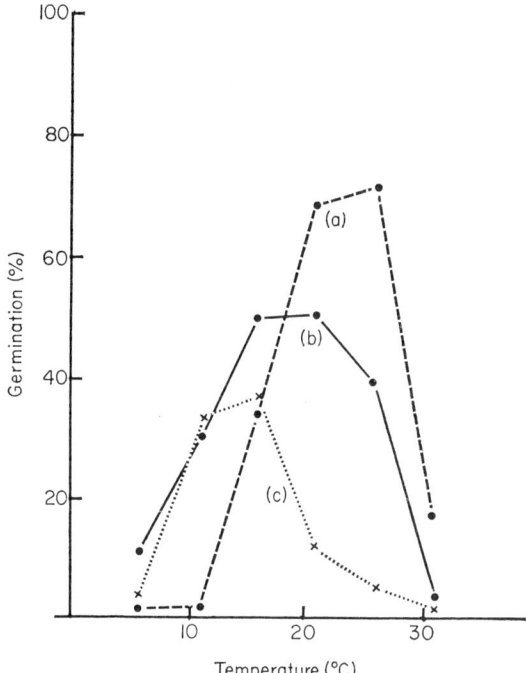

FIG. 9. Germination responses of the three species, detailed in Fig. 8, expressed as means of all the populations of each species. (a) *Lychnis flos-cuculi*, ●- - -●; (b) *Silene alba*, ●——●; (c) *Silene dioica*, ✗- - -✗.

of variance showed the differences to be significant (P = 0·05). Thus, populations of *S. alba* tend to germinate more rapidly at any temperature than populations of *S. dioica*, and are usually capable of germinating at lower temperatures. The relative responses to high and low temperatures of the means of each species were closely similar, resulting in practically parallel regression lines, but in both species, more frequently in *S. alba*, populations occurred in which relative responses at high and low temperatures deviated considerably from the pattern found for the majority.

Examination of differences occurring within a single population may reveal differences between the responses of the progeny of individual plants. Collections of seeds representing the progeny of a number of individuals in a population of *S. dioica* growing in south-eastern England were tested after a period of storage on thermo-gradient bars. The results for three collections shown in Fig. 13 suggested that differences in the characteristics of the temperature responses of the progeny of different plants were quite small, but that

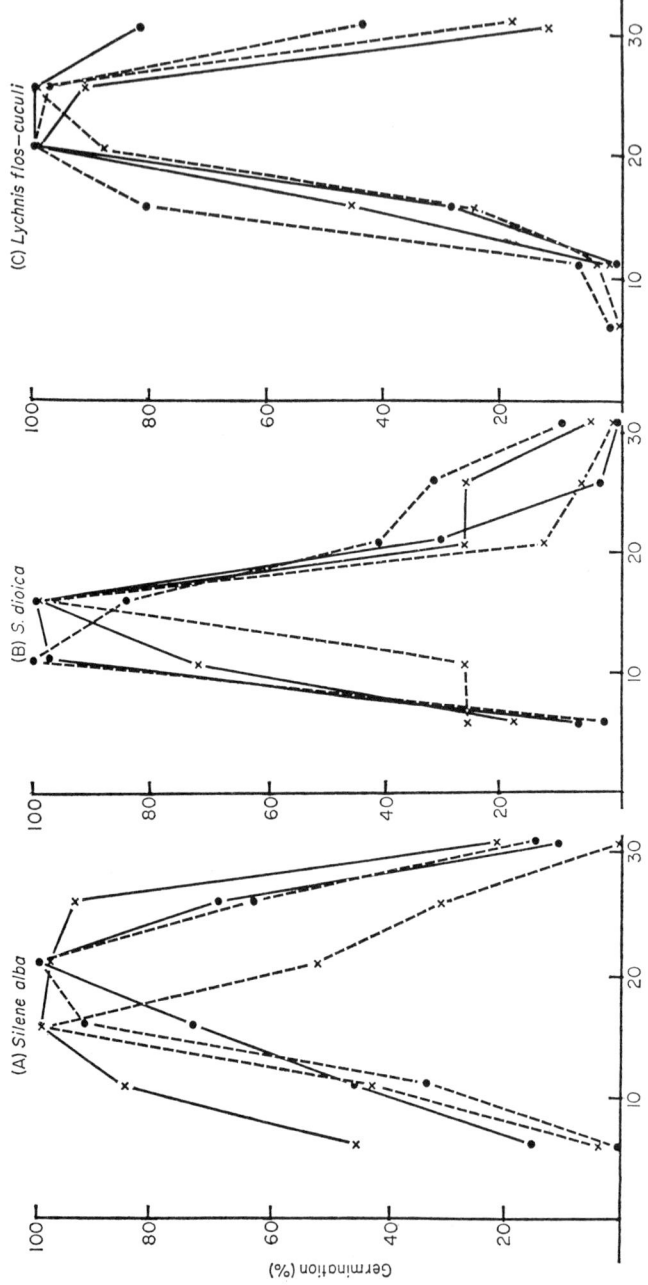

FIG. 10. Relative germination rates at different temperatures of *S. alba*, *S. dioica* and *L. flos-cuculi*. Values are expressed as percentages of the maximum response of each population. (See Fig. 8 for details of the populations tested.)

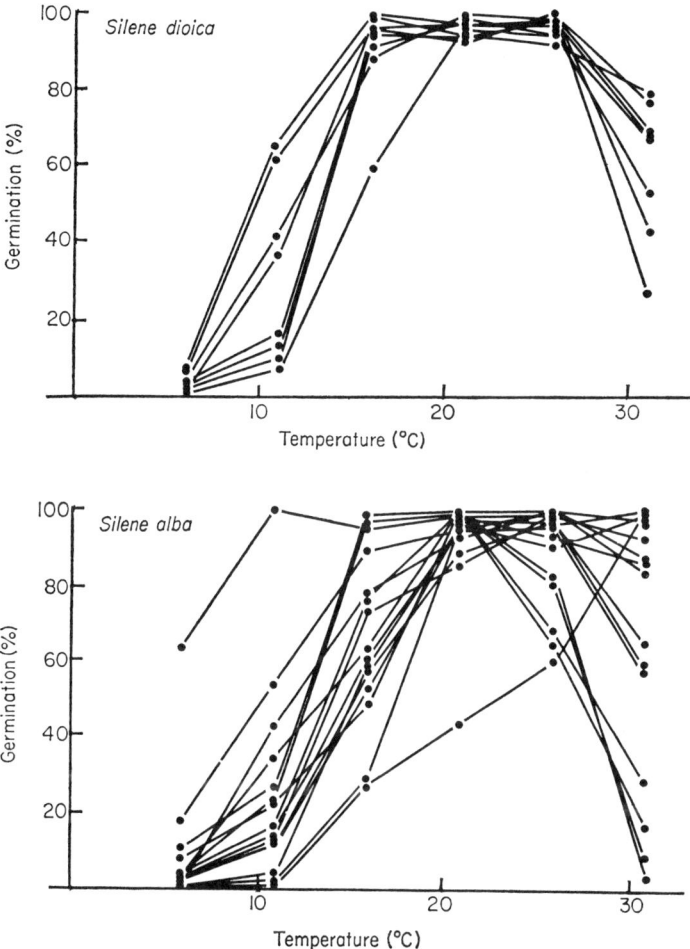

FIG. 11. Responses of stored seed collected from eight populations of *Silene dioica* and fifteen populations of *S. alba* tested in incubators at six different temperatures.

differences between the proportions of dormant seeds were very considerable. The overall range of variation in dormancy was as great as that found previously for differences between populations (see Fig. 8). Subsequent tests involving transfer from the thermo-gradient bars to room temperature (21°±1°C), followed by a chilling treatment at 2°C suggested (see Fig. 14) that there were few, if any, qualitative differences in the temperature responses of different progenies, even in comparisons in which the initial level of dormant seeds had been very different. The major factor affecting reduction in dormancy was

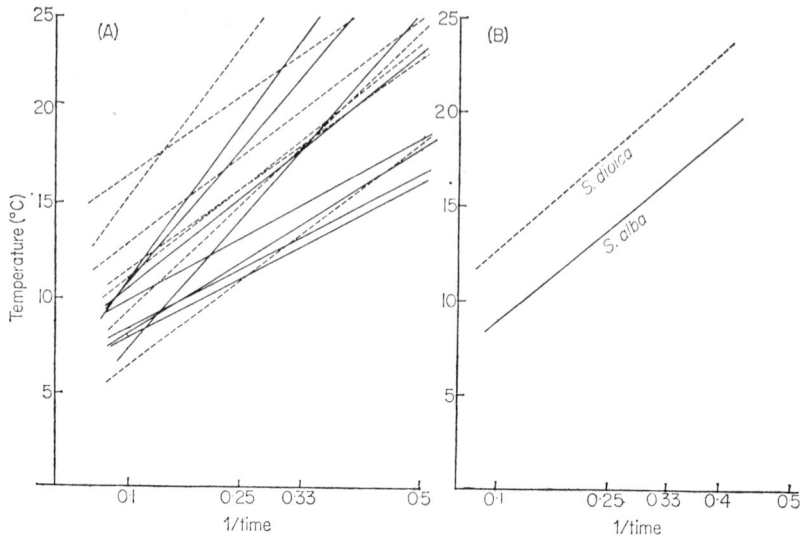

FIG. 12. Germination responses on thermo-gradient bars of stored seed of populations of *Silene alba* and *S. dioica*. Responses are plotted as regressions (Temperature, °C, against the reciprocal of time, days) derived from the minimum temperatures on successive days at which 50% of the final maximum germination rate was reached.
A. Regressions plotted for individual populations of *S. dioica* (- - -) and *S. alba* (———).
B. Regressions plotted as means of all populations.

treatment at low temperatures whether at 2°C in incubators or at the low end of the thermo-gradient bar.

Comparisons between the germination responses of the progeny from different parts of individual plants were made by testing seeds collected separately from different capsules. The results (Fig. 15) showed that there were very considerable differences in the proportion of dormant seeds from one capsule to another. In several of the plants tested in this way germination rates between capsules varied from *ca.* 15 to *ca.* 85%, but in a few plants the overall level of dormant seeds was high in all the capsules examined and maximum germination rates failed to exceed 25–30%. Therefore, although great variations did occur between germination rates of seeds from different capsules on a plant, these variations occurred within a range limited by the overall level of dormancy characteristic of the progeny of each parent plant.

It was suggested earlier that, within an area such as Europe in which topographical barriers to plant distribution are relatively minor, the patterns of distribution of different species are determined largely by the physiological

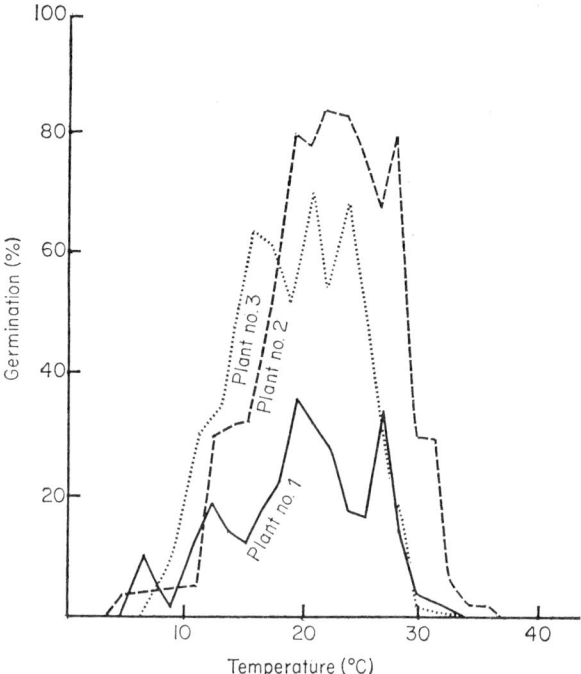

FIG. 13. Germination responses on thermo-gradient bars of the progeny of three separate plants of *Silene dioica* forming part of a small population in southern England.

competence of populations of a species to survive in particular locations. Therefore a study of the factors limiting plant distribution in these areas may largely consist of identification of the components of the physiological responses of different populations, and interpretation of their significance in relation to known climatic, edaphic and biotic parameters of different geographical areas or ecological habitats.

It is usually accepted (e.g. Vegis, 1963, 1965; Harper, 1965) that the germination responses of populations of a species collected from different locations vary. Support for this view may be found in the results presented above and in germination studies with taxa as diverse as *Tsuga canadensis* (Stearns and Olson, 1958), *Typha* spp. (McNaughton, 1966), *Amaranthus retroflexus* (McWilliams et al., 1968), and *Avena* spp. (Thurston, 1957, 1962). Demonstrations of such differences have given rise to the suggestion that germination responses are highly plastic, undergoing rapid selection by environmental factors to produce adaptive changes that distinguish the genetic constitutions of different populations of a species in different locations. If the mechanisms regulating the

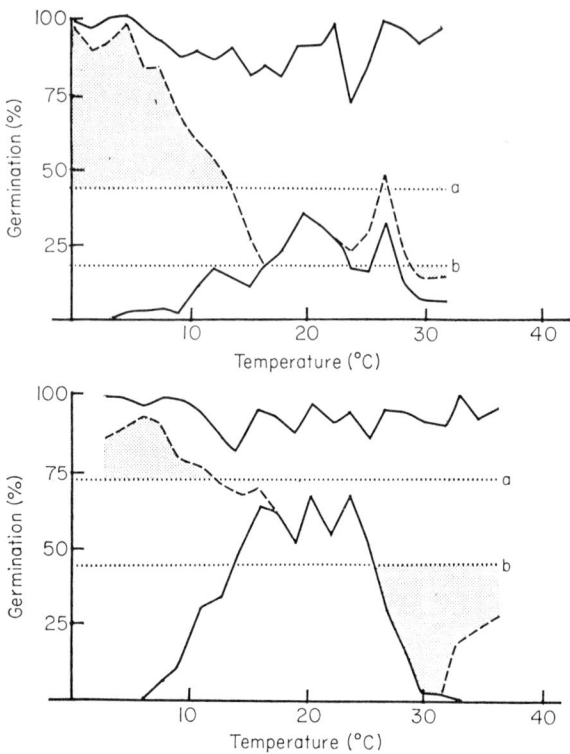

FIG. 14. Germination of the progeny of two plants of *Silene dioica* tested on (a) thermo-gradient bars (———, lower line); (b) after transfer to 21°C (- - -); (c) following a chilling treatment at 2°C for six weeks and final return to 21°C (———, upper line). The dotted lines "a" and "b" represent the upper and lower fiducial limits (P = 0·05) of responses at 21°C on the thermo-gradient bars. Values above or below these fiducial limits indicated by stippling, following transfers to different conditions, indicate promotive or inhibitory effects of the preceding treatment.

pattern and timing of germination were so readily adaptive they would play little part as effective determinants of the geographical distribution of a species. However, Harper (1965) has commented on the wide range of species which are able to complete their life cycles under alien conditions when deliberately sown by man at a favourable season, in comparison with the relatively insignificant proportion which establishes wild populations. Thompson (1970b) has shown that responses to temperature of populations of species of Caryophyllaceae distributed in Europe may vary only slightly, and that, even after the passage of several thousand generations in the environment of central and western Europe,

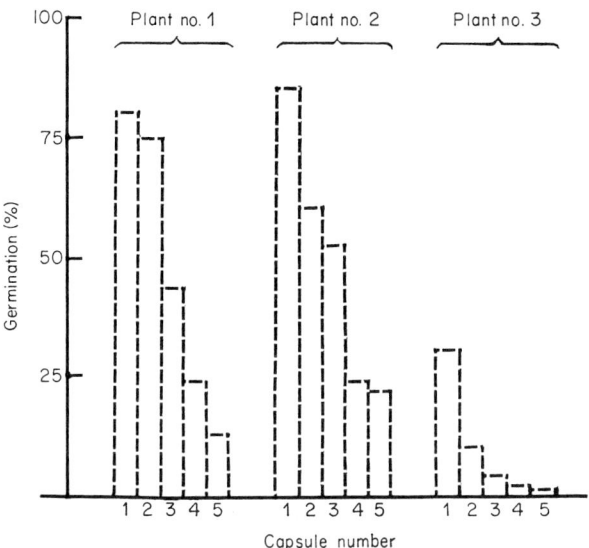

FIG. 15. Germination rates after 28 days of seeds of *Silene dioica* collected separately from different capsules of three different plants. Tests were done in an incubator providing alternate 12 h cycles at 26 and 16°C.

collections of seed of *Agrostemma githago* still retain germination responses characteristic of species growing in its area of origin in the eastern Mediterranean (Thompson, 1973).

It would appear therefore that, whether germination responses are potentially highly adaptive or not, under natural conditions they frequently remain remarkably stable. This suggests that in most species changes in germination responses, and indeed any other physiological change, are constrained by the total genotype of the plant, and that the expression of the adaptive capacity of the germination response will depend on the concurrent adaptive capacity of other physiological and morphological qualities. Thus populations of species such as *Silene dioica*, *Lychnis flos-cuculi* or *Silene cretica* possess relatively consistent responses to temperature which can be closely related to the phytogeographical regions in which the species are distributed. Examination of numbers of populations of such species may reveal none that possesses germination responses to temperature which transgress the range characteristic of plants occurring within the relevant phytogeographical area, even when certain populations are included which have spread as weeds or aliens to other climatic regions. The example of *Agrostemma githago* has been cited already, and similar examples of extension of ranges in the wake of man's activities with no more than minor alterations to the germination responses to temperature occur in *S. gallica*, *S. dichotoma* and *Petrorhagia prolifera*.

Other species, with broader natural distributions, may display less consistent responses from one population to another, and examples shown earlier of the behaviour of *S. alba* and, in particular, *S. vulgaris* illustrate the range of variation that may occur. The latter is notable within the Silenoideae for the extent of its range throughout practically all the phytogeographically distinguishable zones of Europe, and the variety of ecological habitats it occupies, as well as for its morphological variability. Thus these two species which are notable for the variety of their genotypes in other respects also display an unusual degree of variation in their germination responses to temperature.

Temperature responses may be modified by the presence of dormant seed, particularly soon after harvest when a high proportion of the sample may be in a state of relative dormancy similar to that described by Borriss (1940) in *Agrostemma githago*. Dormancy of this type is a widespread protective mechanism amongst the Silenoideae acting, in the case of species from the Mediterranean, to prevent germination at high temperatures during the summer (Thompson, 1970d). A very similar situation has been described by Barton (1936) for species of winter annuals from desert regions of the United States. Almost all populations of Silenoideae on which tests were made immediately after harvest contained a proportion of dormant seeds. This proportion was extremely variable from one population to another, and in the case of *S. dioica* from one plant to another and even between seeds from one capsule to another on the same plant, demonstrating similar plant to plant and within plant variations to those reported by Cavers and Harper (1966) for species of *Rumex*. The variability found in the level of dormancy was therefore in marked contrast to the frequently small differences found between responses to temperature. Similar variations in dormancy from one population to another have been reported for a number of species including *Rosa* spp. (Von Abrams and Hand 1956), *Fraxinus excelsior* (Varasova, 1956), *Lactuca sativa* cultivars (Harrington and Thompson, 1952), *Spergula arvensis* (New, 1958) and *Avena* spp. (Thurston, 1962). In some cases the circumstances of the occurrence of dormancy suggest a strong element of phenotypic control of its expression, but experiments by Harper and McNaughton (1960) have provided evidence for a considerable degree of genetic control of dormancy in species and hybrids of *Papaver*, which also showed great variations in the proportion of dormant seeds depending on the nature and direction of the crosses made. They also suggested that the absence of dormant individuals in the progeny of many of the crosses between different species would result in the germination of a high proportion of the hybrid progeny in the autumn with consequent exposure to adverse weather and cultivations during the winter.

Similarly it would be expected that the proportions of dormant and non-dormant individuals amongst freshly shed seeds of species of Silenoideae would be a critical factor affecting the pattern of autumn and spring germination. The evidence of all the experiments reported, including the results of experiments involving transfers to successive different test conditions, suggests that those species such as *S. dioica*, *L. flos-cuculi* and *S. alba* which are capable of establishing persistent populations in oceanic regions of western Europe all possess germination responses to temperature, and dormancy patterns, which would result in at least a part of the seed germinating in late summer soon after it was shed, and a further part germinating in the following spring after a winter chilling treatment. Polymorphisms in the timing and rate of germination of this sort have been recorded for other species widely distributed in western Europe such as *Rumex* spp. (Cavers and Harper, 1966); and Salisbury (1968) has pointed out the value of such a character in relation to the inconsistent climatic features of the seasons typical of the British Isles.

Consideration of the results discussed above in relation to the model illustrated in Fig. 1 provides some insight into the milieu of factors controlling the season and rate of germination of species in different geographical localities or ecological habitats. Species with a common distribution within a phytogeographically distinguishable area tend to possess germination responses to temperature which are broadly similar, and which can be differentiated qualitatively and quantitatively from the responses of species distributed in other phytogeographical zones. These temperature responses appear, in many species, to be relatively stable, and their stability, however imposed, would be likely to act as a major impediment to the spread of a species to areas with distinctively different climatic patterns. Thus in Fig. 1, the results suggest that common response characteristics would occur in comparisons between the germination behaviour of species A, B, C and D, in spite of differences in their ecological preferences, and common, but different, sets of characteristics would link species F, G and H. A species such as E which possesses populations occurring in different phytogeographical zones and ecological habitats represents a variable species similar to *S. vulgaris*, in which different populations tend to differ considerably from one another, but to resemble other species distributed in their respective phytogeographical zones.

In every population of each species the temperature responses would be modulated in their expression by the presence of a proportion of dormant individuals, and this proportion would be variable between populations, within populations and within the progeny of individual plants as well as between successive generations of a population. It is arguable that such variations in dormancy, allied with relatively small changes in germination responses to temperature, constitute the adaptive mechanisms responsible for the habitat preferences typical of the ecological requirements of different species. Support for this argument can be derived from the differences found in these studies between populations of *S. dioica* and *S. alba* (comparable for example to species B and C in Fig. 1). However, the differences identified in the laboratory were not great, in spite of the considerable differences in the ecology of these two species, and a great deal of overlap occurred between individual populations of the two species. Although the differences found might well be critical in the competitive conditions of the natural environment they do not of themselves suggest that hybrids between the two species would be likely to show marked deviations from the germination responses of the species; as could certainly be the case with hybrids between two species from different phytogeographical zones. This may be a reason for the high level of survival of hybrids between these species and the consequent level of introgression that has been reported for some natural populations.

The patterns of distribution of the Silenoideae illustrated in Fig. 2 are heavily biased toward the evolution of endemic and narrowly distributed species around the Mediterranean, and in steppeland areas of eastern Europe, in contrast to the occurrence in north-western Europe of a limited number of widely distributed species. Germination responses of species from the Mediterranean are typically well defined and related to the winter growing season and the avoidance of germination during periods of transient rainfall in the summer. Species occurring in eastern Europe usually possess relatively few restrictions on the pattern of germination, and this, together with the biennial or perennial nature of their life cycles suggests that much of the seed germinates soon after it is shed in late summer. The progeny then overwinter as young plants, under conditions of winter cold and prolonged snow-cover which are consistent from year to year, and pose relatively few hazards to survival of a species with sufficient hardiness. Those species that occur in north-western Europe are notable for the subtlety of their germination responses and the critical role played both by post-harvest and induced dormancy mechanisms, which result in the division of the population each year into at least two parts, one germinating in late summer, the other in spring.

It is clearly not possible at this stage to state reasons for the high level of speciation of a group of taxa in one particular phytogeographical area compared with another, or even to suggest any means of measuring the contribution that the germination response may make to this situation. It is however possible to define differences that do occur between species with different natural ranges and to define in some measure the similarities and differences that occur between populations of a species within its natural range, or as a consequence of artificial extension of its range by mankind. Such a study, based on established situations, can aim for an increasingly complete description of the nature of particular physiological responses correlated with the physical parameters of particular geographical regions, which contribute to our understanding of the direction and magnitude of the adaptive changes which are required before a species can extend its geographical range or broaden its tolerance to different habitats. Eventually one might hope that this will lead to a much clearer appreciation of the constraints that limit plant distribution and speciation, and the way that these vary in their effectiveness depending on complex interactions between the nature of the genotype of different groups of taxa and exogenous influences characteristic of the parts of the world in which they are distributed.

REFERENCES

BARTON, L. V. (1936). Germination of some desert seeds. *Contr. Boyce Thompson Inst. Pl. Res.* **8**, 7–11.

BORRISS, H. (1940). Über die inneren Vorgänge bei der Samenkeimung und ihre Beeinflussung durch Aussenfaktoren. *Jb. wiss. Bot.* **89**, 254–339.

CAVERS, P. B. and HARPER, J. L. (1966). Germination polymorphism in *Rumex crispus* and *Rumex obtusifolius*. *J. Ecol.* **54**, 367–382.

DATTA, S. C., EVENARI, M. and GUTTERMAN, Y. (1970). The heteroblasty of *Aegilops ovata* L. *Israel J. Bot.* **19**, 463–483.

GREGOR, J. W. (1946). Ecotypic differentiation. *New Phytol.* **45**, 254–270.

HARPER, J. L. (1965). Establishment, aggression and co-habitation in weedy species. In "The Genetics of Colonizing Species" (H. G. Baker and G. L. Stebbins, eds). Academic Press, London and New York.

HARPER, J. L. and MCNAUGHTON, I. H. (1960). The inheritance of dormancy in inter- and intra-specific hybrids of *Papaver*. *Heredity* **15**, 315–320.

HARRINGTON, J. F. and THOMPSON, R. C. (1952). Effect of variety and area of production on subsequent germination of lettuce seed at high temperatures. *Proc. Am. Soc. hort. Sci.* **59**, 445–450.

HESLOP-HARRISON, J. (1964). Forty years of genecology. *Adv. Ecol.* **2**, 159–247.

HIESEY, W. M. and MILNER, H. W. (1965). Physiology of ecological races and species. *Ann. Rev. Pl. Physiol.* **16**, 203–216.

KOLLER, D. (1955). The regulation of germination in seeds. *Bull. Res. Counc. Israel.* **5**, D, 85–108.

KOLLER, D. and NÈGBI, M. (1959). The regulation of germination in *Oryzopsis miliacea*. *Ecology* **40**, 20–36.

LANG, A. (1965). Effects of some internal and external conditions on seed germination. "Encyclopedia of Plant Physiology", Vol. **15/2**, pp. 848–893. (W. Ruhland, ed.) Springer-Verlag, Berlin, Heidelberg and New York.

MAYER, A. M. and POLJAKOFF-MAYBER, A. (1963). The ecology of germination. *In* "The Germination of Seeds". Pergamon Press, Oxford, London, New York and Paris, pp. 179–208.

MCNAUGHTON, S. J. (1966). Ecotype function in the *Typha* community-type. *Ecol. Monogr.* **36**, 297–325.

MCWILLIAMS, E. L., LANDERS, R. Q. and MAHLSTEDE, J. P. (1968). Variation in seed weight and germination in populations of *Amaranthus retroflexus* L. *Ecology* **49**, 290–296.

MORLEY, F. H. W. (1958). The inheritance and ecological significance of seed dormancy in subterranean clover (*Trifolium subterraneum* L.) *Aust. J. biol. Sci.* **11**, 261–274.

NEW, J. K. (1958). A population study of *Spergula arvensis*. *Ann. Bot.* N.S. **22**, 457–477.

SALISBURY, E. J. (1965). The reproductive biology and occasional seasonal dimorphism of *Anagallis minima* and *Lythrum hyssopifolia*. *Watsonia* **7**, 25–39.

SALISBURY, E. J. (1968). Germination experiments with seeds of a segregate of *Plantago major* and their bearing on germination studies. *Ann. Bot.* N.S. **29**, 513–521.

STEARNS, F. and OLSON, J. (1958). Interaction of photoperiod and temperature affecting germination in *Tsuga canadensis*. *Am. J. Bot.* **45**, 53–58.

STOKES, P. (1965). Temperature and seed dormancy. "Encyclopedia of Plant Physiology", Vol. **15/2**, pp. 746–803. (W. Ruhland, ed.) Springer-Verlag, Berlin, Heidelberg and New York.

THOMPSON, P. A. (1970a). Characterization of the germination responses to temperature of species and ecotypes. *Nature, Lond.* **225**, 827–831.

THOMPSON, P. A. (1970b). Germination of species of Caryophyllaceae in relation to their geographical distribution in Europe. *Ann. Bot.* N.S. **34**, 427–449.

THOMPSON, P. A. (1970c). A comparison of the germination character of species of Caryophyllaceae collected in central Germany. *J. Ecol.* **58**, 699–711.

THOMPSON, P. A. (1970d). Changes in germination responses of *Silene secundiflora* Otth. in relation to the climate of its habitat. *Physiol. Pl.* **23**, 739–746.

THOMPSON, P. A. (1973). Effects of cultivation on the germination character of the corncockle (*Agrostemma githago* L.) *Ann. Bot.* **37**, 133–154.

THURSTON, J. M. (1957). Morphological and physiological variation in wild oats (*Avena fatua* L. and *A. ludoviciana* Dur.) and in hybrids between wild and cultivated oats. *J. agr. Sci.* **49**, 261–274.

THURSTON, J. M. (1962). An international experiment on the effect of age and storage conditions on viability and dormancy of *Avena fatua* seeds. *Weed Res.* **2**, 122–129.

TURRILL, W. B. (1946). The ecotype concept: a consideration with appreciation and criticism especially of recent trends. *New Phytol.* **45**, 34–43.

TUTIN, T. G., HEYWOOD, V. H., BURGES, N. A., VALENTINE, D. H., WALTERS, S. M. and WEBB, D. A. (eds) (1964). "Flora Europaea", Vol. 1. Cambridge University Press.

VARASOVA, N. N. (1956). Peculiarities of the seeds of the common ash in relation to different geographic origin (in Russian). *Acta. Inst. bot. Acad. Sci. U.S.S.R.* Ser. IV Bot. exp. **11**, 370–387.

VEGIS, A. (1963). Climatic control of germination, bud break and dormancy. *In* "Environmental Control of Plant Growth" (L. Evans, ed.). Academic Press, London and New York, pp. 265–288.

VEGIS, A. (1965). Ruhezustände bei höheren Pflanzen, Induktion, Verlauf und Beendigung: Ubersicht, Terminologie, allgemeine Probleme. *Handbuch der Pflanzenphysiologie*, Vol. **15/2**, pp. 499–533. (W. Ruhland, ed.) Springer-Verlag, Berlin, Heidelberg and New York.

VON ABRAMS, G. T. and HAND, M. M. (1956). Seed dormancy in *Rosa* as a function of climate. *Am. J. Bot.* **43**, 7–12.

WENT, F. W. (1949). Ecology of desert plants. II. The effect of rain and temperature on germination and growth. *Ecology* **30**, 1–13.

WENT, F. W. (1957). Ecology. *In* "Experimental Control of Plant Growth. Chronica Botanica Co., Waltham, Mass. U.S.A., pp. 248–251.

WENT, F. W., JUHREN, G. and JUHREN, M. C. (1952). Fire and biotic factors affecting germination. *Ecology* **33**, 351–364.

WENT, F. W. and Westergaard, M. (1949). Ecology of desert plants. III. Development of plants in the Death Valley National Monument, California. *Ecology* **30**, 26–38.

WESSON, G. and WAREING, P. F. (1969). The induction of light sensitivity in weed seeds by burial. *J. exp. Bot.* **20**, 414–425.

5 | Phytosociologie et Systématique

M. GUINOCHET

Université de Paris-Sud, 91 ORSAY, France

Abstract: Phytosociology as interpreted in this paper is that of the SIGMA school (Station internationale de Géobotanique méditerranéenne et alpine founded by J. Braun-Blanquet) based on floristic analysis. It is not a branch of ecology as such but is concerned essentially with floristic phenomena so that phytosociological studies are not undertaken as part of an ecological study, but rather ecology is invoked to interpret phytosociological phenomena.

It is evident that inventories listing different species indicate the existence of equally different ecological conditions and conversely two similar lists indicate similar conditions. In view of our knowledge of the rôle of natural selection in the dynamics of populations, one can consider that when a species forms part of several plant associations or other phytosociological units it is represented in each of these by populations of more or less different genotypic composition. Examples are given from *Cardamine pratensis*, *Molinia coerulea*, *Plantago coronopus*, *Festuca ovina*, *Myosotis* spp. and *Centaurea* sect. *Jacea*.

This link between the origin and evolution of taxa and of plant associations is confirmed by studying the comparative distribution and floristic composition of phytosociological units. Examples are discussed from within Europe and between Europe and North America.

Tout d'abord, la Phytosociologie considérée ici sera la Phytosociologie sigmatiste, c'est à dire la branche du savoir fondée sur le concept floristique d'association végétale. Comme je l'ai déjà souvent écrit, ce concept résulte de l'observation banale suivante: pour quelqu'un qui connaît les plantes dans la nature, le simple rappel du nom d'une espèce évoque dans son esprit non seulement son image mais encore celles d'un certain nombre d'autres que l'on trouve ordinairement dans les mêmes endroits qu'elle.

Une association végétale est donc essentiellement une liste idéale d'espèces formée de la réunion de listes établies dans la nature selon une règle concernant la localisation et les dimensions des surfaces analysées. Cette règle est celle de l'homogénéité floristique: nous disons bien *floristique*, c'est-à-dire qu'il s'agit des espèces, autrement dit de concepts explicités par des noms, et non pas des individus par lesquels sont représentées ces espèces et à plus forte raison de leurs densité, dimension, mode de distribution, etc. ... Une surface floristiquement homogène est donc une surface telle qu'en l'explorant on est conduit à n'énoncer qu'un certain nombre de noms d'espèces; dès que l'on est amené à en citer, plus ou moins brusquement, de nouveaux en quantité appréciable il est à peu près sûr que l'on pénètre sur une surface méritant d'être analysée séparément, autrement dit que l'on aborde un autre individu d'association. Pour bien saisir cette notion d'individu d'association, il faut comprendre que la Phytosociologie sigmatiste a été et est encore l'oeuvre de floristes de terrain expérimentés qui ont, par conséquent, en tête les principales associations végétales du territoire auquel ils s'intéressent: ils sont ainsi intuitivement conduits à leur rapprocher, sur le terrain, les éléments de végétation correspondants. La démarche intellectuelle est la même que celle qui consiste à mettre des noms d'espèces sur des individus. C'est la raison pour laquelle on a pris l'habitude d'appeler, par analogie, "individu d'association" une surface de végétation représentative, sur le terrain, d'une association végétale. C'est sur ces individus d'association perçus intuitivement sur le terrain que sont exécutés les **relevés**, c'est-à-dire les listes d'espèces dont il a été question ci-dessus, destinés à établir une description complète, communicable à autrui, de ces associations végétales conçues mentalement.

Pour éviter tout malentendu, il faut préciser qu'en un lieu donné peuvent coexister des individus d'association d'associations végétales différentes: ainsi, un individu d'association végétale forestière constitué d'espèces ligneuses et herbacées implantées dans le sol, héberge des individus d'association d'associations épiphytes, d'associations de champignons, etc. . Cet ensemble d'individus d'association, plus complexe que chacun d'eux pris individuellement, représente une phytocénose. La phytocénose héberge une zoocénose, formée d'un certain nombre d'individus d'association d'associations animales. L'ensemble d'une phytocénose et d'une zoocénose constitue, comme chacun sait, une biocénose.

La description d'une association végétale est présentée sous la forme d'une matrice, dite **tableau d'association**, où les lignes sont consacrées aux espèces, les colonnes aux relevés. Une association végétale est donc bien définie, comme déjà dit, par son «ensemble spécifique normal», c'est-à-dire par une liste synthétique d'espèces formée de la réunion de listes, les relevés, se ressemblant plus entre elles qu'elles ne ressemblent aux autres listes, relevant donc d'autres associations, prises dans le territoire étudié. Un calcul élémentaire montre que la probabilité de trouver deux relevés d'une association végétale rigoureusement identiques est, dans la majorité des cas, si faible qu'elle peut-être considérée comme nulle: c'est cette situation qui empêche de comprendre la phytosociologie les esprits qui ignorent ou ne peuvent assimiler la notion de catégorie polythétique bien qu'ils l'utilisent, sans s'en rendre compte, en systématique (cf. par ex. les familles par enchaînement).

La comparaison des tableaux d'association des associations végétales d'un territoire fait ressortir dans chacune d'elles quelques espèces qui y ont une fréquence* significativement plus élevée que dans les autres: ce sont les **espèces** dites **caractéristiques** de l'association correspondante, les autres étant qualifiées de **compagnes**. Ces notions d'espèces caractéristiques et d'espèces compagnes ont fréquemment donné lieu à contresens. On a, tout d'abord, trop souvent tendance à confondre espèce caractéristique, espèce constante et espèce dominante: une espèce peut être caractéristique d'une association tout en y ayant une très faible fréquence, pourvu que celle-ci soit plus élevée que dans les autres associations; les espèces rares sont toujours, ou presque, des caractéristiques d'association. Quant à la dominance, on lui accorde trop généralement une importance exagérée, du moins en Phytosociologie sigmatiste. Comme l'ont montré de nombreux phytosociologues sigmatistes, la dominance, voir plus généralement l'abondance, ont une valeur d'information bien inférieure à celle du critère présence-absence pour la discrimination des associations végétales. Récemment encore, un auteur qui ne peut-être suspecté d'être un partisan inconditionnel de la phytosociologie sigmatiste a écrit ceci: «On peut admettre *a priori* que les données quantitatives apportent plus d'information que celles ne concernant que la présence-absence des espèces. Cela ne signifie pourtant pas que les données quantitatives représentent le type idéal d'analyse des relations de parenté phytosociologiques. De telles données, étant reliées à des performances variables, peuvent contenir «trop» d'information au sujet de détails obscurcissant par cela d'importantes relations de parenté au niveau de l'établissement et de la survivance dans les divers sites» (Orloci, 1968). Enfin, pour en terminer avec la notion d'espèce caractéristique, notons que, dans des parties distinctes de son aire de distribution géographique, une espèce peut être caractéristique d'associations végétales différentes: ce sont les caractéristiques transgressives. Il va de soi que dans toute cette partie de l'exposé on n'a en vue que des espèces de type linnéen, comprises très largement.

Quant aux espèces compagnes, elles déconcertent curieusement de nombreux esprits. Ainsi, encore récemment, un auteur, adepte de l'école de Wisconsin, comparant ses résultats à ceux obtenus par les méthodes d'autres écoles, dont l'école sigmatiste, a écrit qu' «il se manifeste des chevauchements d'espèces entre

---

* Dans tout cet article, le mot «fréquence» est employé avec son acception mathématique.

les groupes de chacun des autres chercheurs, de sorte que leurs groupes de classification ne peuvent pas être interprétés d'une manière absolument rigide » (Rogers, 1970). Cela dénote beaucoup plus une mentalité de géomètre que de statisticien : or la phytosociologie, comme toute la biologie, relève essentiellement, sinon exclusivement, de cette dernière.

En tout cas, il faut bien comprendre que dans un territoire dont la Flore comporte E espèces, l'ensemble spécifique normal de chacune des A associations végétales qui y sont réalisées correspond à une seule des C combinaisons simples théoriquement possibles de ces E espèces prises N à N, $2 \leq N \leq E-1$: chacun de ces ensembles spécifiques normaux représente donc une combinaison originale d'espèces, même si un nombre appréciable de celles-ci se retrouve dans d'autres combinaisons, d'où l'existence d'espèces compagnes, du moins au niveau ou l'où travaille ordinairement, c'est-à-dire celui des espèces de type linnéen. Mais il faut aussi comprendre qu'un ensemble spécifique normal est formé de la réunion des ensembles spécifiques, de cardinaux toujours très inférieurs au sien, d'un certain nombre d'individus d'association qui correspondent, eux aussi, chacun à l'une des C combinaisons simples théoriquement possibles des E espèces : d'où le caractère de catégorie polythétique de l'association végétale. Dans ce raisonnement, j'ai employé le mot espèce pour ne pas trop dérouter le lecteur : en réalité, il eut été épistémologiquement plus correct d'utiliser celui de binôme, car les relevés et les ensembles spécifiques normaux ne sont, en fait, que des listes de binômes.

Ayant ainsi bien précisé ce qu'en phytosociologie sigmatiste il faut entendre par individu d'association, association végétale et espèce caractéristique, il faut souligner que toutes les considérations mathématiques surajoutées ne sont soit que des tentatives de justification théorique *a posteriori*, soit que de simples techniques destinées, notamment, à faciliter le triage des relevés. En d'autres termes, la mathématisation de la phytosociologie, pourtant très souhaitable, doit être considérée comme un moyen et non une fin.

En opérant, à partir des ensembles spécifiques normaux des associations, comme on l'a fait à partir des relevés des individus d'association, on peut classer les associations végétales en classes plus extensives, les **alliances**. En réitérant le processus, on peut distribuer les alliances entre des **ordres**, ceux-ci entre des **classes**. A chaque niveau, il est possible de reconnaître des espèces caractéristiques, à savoir des espèces caractéristiques d'alliance, d'ordre, de classe. Il va de soi que l'extension d'une catégorie est supérieure à celle des catégories qui lui sont subordonnées : par exemple, l'ensemble spécifique normal d'une alliance est formé de la réunion des ensembles spécifiques normaux des associations dont elle est constituée. Il en résulte qu'une espèce caractéristique de cette alliance peut, si elle y a une fréquence supérieure, être en même temps caractéristique de l'une des associations de ladite alliance.

Bien entendu, l'ensemble spécifique normal de chaque classe, ordre, alliance d'un territoire floristique, correspond, comme dans le cas des associations, à l'une des C combinaisons simples théoriquement possibles des E espèces de la Flore dudit territoire prises N à N, $2 \leq N \leq E-1$, la valeur de N étant de plus en plus élevée en passant des individus d'association aux associations, puis aux alliances, aux ordres et aux classes. Naturellement, parmi les C combinaisons théoriquement possibles, dont le nombre est extraordinairement grand, une partie relativement faible seulement est réalisée, ne fut-ce que parce que certaines sont absurdes : on voit mal une combinaison d'hydrophytes et de chasmophytes, par exemple! Cette remarque conduit tout naturellement à envisager la signification des associations végétales et des unités phytosociologiques en général.

Il tombe sous le sens, encore qu'il semble que cela ne soit pas évident pour tout le monde, et l'on se demande bien pourquoi, que deux listes d'espèces similaires désignent des conditions d'existence équivalentes et que deux listes d'espèces dissemblables signalent des conditions d'existence distinctes. La majorité des travaux de l'école sigmatiste a, d'ailleurs, contribué à le confirmer expérimentalement.

La coexistence d'espèces dans une station est la résultante des influences réciproques des organismes, du sol et du microclimat. Il s'agit toujours d'un enchevêtrement très complexe d'interactions: vouloir tenter de le débrouiller, ce qui fait l'objet de la synécologie, est parfaitement légitime. Mais il faut être très conscient de la difficulté de l'entreprise qui ne peut et ne pourra certainement conduire qu'à des «explications» partielles. Or, il est des circonstances où l'on peut avoir besoin d'une grande rigueur dans la caractérisation des habitats sans qu'il soit pour autant indispensable d'en analyser les composantes.

Tel qu'il est ordinairement défini, l'**habitat** est un type d'environnement; en d'autres termes il y a la même relation entre environnement et habitat qu'entre individu d'association et association végétale.

Comme je l'ai déjà plusieurs fois souligné (Guinochet, 1955, 1967, 1968, 1973), mais il y a des choses qu'il n'est pas mauvais de répéter souvent, l'**environnement**, c'est-à-dire l'ensemble des conditions énergétiques, physiques, chimiques et biologiques qui règnent au voisinage immédiat des organismes, est un concept fondamentalement relatif: s'il n'y avait pas d'environnés il n'y aurait pas d'environnement. On ne peut donc concevoir un environnement que par rapport à un objet déterminé, qu'il s'agisse d'un gène, d'un organite cellulaire, d'une cellule, d'un tissu, d'un organe, d'un organisme, d'une population ou d'un clone, d'un individu d'association, d'une phytocénose, etc. . Par conséquent l'habitat, qui est un type d'environnement, ne peut être, lui aussi, défini qu'en fonction d'un autre concept, soit, pour ce qui nous intéresse ici, une unité systématique ou une unité phytosociologique. Il en résulte qu'il ne peut y avoir d'environnement préexistant, mais seulement des conditions préexistantes : ce n'est qu'à partir du moment où un organisme, par exemple, est introduit dans le système qu'est réalisé son environnement, environnement qu'il contribue, du reste, à façonner par les intéractions qui s'établissent entre lui et les conditions préexistantes.

Cela dit, il faut bien comprendre que si une association végétale—ou autre unité phytosociologique—désigne, de la manière à la fois la plus précise et la plus concise, les habitats de ses espèces constituantes, ceux là ne sont, au moins théoriquement, pas identiques : en effet, l'habitat d'une espèce d'une association végétale est défini par l'ensemble des environnements de chacune des populations par lesquelles elle est représentée dans les individus de ladite association où elle est présente, en entendant, ici, par population un ensemble d'individus c'est-à-dire de phénotypes de génotypes directement issus d'une reproduction sexuée ou de clones. Or, pour un individu d'association, son environnement c'est le sol, l'atmosphère et ce qui la traverse, les individus d'association des autres associations végétales et des associations animales de la biocénose dont il fait partie, ainsi que les biocénoses immédiatement voisines ; tandis que pour une population d'une espèce de cet individu d'association, son environnement c'est tout cela *plus* les populations de toutes les autres espèces présente dans celui-là : c'est à la fois pour cette dernière raison et pour le fait que chacune des espèces participe à l'association pour des aptitudes qui lui sont propres, en fonction, en quelque sorte, de son tempérament, que les environnements des populations des diverses espèces d'un individu d'association ne peuvent pas être considérés comme étant rigoureusement identiques.

Etant donnée cette spécificité des habitats à l'intérieur de l'association végétale, on peut concevoir que pour une espèce donnée des habitats similaires puissent être réalisés dans plusieurs associations végétales par ailleurs assez différentes, ce qui peut être le cas pour de nombreuses espèces compagnes. Cependant, pour aussi similaires que puissent être ces habitats, il est exclu qu'ils soient identiques puisqu'une composante, au moins, de l'environnement, la composition floristique, est différente. En reliant cela à ce que l'on sait de la dynamique génétique des populations et de l'influence de la sélection sur celle-là, il est plus que vraisemblable que les populations par lesquelles une espèce est représentée dans diverses associations végétales n'ont pas des compositions génotypiques tout à fait semblables. Bien que, dans la majorité des cas, les différences doivent être suffisamment faibles pour n'avoir pas de répercussions phénotypiques aisément décelables, on a, cependant, déjà pu rassembler un nombre appréciable d'exemples de nature à corroborer cette hypothèse: ils portent sur *Cardamine pratensis* L. (Guinochet, 1946), *Molinia coerulea* Moench (Guinochet et Lemée, 1950), *Plantago coronopus* (Guinochet et Gorenflot, 1952) *Plantago serpentina* et *P. alpina* (Cartier, 1965) divers taxons de *Festuca ovina* L. sensu lato (Bidault, 1968), divers *Myosotis* (Blaise, 1970, 1972), divers taxons de *Centaurea jacea* L. sensu latissimo (Gardou, 1970).

Il y a donc une concomitance évidente entre l'évolution et la genèse des taxons et celles des associations végétales et autres unités phytosociologiques, ce qui est, d'ailleurs, confirmé par un autre type d'observations fondées sur la chorologie et la composition floristique comparées de ces unités phytosociologiques. Il résulte de ces considérations que les disjonctions géographiques jouent un rôle important dans la différenciation phytosociologique. Ainsi, en Europe, la genèse d'un grand nombre des unités phytosociologiques qui y existent actuellement peut être expliquée, comme celle de nombreux taxons, par les migrations végétales consécutives aux variations climatiques des époques glaciaires et post-glaciaires, événements qui ont provoqué à la fois des disjonctions d'aires et des remaniements génétiques par hybridations résultant de la rencontre de populations isolées les unes des autres pendant plus ou moins longtemps.

Comme exemples d'unités phytosociologiques qui ont pu se différencier par disjonction d'une unité ancestrale au cours des réchauffements climatiques postglaciaires, on peut citer le cas des quatre alliances climatiques de l'étage alpin des massifs montagneux européens, à savoir le Festucion supinae des Pyrénées, le Caricion curvulae des Alpes et des Carpathes, le Seslerion comosae des Balkans, le Juncion trifidi de Scandinavie, qui comportent des taxons vicariants des genres *Festuca*, *Phyteuma*, *Jasione*, etc. . On peut également citer le cas des Festucion eskiae des Pyrénées et Festucion variae des Alpes, ou encore, dans l'ordre des Seslerietalia variae, celui des Festucion scopariae des Pyrénées, Seslerion variae des Alpes et du Jura, Festucion pungentis et Seslerion tenuifoliae balkaniques, Seslerion bielzii des Carpathes. Et l'on pourrait multiplier les exemples.

La comparaison de certaines unités phytosociologiques d'Europe occidentale et d'Amérique du Nord orientale est également très instructive. La première de ces comparaisons est due à Conard (1954) à propos des associations des Ammophiletea, comparaison qui l'a conduit à la conclusion suivante:

«La similitude ou l'identité de la plupart des espèces de ces associations indique l'existence d'un *Ammophila* circumboréal quand l'Amérique du Nord et l'Europe étaient suffisamment en continuité pour posséder une flore uniforme... Il est impossible de savoir quelle espèce contemporaine est la plus proche de la forme ancestrale. Nous pouvons suggérer :

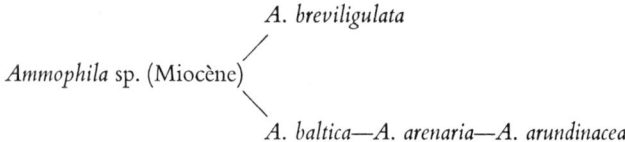

Voilà qui peut donner à penser que *A. arundinacea* est l'espèce la plus proche de celle des rivages chauds du Miocène et qu'elle a survécu seulement sous les climats méridionaux appropriés. Ce qui suggèrerait :

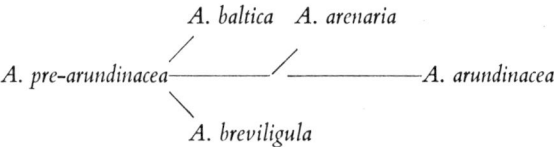

La phylogénie de l'association suit celle de l'espèce caractéristique d'*Ammophila*. Dans l'alliance Ammophilion borealis nous avons les associations :

Miocène : **Ammophiletum pre-arenariae**
Ere contemporaine : **Ammophiletum arundinaceae.** Méditerranée
Euphorbio-Ammophiletum arenariae. Ouest de l'Europe centrale
Elymo-Ammophiletum balticae. Nord-Ouest de l'Europe
Ammophiletum brevilligulatae. Nord-Est de l'Amérique du Nord. »

Puis Nordhagen (1954), Lebrun (1961), Lüdi (1961), Medwecka-Kornas (1961), Grandtner (1962), on comparé diverses unités phytosociologiques forestières. Voici, par exemple, le tableau de concordance établi par Medwecka-Kornas (p. 129).

De telles paires de groupements vicariants, avec des espèces soit identiques, soit vicariantes, résultent certainement de disjonctions de groupements ancestraux préglaciaires.

Il est assez remarquable que dans cette comparaison on ait été conduit à mettre en correspondance des associations d'Amérique avec des alliances d'Europe. On ne peut, en effet, pas s'empêcher de rapprocher cette observation du fait que, jusqu'à sa conquête par les européens, c'est-à-dire une époque fort récente, l'Amérique du Nord n'avait qu'une très faible densité de population nomade ou semi-nomade, vivant essentiellement de chasse et de cueillette, tandis que l'Europe, particulièrement l'Europe occidentale, a été occupée par des peuples cultivateurs, donc défricheurs, depuis plusieurs milliers d'années, peuples qui ont

TABLE I. Groupements Forestiers Correspondants de l'Amérique du Nord Orientale et de l'Europe (d'après Medwecka-Kornas, 1961)

AMÉRIQUE DU NORD ORIENTALE	EUROPE
Cl. **Querco-Fagetea grandifoliae** Knapp 1957	Cl. **Querco-Fagetea silvaticae** Br-Bl. et Vlieger 1937
Ord. **Aceretalia saccharophori** Knapp 1957	Ord. **Fagetalia silvaticae** (Pawl. 1928 n.n.) Tx et Diem. 1936
Ass. **Aceretum saccharophori**	All. du **Carpinion** Oberd. 1953 et **Fagion** Pawl. 1928
Ord. **Gaultherio-Piceetalia** Br-Bl.	Ord. **Vaccinio-Piceetalia** Br-Bl. 1939
Ass. **Piceo-Abietetum balsameae**	All. du **Vaccinio-Piceion** Br-Bl. 1938, 1939
Ass. **Medeolo virginianae-Aceretum**	All. du **Pino-Quercion** Medw.-Korn, 1959

fait preuve d'un accroissement démographique assez rapide, surtout au cours du dernier millénaire. On peut donc penser qu'avant les importants défrichements qu'elle a subi, l'Europe présentait un visage phytosociologique comparable à celui de l'Amérique du Nord moderne : les associations végétales forestières qui formaient l'essentiel de la végétation devaient être réduites à quelques unités représentées chacune par un très petit nombre d'individus d'association de très grandes dimensions. C'est la situation que l'on observe non seulement en Amérique du Nord, mais encore dans toutes les régions du globe peu touchées par l'action humaine, comme l'auteur de ces lignes a pu l'observer en Sibérie méridionale, ou, encore, comme il ressort des travaux de Mangenot (1955) sur la forêt dense intertropicale en Afrique occidentale.

Ainsi, la carte de la végétation de la Côte d'Ivoire publiée par Mangenot (1955) montre que le Turraeantho–Heysterietum n'est représenté que par un seul immense individu d'association, bien qu'il soit entremêlé de savanes, car nulle part elles ne semblent interrompre sa continuité. De même, le Diospyro–Mapanietum, association végétale forestière pélohygrophile, semble ne former qu'un très grand individu d'association dans la partie N.W. de l'arrière pays, ne cédant progressivement la place à d'autres individus d'association que sur les pentes des massifs montagneux, comme ceux du Tonkoui, près de Man, ou des monts Nimba, à la frontière guinéenne.

Or, pour s'appuyer sur un exemple concret, en Europe, une image comparable serait obtenue si, sur la carte phytosociologique de Pontarlier S.W. (Guinochet, 1955), l'on remplaçait les couleurs attribuées au Mesobrometum et à l'Arrhenatheretum, associations manifestement anthropogènes, par celle qui désigne le Fagetum praealpino-jurassicum: on n'y reconnaitrait que deux individus d'association de cette association au maximum, d'ailleurs très peu isolés l'un de l'autre par l'étroite vallée du Drugeon, et qui s'étendraient bien au-delà des limites de la carte si l'on envisageait une plus grande surface, individus d'association dans lesquels ceux des autres associations végétales—Salicetum cinereopentandrae, Mastigobryo-Piceetum, etc.—apparaitraient inclus sans interrompre leur continuité. Il est vraisemblable que l'on aurait là une représentation de ce qui existait avant l'implantation humaine.

Il est bien évident que lorsqu'un individu d'association s'étale sur une vaste étendue, il existe de légères variations de composition floristique entre ses parties extrêmes; cependant, ces variations ne peuvent être que graduelles. Même pour les espèces également réparties sur toute la surface d'un tel immense individu d'association, la composition génotypique potentielle de la population par laquelle chacune d'elle est représentée ne peut pas être homogène: les échanges polliniques, même lorsqu'il s'agit d'une espèce allogame, anémogame et anémochore, sont forcément plus fréquents entre pieds voisins qu'entre plantes situées aux confins de l'aire de la population. Mais, là encore, les variations ne peuvent être que graduelles.

De toute manière, il va de soi que lorsqu'un immense individu d'association est fragmenté, quelle que soit la cause de cette fragmentation, les différences de composition floristique et de composition génotypique potentielle des populations entre les fragments ne peuvent que s'accentuer avec le temps. Et voilà pourquoi l'on peut penser que les importants défrichements accomplis en Europe, en contribuant à la fois à disloquer de grands individus d'association et à accentuer les distances séparant les individus, en réduisant leurs dimensions, d'association d'associations préexistantes, donc leur isolement les uns par rapport aux autres, ne sont peut être pas étrangers à la multiplication de certains groupes d'associations végétales suffisamment apparentées floristiquement pour être réunies en alliances, alliances que nous sommes conduits à mettre chacune en correspondance avec une seule association nord-américaine: ainsi, par exemple, tout en tenant compte aussi des rôles possibles du relief, des variations climatiques, lithologiques, etc., ainsi que du fait que les études phytosociologiques sigmatistes ont été, à ce jour, plus poussées en Europe qu'en Amérique, on pourrait voir là l'une des raisons de la complication de la systématique des associations végétales des Fagetalia européens au regard de la simplicité apparente de celle de l'ordre américain des Aceretalia saccarophori.

Que tout cela reste très conjectural ne fait aucun doute. Il semble cependant intéressant de l'envisager comme hypothèse.

Mais ce qui est, en tout cas, certain, c'est que les défrichements ont contribué à l'extension de certains groupements et même à la création de nouveaux habitats. Prenons, comme premier exemple, le cas du Mesobrometum et de l'Arrhenatheretum du deuxième plateau du Jura, auxquels il a déjà été fait allusion. Le Mesobrometum est une association prairiale qui, après déboisement, s'établit et se maintient, grâce au fauchage répété, sur le sol forestier, qu'elle ne modifie guère; si cette pratique cesse, l'association végétale forestière, le Fagetum praealpino-jurassicum, reprend rapidement ses droits. Ainsi, l'abondance, en certains points de la carte, d'îlots de Convallarieto-Coryletum, association pionnière précédant le Fagetum praealpino-Jurassicum, sont des indices certains de la reconquête en cours par cette association forestière de prairies plus ou moins abandonnées du Mesobrometum.

La composition floristique d'une telle association anthropogène a du se constituer à partir de génotypes d'éléments de celles d'associations préexistantes où ils étaient éliminés. Et l'on peut dire la même chose pour l'Arrhenatheretum qui est manifestement dérivé du Mesobrometum sous l'influence de la fumure. A cet égard, il est intéressant de noter, sur la carte, que les individus d'association les plus typiques d'Arrhenatheretum et de Mesobrometum sont respectivement les plus proches et les plus éloignés des villages. Entre, on trouve tous les intermédiaires; mais il ne s'agit pas, à proprement parler, d'un gradient, comme pourrait le laisser supposer la figuration adoptée, par nécessité de schématisation, sur la carte: en réalité chaque pré est plus ou moins proche, par sa composition floristique de l'une ou de l'autre association suivant la régularité et l'importance des fumures appliquées par le propriétaire.

Des cas similaires peuvent être observés, par exemple à l'étage subalpin des Alpes, où l'on ne peut manquer d'être frappé par le nombre d'espèces qu'ont en commun les prairies du Triseto-Polygonion bistortae et les groupements de l'Adenostylion qui ont dû, à la suite des défrichements, partiellement approvisionner la composition floristique de celles-là, mais certainement toujours par des génotypes différents. Les choses se compliquent cependant, un peu, du fait que dans les Alpes méridionales certains groupements du Triseto-Polygonion bistortae présentent, avec le Festuceto spadiceae-Centaureetum uniflorae des rapports similaires à ceux qui existent, dans le Jura, entre l'Arrhenatheretum et le Mesobrometum (Lacoste, 1971, 1972). Mais cela ne change rien à l'affaire, car il est bien évident que la composition floristique d'une association, ou d'une alliance, ne peut pas dériver que d'une seule autre association ou alliance.

On pourrait multiplier ces exemples de genèse d'associations consécutive à la création d'habitats nouveaux par les défrichements humains. Ainsi, il est très vraisemblable qu'à l'étage méditerranéo-montagnard du midi de la France, les associations du type de celles de l'Ononidion striatae, qui couvrent actuellement d'immenses espaces, se sont constituées et étendues consécutivement à la destruction des groupements forestiers du Quercion pubescenti-petraeae suivie d'un important pâturage ovin et ont tiré une bonne partie de leur composition floristique de génotypes fournis par d'autres, plus spécialisées, vivant dans les parois rocheuses, sur les crêtes impropres à tout boisement, etc., telles que celles des Potentilletalia caulescentis et du Genistion lobelii.

Des suggestions similaires peuvent être formulées à propos de l'origine de certaines associations des Rosmarinetalia, des Cisto-Lavanduletea, des Thero-Brachypodietea, dans le midi de la France, des Calluno-Ulicetea, des Sedo-Scleranthetea, etc., ailleurs.

Les défrichements ont aussi contribué à multiplier les situations favorables à l'établissement et à la diversification des groupements de clairières et de lisières, tels que ceux des Epilobietea, des Prunetalia, des Trifolio-Geranietea, etc., groupements qui sont les lieux d'élection des innombrables taxons hybridogènes de genres complexes tels que les *Rubus* et les *Rosa*, ce qui n'est certainement pas par hasard.

Mais l'exemple le moins conjectural, en l'état de nos connaissances à ce sujet, est assurément celui des associations de mauvaises herbes accompagnant les cultures et des associations, rudérales, associations des Secalinetea, des Chenopodietea, des Plantaginetea majoris, des Paspalo-Heleochloetalia. La diversification et l'extension de ces associations a certainement accompagné celles des cultures et des modes de vie : ceux-ci s'étant modifiés constamment, ou presque, au cours de l'histoire, il en a été de même des types d'habitats créés. Cela n'est certainement pas étranger aux difficultés systématiques qu'offrent certains genres, tels les *Chenopodium*, *Atriplex*, *Rumex*, *Polygonum*, etc., bien représentés dans ces types d'associations, difficultés qui sont un indice de leur diversification récente et, même, en cours. Là aussi, la composition floristique de ces associations s'est constituée à partir de génotypes issus de populations d'espèces d'autres associations dans lesquelles ils étaient éliminés. Cela a été incontestablement démontré par Durand (1963) en ce qui concerne le *Mercurialis annua* L. : il a, en effet, établi que cette espèce, extrêmement commune, caractéristique des Chenopodietalia, qui comporte actuellement de nombreux cytodèmes, est dérivée de *M. hueti* Hanry localisé dans les groupements des Asplenietalia petrarchae des rocailles calcaires dans le midi.

Si l'action humaine est un facteur de diversification phytosociologique et systématique très efficace, il y a cependant un seuil au delà duquel elle aboutit au résultat inverse : ainsi, le surpâturage nomade, tel qu'il a sévi en certaines régions d'Afrique du Nord, par exemple, est certainement un facteur d'appauvrissement floristique et de brassage génétique qui ont contribué à une uniformisation phytosociologique se traduisant par quelques individus d'association de très grande surface d'un nombre réduit d'associations végétales d'ailleurs souvent faiblement individualisées (Guinochet, 1954) et, par conséquent, difficiles, quoique non impossibles, à diagnostiquer: la phytosociologie n'aurait, certainement, jamais pu prendre naissance dans de telles régions et ce n'est sûrement pas par hasard qu'elle a vu le jour dans les pays alpins d'Europe occidentale.

Toutes les suggestions précédentes restent pour le moment très conjecturales, en ce sens qu'elles n'ont pas encore été traduites en des faits grâce à des études biosystématiques délibérément orientées dans cette optique. Cependant, outre l'exemple cité du *Mercurialis annua* L., certaines idées qui se dégagent des travaux de taxonomie expérimentale accomplis durant les dernières décades sont de nature à contribuer à leur justification. Il est, ainsi, bien connu que dans de nombreux complexes polyploïdes eurasiatiques, les types diploïdes ancestraux sont disséminés en colonies, généralement réduites et localisées, fréquemment liées à des associations spécialisées, telles que celles des Asplenietea rupestris à titre d'exemple, à la périphérie de l'aire générale du groupe, aire dont la majeure partie est affectée à des polyploïdes, ordinairement des allopolyploïdes, issus de ceux-là. Or, bien souvent, ces polyploïdes relativement récents sont alliés à des associations anthropogènes plus ou moins jeunes également, telles que celles des Arrhenetheretea. Les exemples offerts par *Dactylis glomerata* L. sensu lato (Stebbins et Zohary, 1959), *Centaurea jacea* L. sensu lato (Gardou, 1970), divers *Achillea* (Ehrendorfer, 1959) sont très significatifs, mais il y en aurait beaucoup d'autres.

Quoiqu'il en soit, en dépit de leur caractère hypothétique, les suggestions exprimées au sujet des modalités d'évolution et de genèse des associations végétales et autres unités phytosociologiques permettent de tirer une conclusion certaine: c'est l'indissolubilité de la différenciation phytosociologique et de la différenciation systématique. Cela, Soukatchev l'a, lui aussi, parfaitement bien senti: «la biogéocénose et la phytocénose doivent être considérées comme l'un des facteurs les plus importants de la formation des espèces, par suite, en premier lieu, de l'interaction spécifique des individus d'une même espèce et d'espèces différentes et aussi par suite des particularités créées par l'ensemble des plantes. Bien que ce problème fasse partie du domaine spécial de la biologie de l'espèce et de la formation des espèces, il ne peut, néanmoins, être étudié avec succès par les évolutionnistes et les généticiens qu'en collaboration avec les phytocénologues» (1954, 326), idée que je défends personnellement depuis 1938. En d'autres termes, l'évolution et la genèse des taxons conditionnent en partie, mais en partie seulement, car il faut tenir compte des facteurs de l'environnement, celles des associations végétales et réciproquement. Le déterminisme de celles-là, et, par conséquent, celui des types d'habitat, si l'on a bien saisi ce qui a été exposé, ne peut donc être entièrement compris en n'invoquant que les facteurs physiques de l'environnement d'une part, et, de l'autre, des espèces considérées comme des objets immuables, ce qui semble, pourtant, à travers la bibliographie, correspondre à un état d'esprit très répandu.

Les données phytosociologiques prendront donc de plus en plus d'importance pour la systématique.

En particulier, j'ai longuement expliqué (1969, 1973) pourquoi je pense fermement, avec quelques autres botanistes (cf., à ce propos, Grant, 1971), qu'au moins en ce qui concerne les végétaux la distinction entre les concepts d'espèce génétique fondée sur le critère d'isolement sexuel et d'espèce unité de classification est, et restera longtemps encore, sinon définitivement, inévitable. Autrement dit, tout en reconnaissant que la notion génétique de l'espèce est incontestablement un concept biologique fondamental, il faut se résigner à admettre que la classification ne peut-être intégralement fondée sur elle. Dans ces conditions, on dispose d'une certaine liberté pour circonscrire les espèces unités de classification désignées par des binômes. Et il est évident qu'il ne peut y avoir que des avantages à utiliser cette liberté pour n'attacher des binômes qu'à des classes, c'est-à-dire des espèces unités de classification, conçues d'une manière très extensive, caractérisables sans ambiguïté, autrement dit telles que tout individu quelque peu entraîné à manier une flore, et non pas le seul botaniste spécialiste de tel ou tel groupe, puisse attribuer sans hésiter un échantillon à une seule d'entre elle. Cela ne dispenserait aucunement d'avoir toujours présent à l'esprit que ces espèces ainsi très largement comprises n'ont pas la même signification biologique, ni de décrire leurs variations internes. Mais il serait désormais certainement plus heuristique de décrire cette variation intraspécifique à l'aide d'une terminologie générale, telle que la terminologie en -*dème* (Gilmour, 1952, 1961; Heslop-Harrison, 1954; Walters et Briggs, 1969, etc.) dont on ne peut que s'étonner qu'elle n'ait pas été plus généralement adoptée, plutôt que de persister à multiplier les épithètes latines de ssp. et var. En d'autres termes, si l'on définit la Systématique comme étant l'étude scientifique de la diversité et de la différenciation des organismes et des relations de parenté qui existent entre eux (Heywood, 1967), la classification devrait être considérée comme une technique d'emmagasinage et de recouvrement des résultats de celle-là. Dire que, du point de vue de la biologie, la classification est «la distribution des organismes en classes hiérarchisées» (*ibid.*) n'est pas tout à fait exact: ce sont, en réalité, leurs descriptions, que l'on distribue en classes hiérarchisées.

Depuis la reconnaissance de la notion d'évolution, on a confondu systématique et classification, en admettant que «la botanique systématique a pour but de représenter en un seul corps hiérarchisé, phylogénétiquement ordonné, les rapports de parenté entre les végétaux. Elle est, en un mot, une généalogie» (Emberger, 1960) ou, encore que «les taxa de tout rang ... ne peuvent pas être uniquement définis par la mention expresse des individus qui en relèvent ou par l'énumération des caractères de ces individus mais seulement en spécifiant implicitement ou explicitement les relations de parenté qui existent entre ces individus» (Simpson, 1967), pour ne citer que deux auteurs récents. Que cette conception de la Systématique, très généralement adoptée depuis plus d'un siècle et demi, ait été très heuristique, cela ne fait aucun doute. Mais, depuis quelques années, de plus en plus nombreux sont ceux qui se demandent si cela n'est pas une erreur que de vouloir obstinément persévérer dans cette voie, où la phylogénie et la classification ont toutes deux plus à perdre qu'à gagner. En effet, une classification aisément maniable par l'esprit humain ne peut, apparemment, être que dichotomique; par conséquent, en voulant l'ajuster sur la phylogénie, cela conduit inconsciemment, mais inévitablement, à n'envisager que des relations phylogénétiques également dichotomiques, ce qui constitue un indéniable «obstacle épistémologique» (Bachelard, 1957). D'autre part, rendre la classification, et, concomitamment, la nomenclature, tributaires des spéculations, voire, parfois, de ratiocinations, phylogéniques, leur confère une très préjudiciable instabilité. Ce que j'ai appris, avec un indicible effarement, lors du dernier Symposium de «*Flora Europaea*», à savoir qu'il a été créé, entre 1965 et 1970, sept mille et quelques noms nouveaux de taxons pour la seule flore vasculaire d'Europe, n'a fait que m'ancrer dans cette idée. Cette inflation de noms, qui ne peut, en grande partie, provenir que de démembrements inconsidérés couverts d'un quelconque prétexte phylogénique—y compris le critère d'isolement sexuel—n'est pas un progrès: c'est, purement et simplement, l'assassinat de la Botanique. A moins que l'on admette qu'elle doive être parfaitement ésotérique, ce à quoi je me refuse personnellement. En écrivant cela, et je l'écris comme je le pense, je m'attends à être accusé de faire fi des données phylogéniques et biologiques, et, d'une manière générale, de tout le travail accompli au cours d'un siècle et demi en souhaitant un retour à l'époque pré-analytique qui a précédé celle-là, que l'on peut qualifier d'analytique. Mais il est aisé de répondre que le travail de Systématique analytique intensif auquel se sont livrés nos prédécesseurs en multipliant les espèces, sous-espèces, variétés, formes, était indispensable pour mettre en évidence la nature et l'importance de la variabilité dans les divers groupes; qu'il n'est pas question de revenir aux espèces extensives conçues superficiellement, par manque d'informations, de la période préanalytique; mais qu'il nous appartient, maintenant, de procéder, grâce aux moyens de traitement des données dont nous disposons et qui ne peuvent que s'améliorer, à des *synthèses motivées, tenant compte de l'ensemble, ou du moins du maximum possible, des connaissances acquises*, ce qui n'a rien à voir avec un retour en arrière.

En parlant de l'ensemble des connaissances, il va de soi que cela comporte, *ipso facto*, les données vérifiées concernant la phylogénèse et l'isolement sexuel, données qui peuvent fournir des caractères, au même titre que celles d'ordre morphologique, anatomique, cytologique, chimique, chorologique, écologique et phytosociologique, etc. Il est, d'autre part, bien évident qu'une classification établie en tenant compte impartialement du maximum possible de données doit, dans la majorité des cas, être un reflet approximatif de la phylogénèse: mais c'est dans ce cas une conclusion et non le critère privilégié de la classification.

Cela dit, il faut souligner que, contrairement à ce que l'on imagine trop souvent, les techniques numériques d'ordination et de classification ne suppriment aucunement l'intervention du jugement: elles permettent essentiellement d'évaluer avec rapidité et sécurité les proximités des objets comparés en fonction d'un nombre de données que notre esprit serait incapable de prendre en considération simultanément. Mais il nous reste la décision finale concernant, entre autres, la délimitation des classes. En ce qui concerne les espèces, cette décision devrait être guidée par deux principes: d'une part, celui qui avait déjà été adopté par Briquet (1899, 1910), dont les exposés sur ce sujet sont trop meconnus des théoriciens de la Systématique, à savoir que l'on ne doit attribuer de binômes spécifiques qu'à des ensembles séparés les uns des autres par des solutions de continuité très nettes; d'autre part, au moins en ce qui concerne les Spermatophytes et Ptéridophytes, elles doivent pouvoir être diagnosticables sans ambiguïté sur le terrain grâce à quelques caractères marqueurs fidèles et aisément repérables: ils ne peuvent, évidemment, être trouvés que parmi ceux de morphologie macroscopique et parmi les aptitudes écologiques, aptitudes qui sont exprimées le plus complètement et le plus précisément par les liaisons phytosociologiques. Certains ne manqueront pas de parler de cercle vicieux en remarquant que l'on prétend diagnostiquer des espèces à l'aide d'unités phytosociologiques elles-mêmes définies par des listes d'espèces. Il s'agit pourtant d'un type de démarche intellectuelle très classique: on commence par définir des espèces sur la base de leur morphologie, à l'aide de celles-ci l'on définit des unités phytosociologiques qui permettent, ensuite, d'en compléter ou rectifier la diagnose. En effet, quel que soit le type de caractère invoqué, qu'il soit d'ordre caryologique, chimique, etc., comment procède-t-on? L'on part d'espèces définies par leur morphologie macroscopique dont on complète la description, voire rectifie la diagnose, à l'aide des nouvelles données. La situation est la même dans le cas du critère phytosociologique, critère qui revient à déterminer, pour une espèce initialement définie sur la base d'autres attributs, quelles sont les espèces avec lesquelles elle est ordinairement associée, ce qui est un caractère qui lui est propre comme les autres. D'ailleurs certaines Flores très classiques, comme celle de Hegi, complètent assez régulièrement la description des espèces par une liste de celles que l'on trouve le plus fréquemment avec elle: remplacer cette liste par un nom d'unité phytosociologique est à la fois plus concis et plus complet, et c'est, d'ailleurs, ce que font certains auteurs de la deuxième édition de la dite Flore de Hegi. C'est également ce qu'ont fait Rothmaler, Soó, Oberdorfer. Dire que remplacer une simple liste d'espèces par un nom d'unité phytosociologique est à la fois plus précis et plus complet est évident pour les phytosociologues. Pour ceux qui en douteraient encore, je citerai un auteur, Whittaker (1962) qui ne peut être suspecté de parti pris favorable à l'égard de la phytosociologie sigmatiste, auteur sous la plume de qui l'on peut lire: «quand l'étude phytosociologique d'un territoire est suffisamment avancée, ses types de communautés peuvent être présentés selon un canevas systématique (e.g. Tüxen, 1937; Braun-Blanquet, 1948-49). Les types de communautés peuvent alors être définis avec concision par leurs espèces diagnostiques, et leurs subdivisions, leurs relations avec l'habitat, leurs successions et leur aménagement peuvent être indiqués. Aucun autre système de classification n'a rendu possible une classification comparable, à la fois dans les grandes lignes et dans le détail, de la végétation d'une région. L'écologiste étudiant une telle classification peut la voir comme une structure formalisée et relativement pauvre, fragmentant les dessins de la végétation en unités artificiellement introduites de force dans une hiérarchie, un exposé squelettique qui lui apporte peu de ce qui l'intéresse. Il apparaît aussi être vrai, pourtant, que, pour l'initié familier avec le système et avec la végétation, de telles classifications réunissent un monde d'informations sur les relations des espèces, les types de communautés, et les relations avec l'habitat, tel qu'il ne peut guère en être réunies avec une égale efficacité par aucun autre système» et encore «pour un travail intensif, détaillé, en écologie appliquée spécialement, l'approche floristique et les concepts d'espèces diagnostiques peuvent s'avérer précieux de compagnie avec des méthodes quantitatives. On peut donc, sans défendre une standardisation universelle de ce système, suggérer que les mérites et ses réalisations justifient un examen renouvelé et favorable de la part des écologistes».

En conclusion, si l'on veut bien y réfléchir, la phytosociologie sigmatiste apparaît comme une floristique améliorée, rationalisée: elle procure, ainsi, d'excellents éléments diagnostiques des habitats, et peut donc contribuer puissamment aux études de biosystématique en permettant, notamment, de mettre en évidence des phénomènes de sélection. Elle est, par conséquent et de surcroit, susceptible de fournir d'excellents caractères systématiques. Par exemple, la distinction des *Brachypodium pinnatum* (L.) P.B. et *Br. phoenicoides* R. et S., essentiellement basée sur des caractères de nervation foliaire, est assez subtile pour que nombre d'excellents auteurs les aient réunis en qualité de sous-espèces d'une même espèce: en France, sur le rivage de la Méditerranée l'on est sûr d'avoir affaire au *Br. phoenicoides*, dans le bassin parisien au *Br. pinnatum*. Mais il est des régions, comme à l'étage montagnard des Alpes maritimes, où les deux espèces entrent en contact, où l'on trouve des formes de transition, où la décision en faveur de l'un ou de l'autre est souvent difficile à prendre sur la base des caractères morphologiques traditionnels; alors je pose la question: peut on sérieusement contester que, dans ces cas litigieux, on a plus de chances de serrer la réalité génétique, donc biologique, de plus près en décrétant que l'on a affaire au *B. phoenicoides* quand on le trouve avec un cortège floristique d'associations franchement méditerranéennes, telles que celles des Thero-Brachypodietalia, Holoschoenetalia, Rosmarinetalia, Quercetalia ilicis, etc., et au Brachypodium pinnatum dans les autres cas?

BIBLIOGRAPHIE

BACHELARD, G. (1957). «La formation de l'esprit scientifique. Contribution à une psychanalyse de la connaissance objective.» (3ème édition) J. Vrin édit., Paris, 256 pp.

BIDAULT, M. (1968). Essai de Taxonomie expérimentale et numérique sur *Festuca ovina* L. s. l. dans le Sud-Est de la France. *Rev. Cytol. et Biol. Vég.* **31**, 217–356.

BLAISE, S. (1970). «Révision systématique de quelques taxons du genre *Myosotis* L.» Thèse Univ. Paris-Sud, Orsay.

BLAISE, S. (1972). Problèmes taxonomiques posés par l'homogénéité apparente du genre *Myosotis*. *Candollea* **27**, 1.

BRIQUET, J. (1899). Observations critiques sur les conceptions actuelles de l'espèce végétale au point de vue systématique. *In* E. Burnat, «Flore des Alpes maritimes», vol. **3**, Genève et Bâle, Georg et Cie, Libraires-éditeurs, Lyon.

BRIQUET, J. (1910). «Prodrome de la Flore Corse.» Genève et Bâle, Georg et Cie, Libraires-éditeurs, Lyon.

CARTIER, D. (1965). Une exemple d'introgression naturelle chez *Plantago alpina* L. et *Plantago serpentina* All. *Bull. Soc. bot. Fr.* **112**, 379–389.

CARTIER, D. (1965). Caryologie des plantains de la section *Oreades* Decne. *C. r. hebd. Séanc. Acad. Sci., Paris* **261**, 4475–4478.

CARTIER, D. (1970). «Etude biosystématique de quelques espèces du genre *Plantago* (Tourn.) L. (sections *Coronopus* DC. et *Oreades* Decne.).» Thèse Univ. Paris-Sud, Orsay.

CONARD, H. S. (1954). Phylogeny and ontogeny in plant sociology. *Vegetatio* **5–6**, 11–16.

DURAND, B. (1963). «Le complexe *Mercurialis annua* L. s. l.: une étude biosystématique.» Masson et Cie, Paris.

EHRENDORFER, F. (1959). Differentiation-Hybridization Cycles and Polyploidy in *Achillea*. *Cold Spring Harbor Symp. Quant. Biol.* **24,** 141–152.
EMBERGER, L. and CHADEFAUD, M. (1960). «Traité de Botanique.» 2 tomes, Masson et Cie, Paris.
GARDOU, C. (1970, 1972). Recherches biosystématiques sur la section *Jacea* Cass. et quelques sections voisines du genre *Centaurea* L. en France et dans les régions limitrophes. *Thèse Univ. Paris-Sud et Fedd. Repert.* **83,** 5–6, 311–472.
GILMOUR, J. S. L. (1952). Taxonomy and philosophy. *In* "The New Systematics" (J. Huxley, ed.). Oxford University Press, pp. 461–474.
GILMOUR, J. S. L. (1961). A decade of nomenclature. *Appl. Gard. Soc. Bull.*
GILMOUR, J. S. L. and HESLOP-HARRISON, J. (1954). The deme terminology and the units of micro-evolutionary change. *Genetica* **27,** 147–161.
GRANDTNER, M. (1962). Sur les forêts du Sud de la Scandinavie et du Québec. *Bull. Soc. r. Forest. Belg.* **69,** 10.
GRANT, V. (1971). "Plant Speciation." Columbia University Press, New York and London. 435 pp.
GUINOCHET, M. (1938). «Etudes sur la végétation de l'étage alpin dans le Bassin supérieur de la Tinée (Alpes-Maritimes).» Thèse de Doctorat es-sciences, Lyon, Bosc. et Riou, et *Comm. S.I.G.M.A.* n° 59.
GUINOCHET, M. (1946). Sur l'existence, dans le Jura Central, de races écologiques aneuploïdes et polyploïdes chez *Cardamine pratensis* L. *C. r. hebd. Séanc. Acad. Sci., Paris* **222,** 1131–1133.
GUINOCHET, M. (1954). Sur les fondements statistiques de la phytosociologie et quelques unes de leurs conséquences. *Veröff. Geobot. Inst. Rübel Zürich* **29,** 41–67.
GUINOCHET, M. (1954). Réflexions sur l'état actuel de nos connaissances phytosociologiques en Afrique du Nord. *Vegetatio* **5–6,** 18–22.
GUINOCHET, M. (1955). «Logique et Dynamique du Peuplement Végétal (Phytogéographie, Phytosociologie, Biosystématique, Applications Agronomiques).» Masson et Cie, Paris.
GUINOCHET, M. (1955). *Carte phytosociologique au* 1/20 000 *Pontarlier Sud-Ouest*, publiée par le Service de la carte des groupements végétaux de la France du C.N.R.S. et par l'institut géo. nat. de France 1948.
GUINOCHET, M. (1967). L'Ecologie végétale: quelques remarques sur ses fondements et ses objectifs. *Mises à jour scient.* **1,** 387–402.
GUINOCHET, M. (1968). Continu ou discontinu en Phytosociologie. *Bot. Rev.* **34,** 273–290.
GUINOCHET, M. (1969). Quelques problèmes de biosystématique méditerranéenne et alpine. "V Simposio de *Flora Europaea*". *Publ. Univ. Sevilla* 177–201.
GUINOCHET, M. (1972). Phytosociologie. *In* «Encyclopedia Universalis».
GUINOCHET, M. (1973). «Phytosociologie.» Masson et Cie, Paris.
GUINOCHET, M. and GORENFLOT, R. (1952). Sur l'existence de formes tétraploïdes chez *Plantago coronopus* L. *C. r. hebd. Séanc. Acad. Sci., Paris* **234,** 2482–2484.
GUINOCHET, M. and LEMÉE, G. (1950). Contribution à la connaissance des races biologiques de *Molinia coerulea* (L.) Moench. *Rev. gén. Bot.* **57,** 565–594.
HEGI, G. (1906 et années suivantes). «Illustrierte Flora von Mitteleuropa.» J. F. Lehmanns Verlag, Munich.
HEYWOOD, V. H. (1967). "Plant Taxonomy." Edward Arnold, London.

LACOSTE, A. (1971). Les groupements à *Festuca spadicea* L. dans les Alpes maritimes et la définition d'un *Festucetum spadiceae* des Alpes austro-occidentales. *Ann. Univ. Besançon. Cahiers de géographie* no. 21, 45–62.

LACOSTE, A. (1972). «La végétation de l'étage subalpin du bassin supérieur de la Tinée (Alpes-Maritimes). Application de l'analyse multidimensionnelle aux données floristiques et écologiques.» Thèse Univ. Paris-Sud, Orsay.

LEBRUN, J. (1961). Quelques remarques sur la Flore et la végétation du Canada (Ontario méridional, Québec, région de Montréal). *Vegetatio* **10**, 25–41.

LÜDI, W. (1961). Botanische Streufzüge durch die Rocky Mountains Nordamerikas. *Ber. Geobot. Inst. E. T. H. Stift. Rübel.* **32**, Festschr. E. Schmid.

MANGENOT, G. (1955). Etude sur les forêts des plaines et plateaux de la Côte d'Ivoire. *Etudes éburnéennes* IV.

MEDWECKA-KORNAS, A. (1961). Some floristically and sociologically corresponding forest associations in the Montreal region of Canada and Central Europe. *Bull. Acad. Pol. Sci. Lett.* **9**, 6.

NORDHAGEN, R. (1954). Vegetation units in the mountain areas of Scandinavia, In "Aktuelle Probleme der Pflanzensoziologie." *Veröff. der Geobot. Inst. Rübel* **29**, 81–95.

OBERDORFER, E. (1958). "Pflanzensoziologische Exkursionsflora für Süddeutschland." (2ème édition.) Verlag E. Ulmer, Stuttgart.

ORLOCI, L. (1968). Information analysis in phytosociology: partition, classification and prediction. *J. theor. Biol.* **20**, 271–284.

ROGERS, D. J. (1970). A preliminary ordination study of forest vegetation in the Kirchleerau area of the Swiss Midland. *Ber. Geobot. Inst. E. T. H. Stift. Rübel* **40**, 28–78.

ROTHMALER, W. (1963). "Exkursionsflora von Deutschland." Volk und Wissen, Volkeigener-Verlag, Berlin.

SIMPSON, G. G. (1967). "Principles of Animal Taxonomy." (3ème edition.) Columbia University Press, New York.

Soó, R. (1964 *et seq.*) "A Magyar Flóra és Vegetáció." Akad. Kiadó, Budapest, 4 vols.

SOUKATCHEV, V. N. (1954). Quelques problèmes théoriques de la phytocénologie. *In* «Essais de Botanique». Ed. Acad. Sc. U.R.S.S., Moscou-Leningrad, **1**, 310–330.

STEBBINS, G. L. and ZOHARY, D. (1959). Cytogenetic and evolutionary studies in the genus *Dactylis*. I. Morphology, distribution and interrelationships of the diploid subspecies. *Univ. Calif. Publ. Bot.* **31**, 1.

WALTERS, M. and BRIGGS, D. (1969). «Les Plantes: Variations et Évolution.» (Trad., J. Kovoor.) L'univers des connaissances, Hachette, Paris.

WHITTAKER, R. H. (1962). Classification of natural communities. *Bot. Rev.* **28(1)**, 1–239.

6 | Taxonomy of Cryptogams and Cryptogam Communities

J. J. BARKMAN
Biologisch Station, Wijster, Netherlands

Abstract: Phytosociology and biocoenology are sciences which study all aspects at one level of integration. Synecology and syntaxonomy are only two of these aspects. Ecology is an "aspect science", studying the relation of living material to the environment at different levels, of which the community level is only one.

Syntaxonomy and idiotaxonomy have many features in common, both being sciences in their own right and auxiliary to many other biological disciplines. Both are hierarchical, and use as many characters as possible. It is pointed out that both sciences sometimes break down (gradual transitions between taxa and between syntaxa, multi-dimensional subdivision of taxa and syntaxa). Classification and ordination are not contradictory, but complementary methods.

There are also differences between syntaxonomy and idiotaxonomy. The individual in syntaxonomy (i.e. the association stand) is a much vaguer notion than the plant individual. Syntaxonomy is essentially a statistical science (but so is modern idiotaxonomy). Syntaxonomy lacks a genetical basis.

Ecotypes and cytological races may be of great value to the phytosociologist, but a warning is given about the risks of circular reasoning and practical difficulties.

Plant sociology may contribute to plant taxonomy in yielding new taxa, new localities of known species, and better knowledge of intra- and inter-specific variability.

Cryptogam communities usually cannot be separated from plant communities, and these again are an integral part of biocoenoses (plant–animal communities). It is pointed out, however, that most plant communities, as they are studied, are "cormophytocoenoses", i.e. they are taxocoenoses just like cryptogam communities. This is regrettable. Bryophytes and lichens are often somewhat neglected although they may outnumber the higher plants in species. Fungi are usually even more numerous and also play a very important rôle in the ecosystem. They are rarely included in phytosociological investigations owing to taxonomic difficulties, the invisibility of the vegetative parts and the often sporadic occurrence of fruit-bodies.

Cryptogam communities are ecologically dependent on the rest of the phytocoenosis, but they may be independent in a syntaxonomic sense. Separate treatment as synusiae and even separate classification of these synusiae may therefore be profitable, in addition to, not as a substitute for, the classification of syntaxa of entire ecosystems (plant communities).

The relations between the taxonomy of cryptogams and of the communities they form constitute only one aspect of a wider problem: the relations between taxonomy as a whole and plant sociology. The study of cryptogam communities is a part of phytosociology which, in turn, is an aspect of biocoenology, i.e. the study of entire ecosystems. In English-speaking countries the word "ecology" has often been used to comprise both the latter science and the study of the interrelations between individual organisms and their environment. This is neither logical nor practical.

The term ecology should be restricted to the study of environmental relations of single organisms (autecology), populations of single species (population ecology) and plant communities or ecosystems (synecology). We may even speak of the ecology of organs (leaves, roots, etc.). Ecology is, therefore, a science spanning many different levels, but at each level dealing with only one aspect. It may be regarded, therefore, as an "aspect science".

Plant sociology and biocoenology on the other hand deal with all aspects at one (highly integrated) level, the plant (or biotic) community, which consists of many different species and life-forms. The aspects studied are: floristic (and faunistic) composition, structure, classification and/or ordination, synecology, succession, area and geographical variation, and also recently, mainly due to the International Biological Programme, synphysiology or the study of the performance of entire ecosystems. The latter aspect includes primary and secondary production, litter production, cycling of energy, water, minerals, etc. Synphysiology, in this sense, should not be confused with ecology. Synecology is the only sub-science which ecology *sensu lato* and phyto(bio)coenology have in common.

From what has been said above, it is obvious that plant sociology should be compared with the whole of organismic botany (idiobotany), and, vice versa, taxonomy of organisms (idiotaxonomy) should be compared only with a part of plant sociology, i.e. the classification of community types or syntaxonomy.

Both taxonomies are indispensable auxiliary sciences to all other biological sciences that deal with their respective objects, viz. individuals and communities. Classical or idiotaxonomy is in addition also auxiliary to syntaxonomy: we need types of plant communities with names and descriptions in order to communicate with other scientists, to compare results, to abstract and to generalize. But we need names of plants first before we can describe plant communities.

However, the two taxonomies are also sciences in their own right. Idiotaxonomy is the basis of our understanding of evolution at work; syntaxonomy serves to understand ecology at work. It has even been said that syntaxonomy is a basis for real understanding of the evolution of species, for all plant and animal species and their ancestors live and always have lived within communities, in which constant competitive pressure as well as specific chemical influences from all surrounding species must have largely affected their course of evolution.

Syntaxonomy and idiotaxonomy share many features. Both are hierarchical, both use, in principle, as many characters as possible, both are successful in many cases and fail in others. We know cases where the one-dimensional system fails in plant sociology, particularly at the infra-associational level: some associations may be subdivided into (ecological) subassociations and also into (geographical) vicariants. The two subdivisions are often quite independent and at the same time equivalent, so that a two-dimensional system would be more natural (for instance in the dry heath association Genisto-Callunetum). But exactly the same applies to idiotaxonomy, especially at the infraspecific level, for instance with regard to the form and colour varieties of the blunt periwinkle, *Littorina obtusata* (Dautzenberg and Fischer, 1915), and with regard to the different series of variants (formae) of the fern *Polypodium vulgare*, based on (1) leaf-shape, (2) dentation of the segments, (3) distance between the lobes, (4) colour (cf. van Ooststroom, 1948).

In some cases one association gradually merges into another, the delimitation is arbitrary and ordination seems more appropriate than classification. But the same phenomenon is known in idiotaxonomy as clinal variation within and between species. Yet, clinal variation is much more frequent in community types than within species. The two aspects, classification and ordination of plant communities, are complementary. This is realized nowadays both by continental European "classifiers" and by the (mainly North-American) "ordinators". The controversy between these groups, marked by bitter verbal disputes some 10–15 years ago, has now almost ceased to exist. "Gradient analysis and classification are alternative approaches to the vegetation of a landscape" (Whittaker, 1967).

However, there are also a number of differences between the two sciences. First of all, species have a genetical basis, associations do not. Syntaxonomy is essentially statistical, its units (syntaxa) being based on large numbers of vegetation records. Species may be based on a single specimen. Yet, this difference is gradual and becoming ever more obscure, since modern idiotaxonomy is mainly based on statistical population studies. The concept of individual (i.e. stand), for that matter, is rather obscure in plant sociology. Stands of the same community type may be well separated or may form a network. Besides, two clear-cut stands may be very different in size and therefore not comparable.

Again, the difference is not fundamental, for we all know how difficult it is to define the "individual" in grasses, sedges, bryophytes, fungi, corals, sponges, etc. Also two individuals of a species forming colonies may differ enormously in size.

Both sciences need a set of nomenclatural rules. But whereas in idiotaxonomy these were established many years ago and have been refined ever since, the first draft of an official code for syntaxa has only just been prepared in Prague (November 1972) and is not yet published. Nomenclature of syntaxa is even more complicated, because their names have to be changed not only for the reasons that apply to taxa (priority and change of taxonomic opinion), but also because the names of the taxa on which they are based are altered. This tends to create a double instability of community names. Another difficulty is that we do not have such a thing as a type-specimen. Conservation of type-stands is impossible, as vegetation is liable to change with time. Type-records cannot be checked by other scientists as to homogeneity, completeness, and correct identification of the component species, once the type-stand is lost or altered. The best thing to do is probably to add a photograph and/or vegetation chart of the plot in question to the type-record and to preserve herbarium specimens of all the species present in the sample plot.

Having dealt so far with similarities and differences between the two taxonomies, we may now ask the question, in which way can they help each other? As noted above, plant sociology is impossible without plant taxonomy. The "new systematics", in particular, is of great value, dealing as it does with small species, ecospecies and ecotypes. These are usually much more indicative of certain habitat factors and community types than the classical Linnaean species. The new trend, however, is still mostly confined to vascular plants in Europe, so that the student of cryptogam communities cannot yet profit by it. On the other hand, a warning is in order here.

Whenever a species occurs in different habitats, taxonomists tend to look for minute morphological differences in order to recognize ecospecies or ecological subspecies. Strictly speaking it is not logical to do this and at the same time not to distinguish taxa within species with morphological differences of a similar magnitude, but occurring in the same type of habitat. Besides, what appears to us as quite a different habitat, may actually be the same to the plant, because of compensation of one factor by another, or because the environmental factors that differ do not affect the plant in question.

If taxa are based on habitat differences only and then used to characterize plant communities, we are trapped in a vicious circle. If ecotypes differ also in chemical or cytological characters, but not in morphology, as sometimes occurs, other problems arise. Sometimes only a few samples have been investigated. If samples from a different habitat happen to differ in chemical reaction or number of chromosomes, it is often assumed that this applies to all individuals occurring in that habitat. But even when large numbers of specimens have been checked and the correlation with habitat has been firmly established, it is rather impractical for the plant sociologist to use such "ecotypes" (cryptic polyploids, for instance) in making relevées, for we can hardly expect him to examine the chromosome number each time he finds these plants in an intermediate or a third type of habitat. And if he finds them in one of the known types of habitat (plant communities) and decides on these grounds which cytological or chemical subspecies he has, he can just as well omit the name of the subspecies, because he does not add any information to the relevée!

One may also ask whether plant sociology can contribute to taxonomy. The answer is certainly yes. A substantial contribution is made through floristics, in particular with regard to cryptogams. These small plants are easily overlooked and one of the best ways to find them still is to search a limited area very carefully. Now this is exactly what the phytosociologist does. Once the phytosociologist has discovered in which plant community a rare species occurs, he can go and look for this species much more efficiently in the field. In this way many new localities have been discovered, and in fungi even new species. The investigation of juniper scrub in the Netherlands by the present author alone yielded, between 1964 and 1969, 14 new species of *Galerina*, as well as many additions to the *Galerina* flora of Europe.

Mycosociologists, in order to get a complete list of fungus species present, have to search their sample plots in all seasons. Thus, not only do they find species not normally found, but also they get a better insight into the morphological variation of the fruit-bodies caused by weather conditions. Species that cannot be identified without a microscope have to be sampled each time in each quadrat. If such species are frequent and widespread, the phytosociologist will eventually have examined many more specimens than the floristic botanist or taxonomist and therefore have a better knowledge of the species' variability. In this way so many intermediate forms between *Mycena vitrea* and *M. iodiolens* were found that the two species had to be united (Barkman, unpublished).

We shall now turn our special attention to cryptogam communities. Both in idiotaxonomy and in syntaxonomy we strive towards a "natural classification". A natural system is a system using as many features of the objects to be classified as possible, but no more. Geographical distribution, for instance, is not an attribute that should be used in classifying species or subspecies, nor is habitat in delimiting plant associations. Both procedures would lead to circular reasoning.

To use all characters of vegetation means using all structural and all floristic and faunistic characters, i.e. a list of all plant and animal species present. For practical reasons this is impossible. Animals are nearly always neglected, and so are bacteria, microscopic algae and soil fungi. Even macrofungi are generally omitted from vegetation records, and as to lichens and bryophytes, I regret to say that the published species lists are often very incomplete. To quote just one example, Schwickerath (1933a, b) made a special study of the chalk grasslands (Gentiano-Koelerietum) near Aix-la-Chapelle (Germany) and recorded 14 species of lichens and bryophytes. A careful investigation of the same association in the adjacent area of the Netherlands yielded no less than 88 species of lichens and bryophytes (Barkman, 1953). The distance between them is no more than 20 km, and the altitude, climate and soil are the same.

Our "plant communities" are in fact vascular plant communities, based on a single, although admittedly large, taxonomic group of plants. Such taxocoenoses (van der Maarel, 1965) are quite normal units to zoocoenologists, who work for instance with mollusc communities or with associations of carabid beetles, neglecting all other animals present in the same biotope. In view of the tremendous number of animal species, the consequent specialization of animal taxonomists and the widely different sizes and behaviour of animals, which necessitates quite different techniques to catch them (due also to the fact that most animals live hidden in soil litter or become active only at night) the zoologists may well be excused for this attitude, but it is nevertheless regrettable. A taxonomist may confine himself to a small group of plants or animals, even to monographing a single genus; a biocoenologist cannot afford to do this. He should have a profound knowledge of a great number of plant and animal groups. In fact he never studies all groups of organisms; he does not use all characters of the community. Yet, there is no fundamental difference with taxonomy of species: no taxonomist has ever used all morphological, anatomical, ethological, physiological, biochemical and cytological characters to describe and classify species. This is poor consolation to the phytosociologist. If he has to limit his field of research, it would be more logical to concentrate on certain strata, niches or life-forms of the ecosystem than to study one or two taxonomic groups only. All biocoenoses consist of three major ecological groups of organisms, viz. producers, consumers and reducers. To study only one of these groups would make more sense than to study only the molluscs, the amphibians, the lichens or the ascomycetes. Cormophytes, it is true, usually form the bulk of the producers, but lichens, algae, chemo- and photoautotrophic bacteria also belong here. On the other hand, parasitic and insectivorous higher plants belong to the consumers, together with parasitic fungi and bacteria and most animals.

Holosaprophytes such as several orchids, *Monotropa* and the liverwort *Cryptothallus mirabilis* belong to the reducers, together with saprophytic fungi, slime-moulds, and bacteria, and with humus-eating small animals like collemboles and worms. In other words, the fundamental division into producers, consumers and reducers does not coincide with taxonomic subdivisions.

Plant sociologists defend their attitude by saying that the flowering plants: (a) form the bulk of the biomass (often 99% or more) of the ecosystem; (b) are the most important organisms, being primary producers of organic material on which all other organisms depend; (c) by their mass, their fixed place and often their longevity constitute the framework of the ecosystem, while other living organisms just have to fit in or are excluded. Biocoenoses or ecosystems, according to their opinion, should therefore be based on the rooting green plants only.

In this connection it is evident that cryptogams cannot be treated as a single group. Bryophytes and lichens may be the dominant life-form (raised bog, tundra) or even the only autotrophic plants present (pioneer communities on rocks, sand, lava, burnt soil, thatched roofs, etc.). Usually, however, they form subordinate, dependent communities, so-called synusiae, within the larger ecosystem. An example is the epiphyte and moss carpet on the forest floor; in acid woods these plants may well outnumber the flowering plants. In a stand of oak wood on poor drift-sand (Dicrano-Quercetum) I found 8 vascular plants and 42 bryophytes and lichens, not even counting the epiphytes. In a wet heath (Ericetum) 14 vascular plants were noted, together with 27 bryophytes and lichens. Both examples are from the Netherlands, where the bryophyte and lichen flora is rather poor, as compared with, for instance, Scandinavia or hyperatlantic regions, not to mention tropical and subtropical mountains. It is evident that if we omit bryophytes and lichens from our vegetation records, we may miss the majority of the green plants.

Another question is whether these cryptogams have to be included in the abstract community types of the phytocoenoses or whether they should be regarded as forming separate communities. They have been treated in different ways (for a detailed discussion of the problem, cf. Barkman, unpublished). Lippmaa (1935) and his Estonian school of phytosociology considered all strata of a phytocoenosis to be representatives of different associations, including not only the moss layer, but also the herb layer, the shrub layer and the canopy of a forest. Du Rietz (1921) was of the same opinion, but later changed his ideas. This extreme viewpoint found support mainly in Northern Europe, but has now become obsolete.

The other extreme position was held by some Eastern European plant sociologists, for instance Sukachev in Russia and Soó and Zolyomi in Hungary, and by Clements in the United States. They considered the whole stand of vegetation in a given biotope to be one community, including the cryptogam communities on tree boles, rotten stumps, ant-hills, rocky outcrops, boulders, etc.

The Braun-Blanquet or Zürich-Montpellier school has always favoured an intermediate opinion, keeping apart (as separate, but dependent associations) the communities on different substrata in a forest (rocks and tree boles), but not the terrestrial cryptogams. The difficulty here is that terrestrial moss communities may also occupy the base of tree trunks (in rainy climates even up to several metres), so that the dividing line is often quite unnatural. Within the Zürich-Montpellier school there is a recent trend to split off more and more of such microcommunities from the main association, even seasonal aspects of the same layer.

The point to be stressed is that most bryophyte and lichen synusiae in a plant community are dependent on the rest of the ecosystem in an ecological sense (the terrestrial, as well as, particularly, the epiphytic, but also the epilithic). On the other hand they are often independent in a syntaxonomic sense, i.e. the abstract types to which they belong may occur in different plant associations. In order to get a natural system of synusiae, they must therefore be treated independently from the system of "whole plant communities", but only as synusiae.

This, fourth, viewpoint has already been proposed by Du Rietz (1932). He reiterated it several times, most recently in 1965 and it was accepted by Wilmanns (1962) and Barkman (1968). It means a return to the whole-stand-communities of Sukachev and Clements, including all microcommunities and synusiae, and those on special substrata, but the delimitation and classification of these communities into types (syntaxa) according to modern principles (based on all species present; use of faithful and differential species and affinity indexes). It also implies the possibility of making a separate hierarchical system, based on the same principles, of all synusiae present within the phytocoenoses, including the herb layer, the shrub layer and the terrestrial moss layer considered separately. The two systems are independent and complementary. The first system is obligatory, the second facultative, i.e. all whole-community-stands must be classified but synusiae can be classified separately, if required for a special investigation.

As to fungi, entirely different problems arise. Being heterotrophic, mushrooms never form communities of their own. Yet they are very important members of the ecosystem, being the main cleansers, not only removing dead organic material, but also recycling it. Research in the International Biological Programme has shown that the energy turnover of the fungi may be four times that of all other lower plants and all animals taken together; their biomass may even be ten times greater. In many cases 90% of the humus is decomposed by fungi, only 5% by bacteria and another 5% by all soil-inhabiting animals. In numbers of species fungi may greatly exceed the other plants. As to microfungi (soil fungi), Apinis (1970) recorded more than 400 species in a total of five littoral communities containing only a few dozen higher plants. As to macrofungi, I may quote an example from my own research. In a single stand of juniper scrub at Sleen, Netherlands, I found 38 species of flowering plants and ferns, 35 bryophytes and lichens and 162 macrofungi. Dense spruce forests are even more extreme: for example, 5 vascular plants, 9 bryophytes and 50 macrofungi were found in one stand. We may safely say that normal vegetation records comprise only a small minority of the plant species present.

In the Zürich-Montpellier school, associations and other syntaxa are based on floristic criteria only, such as faithful and differential species. It is to be expected that fungi, owing to their prolific speciation and often extreme habitat specialization, could yield many faithful species and contribute substantially to a better delimitation of community types.

The reasons why they are usually ignored are manifold. Among these, the scattered and expensive literature, the different interpretations of many species by various authors, the lack of complete descriptions and well preserved type-specimens and, especially, the irregular, not to say sporadic, occurrence of fruit-bodies, are the main obstacles. The vegetative parts of the fungus are hidden. A mycologist is therefore in the same position as many zoologists, who can only catch their animals when they are active. In order to know which species of fungi are present, a sample plot area has to be searched several times a year during at least 5–10 years. A single visit in high season yields about 20% of the species present.

A mycosociologist is also in the position in which a plant sociologist would have been in Linnaean times: he has still to undertake a great amount of taxonomic work. Many species have yet to be described, even in Europe. The genus *Galerina* may serve as an example. A monograph of this genus, published in 1935, mentioned 18 species; another world monograph, issued 1964, described no less than 203 species. As noted earlier, between 1964 and 1969 the investigation of juniper scrub in the Netherlands alone yielded 14 new *Galerina* species, as well as many additions to the *Galerina* flora of Europe.

In the last ten years we have been trying to carry out an integrated research of one vegetation type, i.e. ungrazed juniper scrub on non-calcareous soil, in the

Netherlands and adjacent areas. This association, the Dicrano-Juniperetum, can be divided into 5 terrestrial microhabitats, containing 12 microcommunities or associules. These can be divided into 41 different terrestrial synusiae, 23 of which consist of fungi alone, and there are 11 synusiae of epiphytic lichens, bryophytes and fungi. The 52 synusiae are formed by a total of more than 600 species, namely 80 vascular plants, 80 bryophytes and lichens and more than 450 species of fungi, a number which is still increasing, owing to additions to the mycoflora.

REFERENCES

APINIS, A. E. (1970). Das Verhalten der Pilze in bestimmten Graslandgesellschaften (Engl. summ.). In "Gesellschafts-Morphologie." (R. Tüxen, ed.) *Ber. Int. Symp. Vegetationskunde* **1966**, 172–186.

BARKMAN, J. J. (1953). De Cryptogamen. In Diemont, W. H. and A. J. H. M. van de Ven, De kalkgraslanden van Zuid-Limburg. *Publ. Nat. hist. Gen. Limb.* **6**, 21–30.

BARKMAN, J. J. (1968). Das synsystematische Problem der Mikrogesellschaften innerhalb der Biozönosen. In "*Pflanzensoziologische Systematik*". (R. Tüxen, ed.) *Ber. Int. Symp. Vegetationskunde* **1964**, 21–53.

BARKMAN, J. J. (unpublished). Synusial Approaches to Classification. In "Vegetation Synthesis" (unpublished) (R. H. Whittaker, ed.).

DAUTZENBERG, P. and FISCHER, H. (1915). Étude sur le *Littorina obtusata* et ses variations. *J. Conchyol.* **62**, 87–128.

DU RIETZ, G. E. (1921). "Zur methodologischen Grundlage der modernen Pflanzensoziologie." *Akad. Abh. Uppsala.* Vienna.

DU RIETZ, G. E. (1932). Vegetationsforschung auf soziationsanalytischer Grundlage. *Handb. biol. Arb. Meth.* **11(5)**, 293–480.

DU RIETZ, G. E. (1965). Biozönosen und Synusien in der Pflanzensoziologie. In "Biosoziologie". (R. Tüxen, ed.) *Ber. Int. Symp. Vegetationskunde* **1960**, 23–42.

LIPPMAA, T. (1935). La méthode des associations unistrates et le système écologique des associations. *Acta Inst. hort. Bot. Tartu* **4(1/2)**, 1–7

MAAREL, E. VAN DER (1965). Beziehungen zwischen Pflanzengesellschaften und Molluskenfauna. In "Biosoziologie". (R. Tüxen, ed.) *Ber. Int. Symp. Vegetationskunde* **1960**, 184–198.

OOSTSTROOM, S. J. VAN (1948). Pteridophyta, Gymnospermae. In "Flora Neerlandica" **1(1)** (Weevers *et al.*, eds). Amsterdam, 94 pp.

SCHWICKERATH, M. (1933a). Die Vegetation der Kalktriften (*Bromion erecti*-Verband) des nördlichen Westdeutschland. *Bot. Jahrb.* **65(2/3)**, 212–252.

SCHWICKERATH, M. (1933b). "Die Vegetation des Landkreises Aachen und ihre Stellung im nördlichen Westdeutschland." Mayer, Aachen.

WHITTAKER, R. H. (1967). Gradient Analysis of Vegetation. *Biol. Rev.* **42**, 207–264.

WILMANNS, O. (1962). Rindenbewohnende Epiphytengemeinschaften in Südwestdeutschland. *Beitr. naturk. Forsch. SW-Dlds* **21(2)**, 87–164.

7 | The Relevance of Symbiosis to Taxonomy and Ecology, with Particular Reference to Mutualistic Symbioses and the Exploitation of Marginal Habitats

D. H. LEWIS

Department of Botany, University of Sheffield, England

Abstract: Individual plants and animals as well as other categories in taxonomic and ecological hierarchies are, more often than not, symbiotic systems. After a brief consideration of the features of symbiosis in general and mutualistic symbiosis in particular, mechanisms by which selected mutualistic symbioses (coelenterates, lichens and sheathing and vesicular-arbuscular mycorrhizas) exploit habitats low in nitrogen and phosphorus are discussed. These symbioses tolerate nutrient-poor conditions in contrast to nitrogen-fixing partnerships where nitrogen deficiency in terrestrial and aquatic ecosystems is avoided by utilization of gaseous reserves of the aerial environment. Tolerance is achieved by possession of efficient absorption mechanisms and, with respect to phosphorus in soils, an ability to utilize enzymically complex phosphates, slow metabolic turnover and tight nutrient cycling. Partial elimination of competitors, by production of antibiotics, contributes to the success of mutualistic symbioses in marginal habitats.

INTRODUCTION

Most ecologists are concerned with either terrestrial or aquatic ecosystems. A third group of investigators, who do not normally call themselves ecologists, are concerned with the exploitation of a third environment, living organisms. Their field of study is symbiosis.

The only taxonomists who are explicitly involved in the classification and naming of symbioses are lichenological systematists, but even the binomials they bestow on lichens refer to the fungal partner. All other taxonomists restrict their activity to supposedly single (as distinct from dual) entities.

The purpose of this paper is twofold. Firstly, it emphasizes that individuals, taxa, -demes, biotypes, etc. of animals and plants considered by taxonomists and ecologists are, more often than not, symbiotic systems involving at least two organisms. Secondly and more specifically, it considers mechanisms by which some of these symbioses exploit habitats deficient in particular nutrients.

SYMBIOSIS IN RELATION TO TAXONOMY AND ECOLOGY

1. *Symbiosis*

As originally conceived by de Bary, symbiosis meant a living together of usually dissimilar organisms with no overtones of harm or benefit. For reasons elaborated elsewhere (Lewis, 1973) but in contrast to most botanists, I favour a broad view of symbiosis but, nevertheless, recognize a distinction between parasitic and mutualistic symbioses—albeit a distinction that can be environmentally controlled (Stanier et al., 1971, p. 707; Lewis, 1973). Symbioses are very diverse in kind (Buchner, 1965; Henry, 1966; Read, 1970) involving autotrophs with autotrophs, heterotrophs with heterotrophs and autotrophs with heterotrophs. With the exception of brief mention of other associations, this paper will restrict its consideration to mutualistic symbioses between autotrophic organisms and heterotrophs.

2. *The Relevance of Symbiosis to Ecology and Taxonomy*

Consciously or unconsciously, ecologists, phytosociologists and taxonomists only take note of symbiosis when it suits them. For example, it is clear to ecologists that the competitive ability of an organism will be influenced by its parasites. Since these are normally pathogenic, the effect is usually adverse although sometimes of selective advantage: Bradshaw (1959), for example, showed that infection of *Agrostis tenuis* and *A. stolonifera* by the fungus, *Epichloe typhina*, was of competitive advantage in a grazed situation. With regard to mutualistic symbioses, to mention just two examples, the ecological importance of, on one hand, nitrogen fixing associations to the success of legumes and non-legumes in colonizing habitats deficient in nitrogen (Harley, 1970b) and, on the other, of cellulolytic endosymbionts to the life of wood eating termites (Brooks, 1963) have been well recognized. However, only slowly is the ecological significance of mycorrhizal infection becoming acknowledged (Harley, 1968, 1969b).

From a phytosociological viewpoint, many papers have shown how *Monotropa* and several orchids, especially non-green species, are the end product of a chain in which a fungus forms a link between these species and conventionally autotrophic plants (see Harley, 1969a, pp. 224–226, pp. 316–317). These tripartite associations involve septate fungi, probably all basidiomycetes, and similar infections have been recorded for other so-called saprophytes, e.g. the gametophytes of *Cryptothallus*, and some Gentianaceae. Septate fungi have also been found in the non-green Burmanniaceae and Triuridaceae, but aseptate hyphae have also been reported (Harley, 1969a, pp. 266–267). Non-green pteridophyte gametophytes of the Lycopodiaceae and Ophioglossaceae are also infected by several kinds of fungi but a major one is aseptate and akin to those of the most common type of mycorrhiza of angiosperms—the vesicular-arbuscular group caused by the genus *Endogone* (Harley, 1969a, pp. 251–253). Since all these fungi are "digested" within host cells, the associations probably represent further examples of plants parasitic on fungi (see Lewis, 1973) but the sources of nutrients of the fungi have yet to be established. If, however, tripartite symbioses are a common feature of the biology of heterotrophic angiosperms, the following questions can reasonably be asked. Firstly, in general, how frequent are such mycorrhizal fungal bridges between *two* autotrophic species and, secondly in a more particular, phytosociological context, how far is the ability to recognize phytosociological "associations" a reflection of the existence of mycorrhizal fungi, common to more than one species of the "association"? A corollary to the second question is how far is "fidelity" a feature determined by tripartite associations through fungal bridges? The answers to these questions remain to be elucidated, but passage of metabolites from one higher plant to another via common vesicular-arbuscular mycorrhizal hyphae is more than a remote possibility. This has already been demonstrated for carbon compounds between individuals of *Pinus taeda* via their common sheathing mycorrhizal fungus, *Thelephora terrestris* (Reid and Woods, 1969).

Plant taxonomists, when assessing and describing the characters of a particular taxon, rightly ignore individuals that have been morphologically or otherwise disturbed in a major way by pathogens, i.e. it suits their purpose to reject examples of parasitic symbiosis from their samples. However, many times, particularly in nutrient-poor environments, "atypical" plants, similarly rejected from samples, may have been due to their failure to form effective mycorrhizal or nodular unions. The taxonomists "typical" plants that are described in Floras and monographs are those that manifest a particular range of characters as a result of mutualistic symbioses.

MECHANISMS INVOLVED IN THE TOLERANCE OF NUTRIENT DEFICIENCY
1. Characteristics of Mutualistic Symbioses of Autotrophs with Heterotrophs

Since many different autotrophs and heterotrophs associate in a mutualistic manner (see Table I, below), it is difficult to make universal generalizations about their characteristics. Scott (1969) delimited six features—the association should be a *permanent feature of the life cycles of the organisms*; this involves *physical contact between the participating organisms*; these two cardinal features permit physiological interplay either in the form of *unilateral or bilateral movement of metabolites*, or in the form of *amelioration of environmental status*; physiological interplay can also manifest itself by *morphogenetic effects* and by the *production of metabolites that are not formed by either of the organisms separately.* (Italics here correspond to Scott's heavy type.) Scott considered associations which were characterized by at least four of these attributes to constitute a mutualistic symbiosis. Scott's criteria are re-represented in Fig. 1 from which it can be seen that as a result of the two primary features—permanence and intimacy—bilateral exchange of nutrients is possible. This, either directly or via enhanced morphological and/or biochemical attributes, results in an association of greater ecological amplitude than either partner alone.

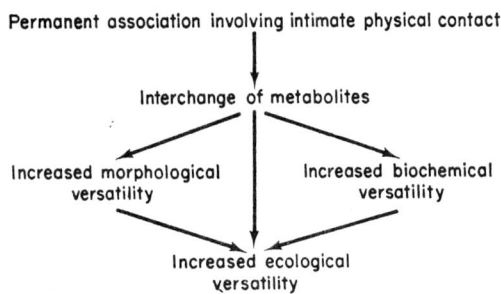

FIG. 1. Characteristics of mutualistic symbioses between autotrophs and heterotrophs. (After Scott, 1969.)

In associations involving autotrophs, the symbiotic unit is photosynthetic and carbohydrates pass from the autotroph to the heterotroph (Smith *et al.*, 1969). This single factor may be used to construct a chequer-board classification of such symbioses (Table I). With respect to carbon, it can be seen that the heterotrophs of these associations *avoid* deficiencies of this element in their environments by exploiting the photosynthetic capacity of their autotrophic partners. Bilateral exchange of nutrients is, however, possible because a further essential feature of these symbioses is their ability efficiently to abstract essential elements from the environment. This is most often, but not exclusively, a special attribute of the heterotroph. In those symbiotic systems where one partner or the dual system is capable of fixing elemental nitrogen, deficiences of this nutrient are also *avoided*. Thus, for all the partnerships of Table I with respect to carbon nutrition and for those with the ability to fix nitrogen with respect to nitrogen nutrition, deficiencies of the terrestrial or aquatic environments are successfully overcome by exploitation of the gaseous reserves of the aerial environment. With respect to phosphorus nutrition for all symbioses and to nitrogen for those that do not fix nitrogen, marginal habitats are successfully exploited by *tolerance* of nutrient deficiency.

Avoidance mechanisms will not be considered further except to illustrate techniques which are also of value in understanding mechanisms by which the second group, those exhibiting tolerance of these environmental conditions, operate. It must be emphasized that the distinction between avoidance and tolerance is only possible when individual elements are being considered. An association may differ with respect to different nutrients.

TABLE I. Mutualistic symbioses between autotrophs and heterotrophs

		DONOR OF CARBOHYDRATE			
		Blue-green algae	Green algae	Dino-flagellates	Green plants other than algae
RECIPIENT OF CARBO-HYDRATE	Bacteria	√	√		ªLegume root nodules ªNon-legume leaf nodules
	Actino-mycetes				ªNon-legume root nodules
	Fungi	ªLichens	Lichens		Sheathing and vesicular-arbuscular mycorrhizas
	Animals	ªProtozoa ªSponges ªEchiuroid worms	Protozoa Sponges Platy-helminths Coelen-terates	Protozoa Platy-helminths Coelen-terates Molluscs	

ª Associations which probably or actually fix nitrogen.

Further discussion will be limited to three kinds of association: symbiotic coelenterates, lichens and some mycorrhizal higher plants (mycorrhizas of the Orchidaceae, etc. are not considered since they have not yet been shown to be mutualistic and they do not function as exploiters of barren habitats: Lewis, 1973). In the first two of these, algae are donors of at least carbohydrates and, in the last two, fungi are recipients of carbon compounds. Of this trio of kinds of association, therefore, lichens occupy an intermediate position with regard to components.

2. *The Exploitation of Marginal Habitats by Selected Mutualistic Symbioses*

(a) *Coelenterates.* The incidence of symbiosis between dinoflagellates (zooxanthellae) and coelenterates increases markedly in nutrient-poor tropical seas (Yonge, 1968) suggesting that algae are important in the nutrition of their animal hosts. Goreau *et al.* (1971) have recently even posed the question—are symbiotic coelenterates autotrophic or heterotrophic? They carefully assess the available evidence and come to the conclusion that some species, e.g. some zoanthids, never feed in a typical holozoic manner and derive all their organic compounds other than can be obtained from solution directly from their algae. Other groups, e.g. most scleractinian corals, are capable of and do indulge in holozoic nutrition by capture and digestion of plankton. As well as their effects on carbon nutrition, it must also be emphasized that zooxanthellae also greatly enhance rates of calcification and, therefore, rates of reef building (Goreau, 1959; Goreau and Goreau, 1959; Pearse and Muscatine, 1971). Less attention seems to have been paid to nitrogen and phosphorus nutrition of symbiotic coelenterates, but, as will be discussed below, special features of their nitrogen metabolism play a significant role in their ecological success.

(b) *Lichens.* Smith (1963) has emphasized that, among the most interesting problems in the biology of lichens is their persistence in habitats where environmental conditions do not support the growth of most other kinds of plants. Despite this, little attention has been paid to their mineral nutrition compared with their carbon metabolism. Smith (1962) and subsequent reviewers of lichen physiology (see Lewis, 1970) have made the point that, since the supply of nutrients is likely to be not only meagre but erratic, it would not be surprising to find efficient mechanisms of utilization. Indeed, mechanisms for absorption of substances may be "too efficient", resulting in accumulation of radioactive fall-out and other pollutants to such levels that the lichen "species" can no longer survive.

(c) *Mycorrhizas.* Harley (1969a) and many others have stressed that infection of forest trees by sheathing ("ectotrophic") mycorrhizal fungi is greatest under nutrient-poor conditions. Although less is known about incidence of infection in the field by vesicular-arbuscular mycorrhizal fungi, both types of infection, in studies of both natural and experimentally induced mycorrhizas, enhance dry matter production. Most data to 1970 have been summarized by Harley (1969, Tables 22, 24–26 and 49; 1970, Table 1; 1971, Figs 4 and 15). More recently Hayman and Mosse (1971), Mosse and Hayman (1971) and Mosse (1972) have not only shown that different strains of *Endogone*, the fungal genus responsible for most vesicular-arbuscular mycorrhizas, have different effects on dry matter production in differing soils, but also that other edaphic factors such as lime content have significant effects on the response to infection. Harley (1969, Tables 22 and 24–26; 1970, Tables 3 and 14; 1971, Fig. 7) has also brought together and assessed the evidence that mycorrhizal plants are richer in nutrients such as nitrogen, phosphorus, potassium, calcium and magnesium than non-mycorrhizal controls.

(d) *Conclusions*. The abundant evidence that has been briefly summarized in the above three paragraphs clearly establishes that these kinds of mutualistic symbioses are efficient exploiters of habitats poor in inorganic nutrients. The final section of this paper will, with special reference to nitrogen and phosphorus, examine how far common mechanisms are responsible for this efficiency.

3. Mechanisms
(a) *Nitrogen*

(i) *Coelenterates*. Since some symbiotic coelenterates have not been observed to feed in a holozoic manner (Goreau et al., 1971), they must depend on soluble, exogenous sources of nitrogen. Dissolved organic matter potentially can be absorbed directly by animal tissues, but direct utilization of inorganic ions is unlikely. At the pH of sea-water (approx. 8), almost all inorganic nitrogen will be present as nitrate since ammonia is volatile at this pH. As nitrate cannot be reduced and incorporated into organic form by almost all animal tissues, this ion must be initially metabolized by the zooxanthellae. Nitrate reductase has not been explicitly demonstrated in these but is common to most algae (Syrett, 1962) and zooxanthellae in culture can utilize nitrate (McLaughlin and Zahl, 1966). It follows that, in addition to the well documented supply of carbohydrates from algae to hosts (Smith et al., 1969), algae must supply organic nitrogen compounds also. These two alternative supplies of organic nitrogen—dissolved in sea-water and from symbiotic algae—potentially can supplement the "normal" nutrition of carnivorous coelenterates.

Evidence that this occurs has come from studies of both isolated zooxanthellae and intact symbioses. Trench (1971) has pointed out some of the possible pitfalls of working with isolated algae. Especially important among these are the following. Symbiotic algae in culture differ from their behaviour in symbiosis particularly with regard to release of metabolites. These differences occur soon after isolation. Also, since isolated algae are fragile, it is important to demonstrate that extracellular products are derived from intact cells and are not due to cell lysis. To ensure that the latter is not occurring, it is important to compare the distribution of fixed ^{14}C between extracellular and intracellular compounds. With these problems of interpretation in mind, Trench clearly demonstrated that freshly isolated zooxanthellae from seven species, representing four orders of the Coelenterata, selectively liberated the endogenous metabolites which incorporated ^{14}C during photosynthesis in $NaH^{14}CO_3$. Although carbohydrates and organic acids were the major compounds released, significant amounts of the amino acid, alanine (1–7% of total radioactivity in the media), were liberated from all isolates of zooxanthellae. That the release of amino acids in particular was selective was shown by the fact that, although ^{14}C-glutamate accounted for up to 16% of the radioactivity in the soluble intracellular compounds, no glutamate was detected in the media. ^{14}C-alanine was also selectively released following dark fixation of $^{14}CO_2$.

Lewis and Smith (1971) investigated the release of nitrogen compounds from intact symbioses by using a technique initially developed for studying carbon movement in lichens (Drew and Smith, 1967). This involved incubating lichens in solutions of $NaH^{14}CO_3$ in the light in the presence of a high concentration of the non-radioactive (^{12}C) form of the carbohydrate which moves between the symbionts. The ^{14}C form of the mobile carbohydrate then accumulates in the medium. Since algal products were prevented from reaching the fungus, the experimental procedure was termed the "inhibition technique". It was originally thought that the radioactive carbohydrate released from the alga was unable to compete for entry to the fungus with the much higher concentration of the non-radioactive form and so accumulated in the medium (see Smith et al., 1969, Fig. 1). More detailed studies with lichens (Hill and Smith, 1972) show that this explanation is too simple and that competition or exchange between ^{12}C and ^{14}C carbohydrates may begin at stages before the fungal uptake sites. Notwithstanding our ignorance of the fine details of mechanisms, the technique has proved most valuable in the identification of mobile carbohydrates.*

---

* Although not directly relevant to the particular theme of this paper, the following observations concerning carbohydrate movement are relevant to the general theme of the symposium. It is of chemotaxonomic interest that, in all lichens with eukaryotic algae and all coelenterates with zooxanthellae that have been examined, the mobile carbohydrate is an acyclic polyhydric alcohol (polyol)—glycerol in the case of zooxanthellae, and erythritol, ribitol or sorbitol in the green algae (Smith et al., 1969). These compounds appear to be especially prevalent in free-living organisms and other symbioses which potentially or actually endure osmotic stress (Lewis and Smith, 1967). By possessing the capacity to produce polyols, these algae may be considered to be pre-adapted to the formation of symbioses under ecological conditions likely to produce osmotic stress, e.g. the dry habitats of lichens and the saline environment of coelenterates. It is significant that the symbiotic algae of fresh-water coelenterates do not release polyols but maltose or glucose instead (Smith et al., 1969). Few green algae form symbioses in the marine environment (Droop, 1963, Table I). However, in the present context, it is significant that, in *Platymonas convolutae*, the green algal associate of the acoelous turbellarian flatworm, *Convoluta roscoffensis*,

Basing their experimental approach on the observations of Trench (1971), Lewis and Smith (1971) applied the "inhibition technique" using alanine to seven species of coelenterate. Three of these were stony corals and the other four were single species representing some other orders of the Coelenterata. In all cases, inclusion of 1% alanine into sea-water containing $NaH^{14}CO_3$ in the light stimulated the release of ^{14}C-alanine into the medium. The proportion of net ^{14}C fixed released into alanine media varied from species to species (0·5–4·0% compared with 0·1–1·1% in sea-water controls). Since this was also true of the release of carbohydrates (Lewis and Smith, 1971), this may reflect the fact that zooxanthellae with differing physiology enter into symbiosis with different coelenterates. This conclusion agrees with that of Trench (1971) who showed marked differences in the patterns of incorporation of ^{14}C from ^{14}C-bicarbonate into the intracellular metabolites of zooxanthellae isolated from a range of coelenterates different from that of Lewis and Smith. The systematics of zooxanthellae has not been extensively studied, but the existing literature is confused and unhelpful from both taxonomic and nomenclatural viewpoints. Taylor (1971) has reviewed and clarified the situation by including symbiotic dinoflagellates in free-living genera, which brings the taxonomic treatment of symbiotic dinoflagellates into line with that of the algae of lichens. Two species are presently recognized—*Amphidinium chattonii* (Hovasse) Taylor and *Gymnodinium microadriaticum* (Frendenthal) Taylor. The metabolic data of Trench (1971) and Lewis and Smith (1971) suggest that certainly different strains of *G. microadriaticum* exist and further information may even warrant the delimitation of further symbiotic species within the genus.

In the preliminary species survey of Lewis and Smith (1971), *Porites divaricata* was not only the most readily sampled species of coral, but also responded to the "inhibition technique" best (e.g. in four separate experiments, photosynthesis of ^{14}C-bicarbonate in the presence of alanine resulted in a 12–40-fold stimulation of release of ^{14}C over sea-water controls). It was therefore used for further studies of nitrogen nutrition. In addition to alanine, other ^{14}C amino acids, e.g. glycine and serine, may also be released in the light, but stimulation of release of ^{14}C by these compounds was less than 5 times the accumulation of ^{14}C into sea-water alone. Significantly, in relation to Trench's results discussed above, glutamic acid was almost without effect showing that this amino acid is not released from zooxanthellae *in vitro* or *in vivo*.

---

the polyol, mannitol, accounts for more than 99% of the soluble carbohydrates (Gooday, 1970), and the green symbiont of the echinoderm *Ophioglypha* belongs to the genus *Coccomyxa* (Droop, 1963). The lichenized members of this genus contain ribitol (Smith *et al.*, 1969).

It is nearly a hundred years since the suggestion was first made that one of the roles of the algae of symbiotic animals is removal of nitrogenous wastes (Geddes, 1882). The studies of McLaughlin and Zahl (1966) clearly demonstrate that zooxanthellae in culture are capable of utilizing a range of nitrogen sources including ammonia, urea, uric acid, guanine, adenine and many amino acids. In order to investigate possible effects of excretory substances on algal nitrogen metabolism *in situ* within a coral, Lewis and Smith (1971), studied the effect of exogenous ammonia on the release of ^{14}C into alanine solutions in both light and dark by *P. divaricata*. Their results may be summarized as follows: (1) During incubation in alanine, a larger proportion of the net fixed ^{14}C was released in the dark than in the light (this was also true for the sea anemone, *Condylactis gigantea*); (2) in the light, a greater percentage of the fixed ^{14}C was released in the presence of ammonia; (3) when *P. divaricata* was pretreated in ammonia and ^{14}C bicarbonate in the light, the coral released more ^{14}C during subsequent incubation in alanine than when pretreated in sea-water; (4) both ammonia and alanine increased net dark fixation, but ammonia did not increase the proportion of dark-fixed ^{14}C released in the presence of alanine. From these experiments, it was concluded that ammonia increases the proportion of fixed ^{14}C incorporated by the zooxanthellae into alanine, i.e. in the presence of a potential animal waste product, the algae incorporate more fixed carbon into the principal amino acid that is subsequently released by the zooxanthellae to their hosts.

In their studies of the trophic structure of the coral reef community on Eniwetok Atoll, Odum and Odum (1955) concluded from their estimates of productivity and available nutrient levels that it was probable that there was cyclic re-use of nitrogen. Droop (1963, p. 187) has reviewed the evidence that the nitrogenous waste products of coelenterates are completely taken up by their symbiotic algae. These data and those of Trench (1971) and Lewis and Smith (1971) clearly show that a short-circuited biogeochemical cycle (Fig. 2) operates so that nitrogen-containing compounds extracted from the sea by symbiotic coelenterates mostly remain held in the standing biomass.

(ii) *Lichens*. As indicated in Table I, lichens with blue-green algae fix nitrogen and, by so doing, exhibit avoidance of nutrient deficiency rather than tolerance. That nitrogen obtained in this manner is available to the fungal partner has recently been elegantly demonstrated by Kershaw and Millbank (1970) who have reviewed the literature on N-fixing lichens in general. Although lichens with green algae must obtain their nitrogen in combined form, such as nitrate or ammonium ions or organic compounds, these are also available to N-fixing "species". Unlike the heterotrophic partner of coelenterate symbioses, fungi of lichens are potentially capable of utilizing oxidized forms of nitrogen.

Like fungi in general, this capacity may be constitutive or adaptive although there have been no specific studies of nitrate reductase in lichen fungi.

Most investigations of the utilization of nitrogen by lichens have concerned the isolated symbionts in culture and is a well reviewed topic (e.g. Ahmadjian, 1966; Hale, 1970). Both fungi and algae can utilize oxidized and reduced forms of inorganic nitrogen as well as a range of organic compounds. However, such studies merely indicate the potential behaviour of the symbionts and, on the whole, this has not been correlated with the actual behaviour of lichen thalli in the field. The limited information on the latter subject and laboratory experiments of direct relevance have been discussed by Smith (1962) and Ahmadjian (1966).

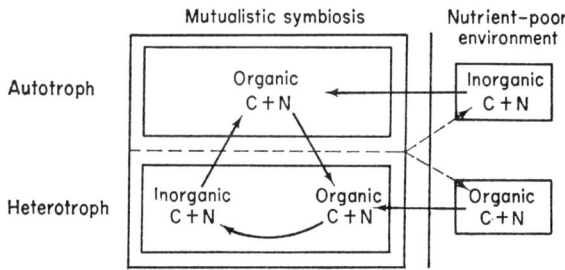

FIG. 2. The role of autotrophic mutualistic symbionts in recycling of carbon and nitrogen in coral reef ecosystems. (From Lewis and Smith, 1971.)

With regard to availability, organic nitrogen is present in the environment of most lichens. This is especially evident in habitats such as roof and stonework of farms, "ornithicoprophilous" rocks, etc. (Barkman, 1958), but Smith (1960b) also found considerable organic nitrogen in run-off and soil in disused clay pits where *Peltigera polydactyla* grew. Smith emphasized that a more detailed study of the range of nitrogen compounds in the environment is desirable. With regard to the utilization of compounds likely to be present in habitats rich in the excreta of birds, etc., Galinou (1954) found the enzymes, allantoicase, allantoinase and urease in several lichens of these habitats and the utilization of some amino acids and amides, especially asparagine by *P. polydactyla* has been described by Smith. Intact lichen thalli can absorb both nitrate and ammonium ions (Salomon, 1914; Smith, 1960a), but the rate of assimilation of these compounds is slow. Smith is of the view that the twin processes of rapid uptake of nutrients and their slow metabolic turnover are adaptations to unfavourable environmental conditions. Salomon's studies indicated that lichens with green algae absorbed more nitrate than those with blue-green algae, but that their capacities to absorb ammonia were not very different. It should be emphasized that absorption and utilization are not synonymous. Smith (1960a), for example, showed that, of the nitrate absorbed by *P. polydactyla* in 24 h, 90% could be recovered unchanged from the lichen. In view of the differing distribution of nitrogen as nitrate or ammonia in acid and alkaline habitats, it would be especially valuable to know more of the capacity of lichens to utilize nitrate. In this context, it may be significant that the proportion of littoral lichens that contain blue-green algae increases seawards and in some genera, e.g. *Arthopyrenia*, where terrestrial species contain green algae, marine species have blue-green algae (Duncan, 1959). As sea-water has a pH where ammonia is volatile, the available inorganic nitrogen is nitrate. The increasing proportion of blue-green algal lichens in the marine habitat with their capacity to fix nitrogen may be an evolutionary mechanism by which the necessity to utilize nitrate is circumvented.

Lichenized blue-green algae are presumably potentially self-sufficient with regard to nitrogen nutrition. Little attention has been paid to the nitrogen compounds available to lichenized green algae. Unless concentrations in the environment are high as in the habitats rich in excreta mentioned above, it is unlikely that compounds reach them directly from the environment owing to the avidity of fungal uptake mechanisms. Admittedly, Smith (1960c) demonstrated that the algal zone of *P. polydactyla* was more active in uptake on a dry weight basis than the entirely fungal medulla. However, this cannot be attributed to algae until the relative activities of medullary and algal zone hyphae have been assessed. Nothing is known of the passage of nutrients from fungus to alga. This is a topic worthy of study to determine the validity of the criterion of bilateral exchange of nutrients as an essential feature of mutualistic symbiosis (Fig. 1).

(iii) *Mycorrhizas*. As far as absorption of nitrogen compounds is concerned, mycorrhizas have much in common with lichens. Morphologically, in both, the autotroph is at least partially separated from, but also connected to the terrestrial environment by fungal hyphae; physiologically, both partners are capable of absorbing inorganic and organic nitrogen compounds and, experimentally, much time has been spent studying the nutrient physiology of the separately cultured symbionts and less on that of the dual system. (Study of cultured *Endogone* of vesicular-arbuscular mycorrhizas is still not possible.)

Very little is known about the nitrogen nutrition of vesicular-arbuscular mycorrhizas and the following discussion will be confined to sheathing mycorrhizas. Since these develop most abundantly on acidic, mor soils, ammonia is the most abundant inorganic form of available nitrogen and, especially in the subsurface layers where mycorrhizas are most abundant, nitrate is present in trace amounts only or absent altogether (Carrodus, 1966). Studies of the ability of intact mycorrhizas or mycorrhizal fungi in culture to utilize nitrate are therefore of more academic interest than ecological relevance. Carrodus (1966, 1967) and Smith (1972) showed by several independent techniques that beech mycorrhizas have a very low capacity to absorb and utilize nitrate. Since most higher plant roots can readily utilize this ion, this inability by mycorrhizas is probably a reflection of fungal behaviour. Cochrane (1958, p. 243) pointed out that higher basidomycetes in general have restricted powers to utilize nitrate and, since species within genera differed, concluded that nitrate utilization is probably governed by ecological rather than taxonomic criteria. With regard to mycorrhizal fungi in particular, this problem has been re-examined by Lundeberg (1970) whose results showed good agreement with those of previous workers, namely that nitrate was unavailable to many isolates (11 out of 26 mycorrhizal strains of various species) and, when utilized, was a much poorer nitrogen source than ammonia (7 out of 26). Eight isolates grew at comparable rates on nitrate or ammonium sources. These 26 isolates were also incapable of significantly releasing organically bound nitrogen existing in raw humus, but, in common with earlier studies reviewed by Lundeberg, some were capable of utilizing some soluble organic nitrogen compounds, especially asparagine.

The principal sources of nitrogen for mycorrhizas in the field are therefore ammonia and any available soluble organic nitrogenous compounds. Both are efficiently absorbed by excised beech mycorrhizas. Ammonia is rapidly converted to organic form, especially glutamine, and this transformation is effected mostly in the fungal sheath (Carrodus, 1966, 1967). In excised mycorrhizas, the uptake of ammonia is stimulated by exogenous supply of some acids of the Krebs Cycle (Carrodus, 1966), and Melin (1963) has reviewed previous work in his laboratory which showed that ammonia and α-ketoglutarate could at least partially substitute for glutamate during growth of several mycorrhizal fungi in culture. Under natural conditions these carbon skeletons would be derived either following metabolism of host sugars by the fungus (Carrodus, 1966) or, perhaps, directly from the host.

Although the tracer studies of Melin and Nilsson (1952, 1953) clearly demonstrated that nitrogen absorbed by mycorrhizal fungi from either (^{15}N-ammonium) nitrate or ^{15}N-glutamate could be translocated to pine seedlings, there has been no investigation of what compounds actually pass from fungus to host. The evidence from the ^{15}N-glutamate work of Melin and Nilsson and the study of incorporation of ammonia into organic compounds by Carrodus (1966) show that the compound(s) moving must be organic. Mycorrhizal higher plants therefore act as though heterotrophic for nitrogen compounds in a manner comparable to nodulated legumes, except that the ultimate source of nitrogen for the latter is the atmosphere (Pate, 1968). Implicit in this conclusion is that carbon initially derived from the host as carbohydrate can return to the host as an amino compound. This does not conflict with the hypothesis of Lewis and Harley (1965) that flow of carbohydrates is one-way—from host to symbiont. What may be true of carbohydrates need not apply to amino compounds. Movement of such nitrogenous organic compounds may account for the observation of Reid and Woods (1969) mentioned above that carbon can move from one higher plant to another via its common mycorrhizal fungus.

(b) *Phosphorus*

(i) *Coelenterates*. Like nitrogen, phosphorus is deficient in tropical seas, although Odum and Odum (1955) concluded from their estimate that, relative to nitrogen, phosphorus was more abundant. Comparatively little is known of the phosphorus nutrition of symbiotic coelenterates but, as for nitrogen, efficient re-cycling is probable. Zooxanthellae, in addition to being able to utilize orthophosphate, can metabolize organic phosphates such as glycerophosphate and cytidylic, adenylic and guanylic acids (McLaughlin and Zahl, 1966) and Droop (1963) has reviewed the evidence that algae not only can absorb all the phosphorus compounds excreted by coelenterates but can also take them up when added to the medium.

(ii) *Lichens*. Even less is known of the phosphorus nutrition of lichens and their components than that of symbiotic coelenterates and theirs. Epiphytic lichens of oak contain a higher proportion of phosphorus than oak bark (Hale, 1970, p. 60) which is perhaps to be expected. Ahmadjian (1966) noted that, apart from nitrogen, the specific inorganic requirements of both lichen algae and fungi in culture were not known and Hale (1971) commented, "Beyond the recognition of their unusual capacity for absorbing metallic ions, we know nothing of the mineral requirements of lichens". Smith (1960b) has, however, shown that this capacity does extend to orthophosphate.

(iii) *Mycorrhizas*. In contrast to coelenterates and lichens, the phosphorus nutrition of both sheathing and vesicular-arbuscular mycorrhizal plants has been both extensively studied and extensively reviewed. Among reviews that have considered this topic in the last few years are those of Nicolson (1967), Gerdemann (1968), Harley (1969a, 1970, 1971), Harley and Lewis (1969), Bowen and Rovira (1969), Bowen (1973) and Khan (1972). Those of Harley (1969a) and Bowen, in particular, have considered kinetic and mechanistic aspects of orthophosphate absorption and utilization and no more will be said here other than to emphasize that mycorrhizal plants, by virtue (1) of extremely efficient uptake mechanisms, (2) of the fact that they exploit a larger volume of soil than non-mycorrhizal plants, and (3) of their longevity, are very much more successful in utilizing soluble phosphate from soils low in nutrients than uninfected controls.

Since the proportion of soil phosphorus that is present as orthophosphate is low in many soils, especially the humus-rich soils exploited by sheathing mycorrhizas (Cosgrove, 1967), it is relevant to consider utilization of both complex inorganic and organic phosphates by mycorrhizas. Several workers have proposed that excretion of various substances by mycorrhizas can effect solubilization of these sources of phosphorus, but this has not been experimentally tested (Bowen and Theodorou, 1967). These workers did show that several mycorrhizal fungi in culture could release orthophosphate from rock phosphate and that, when mycorrhizas were synthesized between these species and *Pinus radiata*, rock phosphate was more efficiently utilized than by uninoculated controls. Similar results for vesicular-arbuscular infections have been obtained by Daft and Nicolson (1966) and Murdoch, Jackobs and Gerdemann (1967). That enzymic degradation of these compounds is probable was indicated by Bartlett and Lewis (1973) who demonstrated the presence of a surface inorganic pyrophosphatase on mycorrhizal roots of beech.

Organic phosphates such as inositol phosphates, sugar phosphates and nucleotides can be used as phosphate sources for mycorrhizal fungi in culture (Theodorou, 1968; Bowen, 1973). That the first of these groups can be utilized is significant for these compounds are commonly the major form of organic phosphates in soil, usually as insoluble calcium, aluminium or iron salts (Anderson, 1967). Various organic phosphates can also be hydrolyzed by surface phosphatases present on mycorrhizas of beech (Woolhouse, 1969; Bartlett and Lewis, 1973) and pine (Bowen, 1973). Bartlett and Lewis demonstrated activity towards *p*-nitrophenyl phosphate and the sodium salts of glucose-6-phosphate, β-glycerophosphate, inositol triphosphate and inositol hexaphosphate. The products of the phytase reaction were investigated and found to be different from those of the mycorrhizal fungus, *Rhizopogon luteolus*, in culture (Theodorou 1971). Mosse and Phillips (1971) also demonstrated that pure cultures of *Trifolium parviflorum* with *Endogone* could utilize several organic phosphates including phytate, hexosephosphates, DNA and phospholipids. Taken together, these results indicate that mycorrhizal plants short-circuit normally accepted biogeochemical cycles as shown for sheathing mycorrhizas in Fig. 3, which is taken from Bartlett and Lewis (1973). This idea has also been discussed by Bowen and co-workers (Bowen and Theodorou, 1967; Bowen and Rovira, 1969).

If mycorrhizal plants utilize phosphate compounds in soil which are not available to uninfected plants, the specific activity of ^{32}P absorbed by mycorrhizas from soils supplemented with radioactive orthophosphate should be less than that absorbed by uninfected controls. The studies of Sanders and Tinker (1971) and Hayman and Mosse (1972) do not show this difference and both sets

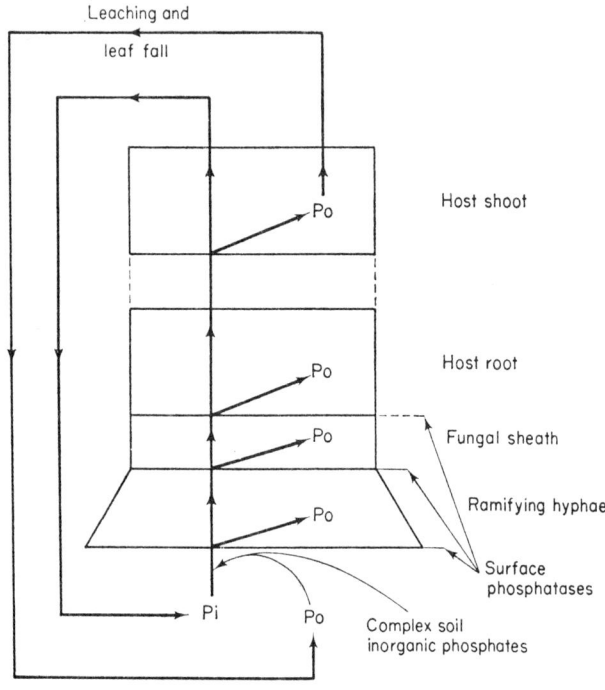

Fig. 3. Phosphate cycle operating in the presence of sheathing mycorrhizal plants. (Pi = orthophosphate, Po = organic phosphates.) (From Bartlett and Lewis, 1973.) The broader section of the diagram representing hyphae ramifying in the soil indicates that these exploit a greater volume of soil than roots. With the omission of fungal sheath, a similar cycle may operate in the presence of vesicular-arbuscular plants.

of authors conclude that the same sources of phosphorus—orthophosphate— were being tapped by both kinds of plant. These results are surprising in view of the utilization of complex phosphates and the presence of surface active phosphatases as described above and only further research will resolve the anomaly. However, among points that should be considered in this problem are the following. Firstly, since orthophosphate is known to be an inhibitor of organic phosphatases, a diauxie phenomenon may occur, i.e. orthophosphate has to be reduced to a very low level before utilization of complex phosphates occur. Secondly, it must be demonstrated in experiments of the kind conducted by Sanders and Tinker and Hayman and Mosse that there are sufficient alternative sources of phosphorus to orthophosphate present. Thirdly, perhaps further experiments should be extended in time. As Bowen and Rovira (1969) pointed out, the longevity of mycorrhizas permits a prolonged exploitation of solid phase phosphate close to the mycorrhizal surface since solution phase phosphate absorbed by the mycorrhizas is replenished from the solid phase. Certainly, repetition of the vesicular-arbuscular mycorrhizal experiments using ^{32}P over a longer period in soils of very low phosphate status with and without added complex phosphate would be instructive.

(c) *Antibiosis*

In habitats poor in nutrients, competition for these nutrients will be intense. It might be expected that natural selection would encourage the development of effective mechanisms for chemical elimination of competitors by antibiotic production and chemical dissuasion of predators by exogenous or endogenous production of toxins or unpalatability (cf. papers by Janzen and Jones in this volume). Certainly, coelenterates in general, but also including species symbiotic with algae, secrete biologically active substances, but little is known of the chemistry of these toxins (Ciereszko, 1970). Also, whether they are to be regarded as agents of offence (antibiotics) or defence cannot readily be assessed. In contrast to this, antibiotic production by both lichens and mycorrhizas has been investigated considerably. Hale (1970) has pointed out that lichens are seldom attacked by bacterial or fungal pathogens and he, Smith (1962) and Ahmadjian (1966) have reviewed the available evidence concerning the production, identity and activity of antibiotics by lichens.

More is known of the biological effects of the antibiotics produced by sheathing mycorrhizas. They suppress both soil saprophytes such that litter decomposition in the vicinity of mycorrhizas is much reduced and they also eliminate pathogens of the host tree (Zak, 1964; Gadgil and Gadgil, 1971; Krywolap, 1971). This role of mycorrhizal fungi as deterrents to pathogenic root infections is attracting increasing attention (Marx, 1972). The net effect to the fungi of this partial elimination of competitors is that they are then in a position to exploit maximally available nutrient reserves and, as shown above for nitrogen and phosphorus, to act as crucial links in tight biogeochemical cycles. It has recently been shown by Marks and Bormann (1972) that clear felling of temporate forest and subsequent herbicidal spraying to prevent regeneration is followed by significant increases in decomposition, nitrification and run-off of nitrate. Clear felling will also eliminate active mycorrhizal fungi, most of which cannot survive saprophytically (Lewis, 1973). The removal of mycorrhizal antibiotics may therefore be a significant factor leading to the microbial changes evident as increased decomposition and nitrification. In nutrient-poor conditions, inhibition of nitrifying organisms may be especially important since, as positively charged ammonium ions will be bound to the abundant negatively charged sites in soils, they are more resistant to leaching than negatively charged nitrite and nitrate. The potential importance of mycorrhizas in nutrient cycling and overcoming problems of leaching in tropical forests has also been pointed out by Stark (1971) and the involvement of antibiotics in soil ecosystems in general has been discussed by Alexander (1971).

CONCLUSIONS

Although only three kinds of association have been considered in detail above, it is probable that the following conclusions from studies of these are applicable to other mutualistic symbioses of Table I that tolerate low levels of nutrients in their environment. Efficient mechanisms of abstracting essential minerals from dilute solution are present and absorbing organs, whether the whole symbiosis as in lichens and reef corals or a special organ as in sheathing mycorrhizas, are commonly long-lived. In the case of exploitation by terrestrial symbioses, this permits slow, probably enzymic, mineralization of solid phase reserves. Once absorbed, utilization of elements is slow so that compounds or ions absorbed in times of comparative plenty can be metabolized over a long period (Harley and Lewis, 1969). The further way in which mutualistic symbioses can be more productive than free-living organisms in nutrient-poor environments is by the elaboration of systems for tight nutrient cycling as shown in Figs 2 and 3—a tightness which is assisted by antibiotic production and the partial elimination of competitors.

ACKNOWLEDGEMENT

I am especially grateful to Professor H. G. Baker for drawing my attention to the paper by Marks and Bormann.

REFERENCES

AHMADJIAN, V. (1966). Lichens. In "Symbiosis" (S. M. Henry, ed.), Vol. **1**, pp. 35–98. Academic Press, New York and London.

ALEXANDER, M. (1971). "Microbial Ecology." John Wiley, New York.

ANDERSON, G. (1967). Nucleic acids, derivatives and organic phosphates. In "Soil Biochemistry" (A. D. McLaren and G. H. Peterson, eds), pp. 67–90. Marcel Dekker, New York.

BARKMAN, J. J. (1958). "Phytosociology and Ecology of Cryptogamic Epiphytes." Van Gorcum, Assen.

BARTLETT, E. M. and LEWIS, D. H. (1973). Surface phosphatase activity of mycorrhizal roots of beech. *Soil Biol. Biochem.* **5**, 249–257.

BOWEN, G. D. (1973). Mineral nutrition of mycorrhizas. *In* "Physiology and Ecology of Ectotrophic Mycorrhizae" (D. H. Marx and T. T. Kozlowski, eds). Academic Press, New York (in press).

BOWEN, G. D. and ROVIRA, A. D. (1969). The influence of micro-organisms on growth and metabolism of plant roots. *In* "Root Growth" (W. J. Whittington, ed.), pp. 170–201. Butterworths, London.

BOWEN, G. D. and THEODOROU, C. (1967). Studies on phosphate uptake by mycorrhizas. *14th Congr. Int. Un. Forest Res. Org.*, Munich, Vol. **5**, pp. 116–138.

BRADSHAW, A. D. (1959). Population differentiation in *Agrostis tenuis* Sibth. II. The incidence and significance of infection by *Epichloe typhina*. *New Phytol.* **58**, 310–315.

BROOKS, M. A. (1963). Symbiosis and aposymbiosis in Arthropods. *In* "Symbiotic Associations" (P. S. Nutman and B. Mosse, eds). *13th Symp. Soc. gen. Microbiol.*, pp. 200–231. Cambridge University Press.

BUCHNER, P. (1965). "Endosymbiosis of Animals with Plant Micro-organisms." Interscience, New York.

CARRODUS, B. B. (1966). Absorption of nitrogen by mycorrhizal roots of beech. I. Factors affecting the assimilation of nitrogen. *New Phytol.* **65**, 358–371.

CARRODUS, B. B. (1967). Absorption of nitrogen by mycorrhizal roots of beech. II. Ammonium and nitrate as sources of nitrogen. *New Phytol.* **66**, 1–4.

CIERESZKO, L. S. (1970). Nitrogen compounds in Porifera and Coelenterata. *In* "Comparative Biochemistry of Nitrogen Metabolism—The Invertebrates" (J. W. Campbell, ed.), Vol. **1**, pp. 57–66. Academic Press, New York.

COCHRANE, V. W. (1958). "Physiology of Fungi." John Wiley, New York.

COSGROVE, D. J. (1967). Metabolism of organic phosphates in soil. *In* "Soil Biochemistry" (A. D. McLaren and G. H. Peterson, eds), pp. 216–228. Marcel Dekker, New York.

DAFT, M. J. and NICOLSON, T. H. (1966). Effect of *Endogone* mycorrhiza on plant growth. *New Phytol.* **65**, 343–350.

DREW, E. A. and SMITH, D. C. (1967). Studies in the physiology of lichens. VIII. Movement of glucose from alga to fungus during photosynthesis in the thallus of *Peltigera polydactyla*. *New Phytol.* **66**, 389–400.

DROOP, M. R. (1963). Algae and Invertebrates. *In* "Symbiotic Associations" (P. S. Nutman and B. Mosse, eds). *13th Symp. Soc. gen. Microbiol.*, pp. 171–199. Cambridge University Press.

DUNCAN, U. K. (1959). "A Guide to the Study of Lichens." Buncle, Arbroath.

GADGIL, R. L. and GADGIL, P. D. (1971). Mycorrhiza and litter decomposition. *Nature, Lond.* **233**, 133.

GALINOÙ, M. A. (1954). Sur la mise en évidence de quelques biocatalyseurs chez les lichens. *8th Int. bot. Congr. Paris*, Sect. 18, 1–4.

GEDDES, P. (1882). The yellow cells of radiolarians and coelenterates. *Proc. R. Soc. Edinb.* **11**, 377–396.

GERDEMANN, J. W. (1968). Vesicular-arbuscular mycorrhiza and plant growth. *A. Rev. Phytopathol.* **6**, 397–418.

GOODAY, G. W. (1970). A physiological comparison of the symbiotic alga, *Platymonas convolutae*, and its free-living relatives. *J. mar. biol. Ass. U.K.* **50,** 199–208.

GOREAU, T. F. (1959). The physiology of skeleton formation in corals. I. A method of measuring the calcium deposition by corals under different conditions. *Biol. Bull. mar. biol. Lab., Woods Hole* **116,** 59–75.

GOREAU, T. F. and GOREAU, N. I. (1959). The physiology of skeleton formation in corals. II. Calcium deposition by hermatypic corals under various conditions in the reef. *Biol. Bull. mar. biol. Lab., Woods Hole* **117,** 239–250.

GOREAU, T. F., GOREAU, N. I. and YONGE, C. M. (1971). Reef corals: autotrophs or heterotrophs. *Biol. Bull. mar. biol. Lab., Woods Hole* **141,** 247–260.

HALE, M. E. (1970). "The Biology of Lichens." Edward Arnold, London.

HARLEY, J. L. (1968). Review of "Les Mycorrhizas" by B. Boullard. *New Phytol.* **67,** 979.

HARLEY, J. L. (1969a). "The Biology of Mycorrhiza." (2nd edition) Leonard Hill, London, pp. 334.

HARLEY, J. L. (1969b). A physiologist's viewpoint. *In* "Ecological Aspects of the Mineral Nutrition of Plants" (I. H. Rorison, ed.). *9th Symp. Br. ecol. Soc.*, pp. 437–447. Blackwell, Oxford and Edinburgh.

HARLEY, J. L. (1970a). Mycorrhiza and nutrient uptake in forest trees. *In* "Physiology of Tree Crops" (L. C. Luckwill and C. V. Cutting, eds), pp. 163–179. Academic Press, London.

HARLEY, J. L. (1970b). The importance of micro-organisms to colonizing plants. *Trans. bot. Soc. Edinb.* **41,** 65–70.

HARLEY, J. L. (1971). "Mycorrhiza." Oxford University Press.

HARLEY, J. L. and LEWIS, D. H. (1969). The physiology of ectotrophic mycorrhizas. *Adv. Microbial Physiol.* **3,** 53–81.

HAYMAN, D. S. and MOSSE, B. (1971). Plant growth responses to vesicular-arbuscular mycorrhiza. I. Growth of *Endogone*-inoculated plants in phosphate-deficient soils. *New Phytol.* **70,** 19–28.

HAYMAN, D. S. and MOSSE, B. (1972). Plant growth responses to vesicular-arbuscular mycorrhiza. III. Increased uptake of labile P from soil. *New Phytol.* **71,** 41–48.

HENRY, S. M. (ed.) (1966). "Symbiosis." Academic Press, New York and London.

HILL, D. J. and SMITH, D. C. (1972). Lichen physiology. XII. The "inhibition technique". *New Phytol.* **71,** 15–30.

KERSHAW, K. A. and MILLBANK, J. W. (1970). Nitrogen metabolism in lichens. II. The partition of cephalodial-fixed nitrogen between the mycobiont and phycobionts of *Peltigera aphthosa*. *New Phytol.* **69,** 75–80.

KHAN, A. G. (1972). Mycorrhizae and their significance in plant nutrition. *Biologia, Lahore* (Special Supplement), pp. 42–78.

KRYWOLAP, G. N. (1971). Production of antibiotics by certain mycorrhizal fungi. *In* "Mycorrhizae" (E. Hacskaylo, ed.). *Proc. 1st N. Amer. Conf. Mycorrhizae*, pp. 219–221. U.S. Government Printing Office, Washington.

LEWIS, D. H. (1970). Physiological aspects of symbiotic associations between fungi and other plants. *Lichenologist* **4,** 326–336.

LEWIS, D. H. (1973). Concepts in fungal nutrition and the origin of biotrophy. *Biol. Rev.* **48,** 261–278.

Lewis, D. H. and Harley, J. L. (1965). Carbohydrate physiology of mycorrhizal roots of beech. I. The identity of endogenous sugars and utilization of exogenous sugars. *New Phytol.* **64,** 224–237.

Lewis, D. H. and Smith, D. C. (1967). Sugar alcohols (polyols) in fungi and green plants. I. Distribution, physiology and metabolism. *New Phytol.* **66,** 143–184.

Lewis, D. H. and Smith, D. C. (1971). The autotrophic nutrition of symbiotic marine coelenterates with special reference to hermatypic corals. I. Movement of photosynthetic products between the symbionts. *Proc. R. Soc. B.* **178,** 111–129.

Lundeberg, G. (1970). Utilization of various nitrogen sources, in particular bound soil nitrogen, by mycorrhizal fungi. *Studia Forestalia Suecica* **79,** 1–95.

McLaughlin, J. J. A. and Zahl, P. A. (1966). Endozoic Algae. *In* "Symbiosis" (S. M. Henry, ed.), Vol. **1,** pp. 257–297. Academic Press, New York and London.

Marks, P. C. and Bormann, F. H. (1972). Revegetation following forest cutting: Mechanisms for returning to steady-state nutrient cycling. *Science, N.Y.* **176,** 914–915.

Marx, D. H. (1972). Ectomycorrhizae as biological deterrents to pathogenic root infections. *A. Rev. Phytopathol.* **10,** 429–454.

Melin, E. (1963). Some effects of forest tree roots on mycorrhizal basidomycetes. *In* "Symbiotic Associations" (P. S. Nutman and B. Mosse, eds.). *13th Symp. Soc. gen. Microbiol.*, pp. 125–145. Cambridge University Press.

Melin, E. and Nilsson, H. (1952). Transfer of labelled nitrogen from an ammonium source to pine seedlings through mycorrhizal fungi. *Svensk. bot. Tidskr.* **46,** 281–285.

Melin, E. and Nilsson, H. (1953). Transfer of labelled nitrogen from glutamic acid to pine seedlings through the mycelium of *Boletus variegatus* (S.W.) Fr. *Nature, Lond.* **171,** 134.

Mosse, B. (1972). Effects of different *Endogone* strains on the growth of *Paspalum notatum*. *Nature, Lond.* **239,** 221–222.

Mosse, B. and Hayman, D. S. (1971). Plant growth responses to vesicular-arbuscular mycorrhiza. II. In unsterilized field soils. *New Phytol.* **70,** 29–34.

Mosse, B. and Phillips, F. M. (1971). The influence of phosphate and other nutrients on the development of vesicular-arbuscular mycorrhiza in culture. *J. gen. Microbiol.* **69,** 157–166.

Murdoch, C. L., Jackobs, J. A. and Gerdemann, J. W. (1967). Utilization of phosphorus sources of different availability by mycorrhizal and non-mycorrhizal maize. *Pl. Soil.* **27,** 329–334.

Nicolson, T. H. (1967). Vesicular-arbuscular mycorrhiza,—a universal plant symbiosis. *Sci. Prog. Oxf.* **55,** 561–581.

Odum, H. T. and Odum, E. P. (1955). Trophic structure and productivity of a windward coral reef community on Eniwetok Atoll. *Ecol. Monogr.* **25,** 290–320.

Pate, J. S. (1968). Physiological aspects of inorganic and intermediate nitrogen metabolism (with special reference to the legume, *Pisum sativum* L.). *In* "Recent Advances of Nitrogen Metabolism in Plants" (E. J. Hewitt and C. V. Cutting, eds), pp. 214–240. Academic Press, London.

Pearse, V. B. and Muscatine, L. (1971). Role of symbiotic algae (zooxanthellae) in coral calcification. *Biol. Bull. mar. biol. Lab., Woods Hole* **141,** 350–363.

Read, C. P. (1970). "Parasitism and Symbiology." Ronald Press, New York.

REID, C. P. P. and WOODS, F. W. (1969). Translocation of C^{14}-labelled compounds in mycorrhizae and its implications in interplant nutrient cycling. *Ecology* **50**, 179–187.

SALOMON, H. (1914). Über das Vorkommen und ·die Aufnahme einiger wichtigen Nahrsalze bei den Flechten. *Jb. wiss. Bot.* **54**, 309–354.

SANDERS, F. E. and TINKER, P. B. (1971). Mechanism of absorption of phosphate from soil by *Endogone* mycorrhizas. *Nature, Lond.* **233**, 278–279.

SCOTT, G. D. (1969). "Plant Symbiosis." Edward Arnold, London.

SMITH, D., MUSCATINE, L. and LEWIS, D. (1969). Carbohydrate movement from autotrophs to heterotrophs in parasitic and mutualistic symbiosis. *Biol. Rev.* **44**, 17–90.

SMITH, D. C. (1960a). Studies in the physiology of lichens. I. The effects of starvation and of ammonia absorption upon nitrogen content of *Peltigera polydactyla*. *Ann. Bot.* **24**, 52–62.

SMITH, D. C. (1960b). Studies in the physiology of lichens. II. Absorption and utilization of some simple organic nitrogen compounds by *Peltigera polydactyla*. *Ann. Bot.* **24**, 172–185.

SMITH, D. C. (1960c). Studies in the physiology of lichens. III. Experiments with dissected discs of *Peltigera polydactyla*. *Ann. Bot.* **24**, 186–199.

SMITH, D. C. (1962). The biology of lichen thalli. *Biol. Rev.* **37**, 537–570.

SMITH, D. C. (1963). Experimental studies of lichen physiology. *In* "Symbiotic Associations" (P. S. Nutman and B. Mosse, eds). *13th Symp. Soc. gen. Microbiol.*, pp. 31–50. Cambridge University Press.

SMITH, F. A. (1972). A comparison of the uptake of nitrate, chloride and phosphate by excised beech mycorrhizas. *New Phytol.* **71**, 875–882.

STANIER, R. Y., DOUDOROFF, M. and ADELBERG, E. A. (1971). "General Microbiology" (3rd edition). Macmillan, London.

STARK, N. M. (1971). Mycorrhizae and nutrient cycling in the tropics. *In* "Mycorrhizae" (E. Hacskaylo, ed.). *Proc. 1st N. Amer. Conf. Mycorrhizae*, pp. 228–229. U.S. Government Printing Office, Washington.

SYRETT, P. J. (1962). Nitrogen assimilation. *In* "Physiology and Biochemistry of Algae" (R. A. Lewin, ed.), pp. 171–188. Academic Press, New York.

TAYLOR, D. L. (1971). Ultrastructure of the "zooxanthella", *Endodinium chattonii, in situ*. *J. mar. biol. Ass. U.K.* **51**, 227–234.

THEODOROU, C. (1968). Inositol phosphates in needles of *Pinus radiata* D. Don and the phytase activity of mycorrhizal fungi. *Proc. 9th Int. Cong. Soil Sci., Adelaide* **3**, 483–493.

THEODOROU, C. (1971). The phytase activity of the mycorrhizal fungus *Rhizopogon luteolus*. *Soil Biol. Biochem.* **3**, 89–90.

TRENCH, R. K. (1971). The physiology and biochemistry of zooxanthellae symbiotic with marine coelenterates. II. Liberation of fixed ^{14}C by zooxanthellae *in vitro*. *Proc. R. Soc. Lond.* B. **177**, 237–250.

WOOLHOUSE, H. W. (1969). Differences in the properties of the acid phosphatases of plant roots and their significance in the evolution of edaphic types. *In* "Ecological Aspects of the Mineral Nutrition of Plants (I. H. Rorison, ed.), pp. 357–380. Blackwell, Oxford and Edinburgh.

YONGE, C. M. (1968). Living Corals. *Proc. R. Soc. Lond.* B. **169**, 329–344.

ZAK, B. (1964). Role of mycorrhizae in root disease. *A. Rev. Phytopathol.* **2**, 377–392.

8 | Snails, Schistosomes and Systematics: some Problems Concerning the Genus *Bulinus*

A. D. BERRIE

Department of Zoology, University of Reading, England

Abstract: The taxonomic history of the snail genus *Bulinus* is described and the inherent variability of the snails is stressed. Interactions between the snails and *Schistosoma haematobium* are discussed and the problems are considered in relation to the epidemiology of schistosomiasis. The *truncatus/tropicus* species complex presents particular difficulty and the occurrence of polyploidy in this group is considered. Geographical problems involved in the ecology of the *africanus* group are described and related to associated taxonomic problems. The importance of the River Nile as a link between the species to the north and south of the Sahara is assessed. Great taxonomic and systematic difficulties exist within *Bulinus* and these are closely linked with the ecology of the snails.

INTRODUCTION

The schistosomes are blood flukes (Trematoda: Digenea: Schistosomatidae) that parasitize mammals and birds. They have a complex life cycle involving an aquatic or semi-aquatic snail as an intermediate host and they show a high degree of specificity in their choice of snail. The definitive host and the snail host are infected by appropriate larval stages which swim freely in fresh water. Three species are important as human parasites causing the disease known as schistosomiasis or bilharzia which is estimated to affect at least 150 million people. *Schistosoma haematobium* (Bilharz, 1852) occurs through most of Africa and the Near East and is transmitted by planorbid snails of the genus *Bulinus* Müller, 1781. *S. mansoni* Sambon, 1907 occurs through most of Africa, in northern South America and some Caribbean islands and is transmitted by planorbid snails of the genus *Biomphalaria* Preston, 1910. *S. japonicum* Katsurada, 1904 occurs in the West Pacific region and is transmitted by hydrobiid snails of the genus *Oncomelania* Gredler, 1881. During the past 25 years increased efforts have been devoted to studying various aspects of schistosomiasis. It is clear that both the snails and the parasites show a considerable amount of subspecific variation and that this is associated with the ecology of the animals and the epidemiology of the disease.

THE GENUS *BULINUS*

1. Taxonomic History

"Le Bulin" was described from Senegal by Adanson in 1757 and in 1781 it was designated, *Bulinus senegalensis* O. F. Müller. It remained the only known species in this genus for 70 years. By the middle of the nineteenth century several more species had been described from material brought to Europe from Egypt and South Africa. As the exploration of central Africa progressed during the second half of the nineteenth century, more shells reached Europe and a number of conchologists became interested in naming African shells. This gave rise to a proliferation of species that were described inadequately and based on characters that could not be related to each other.

Within the last 50 years the bodies of the snails have been examined also and organs such as the radula and parts of the reproductive system have come into use as taxonomic characters. As more suitable material became available efforts were made to resolve the taxonomic chaos. The first major attempt to reduce the number of valid species was made by Pilsbry and Bequaert (1927). The next attempt by Amberson and Schwarz (1953) was over-enthusiastic and added more confusion to the situation. Mandahl-Barth (1957) revised *Bulinus* into 20 species, 3 of which he subdivided into 10 subspecies. This has proved a reliable basis for taxonomic and ecological studies although it is continually subject to modification. The pattern of change in the number of recognized species of *Bulinus* in Africa is illustrated in Fig. 1.

2. The Species Groups

Mandahl-Barth (1957) arranged the species of *Bulinus* into four groups of closely related species.

(a) The *africanus* group contains the members of the former genus *Physopsis* Krauss, 1848 and this name is retained sometimes as a subgenus. These snails are divided reasonably clearly from the other groups on morphological characters and are the intermediate hosts of *S. haematobium* south of the Sahara.

(b) The *forskali* group represents the former genus *Pyrgophysa* Crosse, 1879. These are mostly small, high-spired snails which are reasonably distinctive but their range of variation raises some problems in separating them from other groups. Their maintenance as a separate genus was doubtful on morphological grounds and became untenable when it was found that *B. senegalensis*, the type-species of *Bulinus*, belonged to this group.

(c) The *truncatus* group includes the intermediate hosts of *S. haematobium* in northern Africa and the Near East. A curious feature of the group is that aphallic specimens are very common. They are very variable in appearance and difficult to distinguish from the following group on shell characters or internal anatomy.

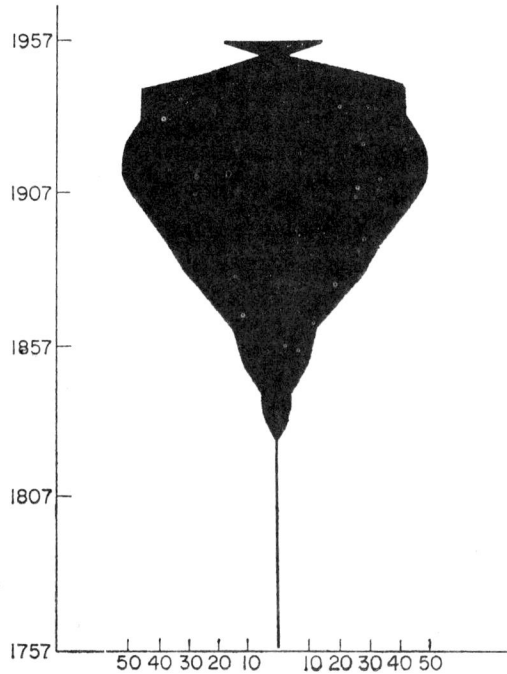

FIG. 1. Diagrammatic representation of the changes which have occurred in the numbers of recognized, named species of *Bulinus* in Africa. (From Wright, 1972. Reproduced with permission from "Flukes and Snails", p. 138. Allen and Unwin, London.)

(d) The *tropicus* group occurs south of the Sahara and is not known to transmit *S. haematobium*. The species show a great range of variation and are difficult to distinguish from the *truncatus* group. The fact that five of the subspecies retained by Mandahl-Barth are in this group, and three more are in the *truncatus* group, is indicative of the difficulties encountered with these groups.

3. Cytological and Immunological Studies

The chromosomes of several species of *Bulinus* have been examined in recent years (Burch, 1960, 1964; Brown et al., 1967; Brown and Burch, 1967).

Most species have a diploid number of 36 which is typical of the family Planorbidae. A few minor variations have been recorded, but the only major exception is that all species in the *truncatus* group are polyploid. They are usually tetraploid, having 72 chromosomes, but higher multiples also occur.

Biochemical and immunological investigations have been carried out and these techniques have been reviewed recently elsewhere (Wright, 1972). The most useful results have come from electrophoretic studies of egg proteins and esterase iso-enzymes which have shown differences at the species and the population levels.

These techniques have already contributed to the solution of some taxonomic problems. Wider application and further development will provide more information about the snails and, it is hoped, may lead to a better understanding of the species.

4. Variability

All the above studies have confirmed the variability of the freshwater planorbids and various reasons have been advanced to explain this. Lakes and river systems are more or less isolated from each other and their snail populations have limited powers of dispersal and cannot readily mix. Nearly all bodies of fresh water are transient in geological time and some snail habitats dry out each year. The snails are hermaphrodite. They can maintain themselves by self-fertilization and a population can be established from one individual. The snails are influenced by their environment and some characters, notably the shape of the shell, show considerable ecophenotypic variation. The proportions of the shell and of the reproductive organs change with age and sexual maturity. Thus, ecological and genetic factors are constantly interacting so that individual populations may develop quite distinctive features but are not likely to remain undisturbed for long enough to allow speciation to take place. Clarifying the systematics of such species distributed across the African continent has proved difficult and the situation changes frequently as more and more material becomes available for examination and new techniques are applied to studying the snails.

HOST–PARASITE RELATIONS

1. Susceptibility to Infection

Numerous records have been published of the results of exposing snails from various localities to infection with different geographical strains of *S. haematobium*. The data demonstrate again the great variation that exists among the snails and the parasites. Many factors influence the development of an infection in a snail (Berrie, 1970) and what happens in the confines of a laboratory may not happen under natural conditions. In an area where the disease is common there is usually at least one intermediate host that is highly susceptible to the local strain of parasite, but the same snails may be totally refractory to another strain of parasite. Infection rates up to 100% are possible in the laboratory and in the field. Many parts of Africa have a dry season during which the human population may be dependent on water from a limited number of pools. These often provide good habitats for host-species of *Bulinus* and they may be contaminated by people bathing. This can produce high levels of infection in the snails and such pools then become dangerous foci of disease transmission.

2. Two Forms of Schistosoma haematobium

S. haematobium can be divided into two ecologically distinct forms. One occurs in the Near East, North and West Africa and is transmitted by *Bulinus* belonging to the *truncatus* species group. The other is confined to Africa south of the Sahara and is transmitted by *Bulinus* of the *africanus* group. These distributions conform with the general distribution of the two groups of snails (Fig. 2). Species of the *truncatus* group are present in East Africa but do not act as intermediate hosts of human schistosomes in this area. Both forms of *S. haematobium* occur sympatrically in parts of West Africa where the ranges of the two groups of snails overlap. In most of Ghana *S. haematobium* is transmitted by *B. globosus* (Morelet, 1866), a member of the *africanus* group. In the Ke district it is transmitted by *B. truncatus rohlfsi* (Clessin, 1886). Each snail is associated with its own strain of parasite. Attempts to obtain cross-infections failed except in a few cases when the snails were exposed to an extremely large number of miracidia larvae (McCullough, 1959). In addition there is some evidence that the pathological effects produced in man by the *truncatus*-borne form of *S. haematobium* are more severe than those produced by the *africanus*-borne parasites.

S. haematobium was described from Egypt where the intermediate host is *B. truncatus truncatus* (Audouin, 1827). It has been proposed that the southern form is a separate species which should be regarded as *S. capense* (Harley, 1864) (Le Roux, 1958; Wright, 1962). However, no morphological differences have been established and this has given rise to opposition to such a change (Nelson *et al.*, 1962; Pitchford, 1965). The position remains unresolved.

3. Selection Pressures

Several investigations have indicated that larval schistosomes have a number of effects on the snail which they parasitize (Berrie, 1970). Infected snails do not survive as long as uninfected ones although the reduction in survival may be proportional to the incompatibility between the strains of snail and parasite, and

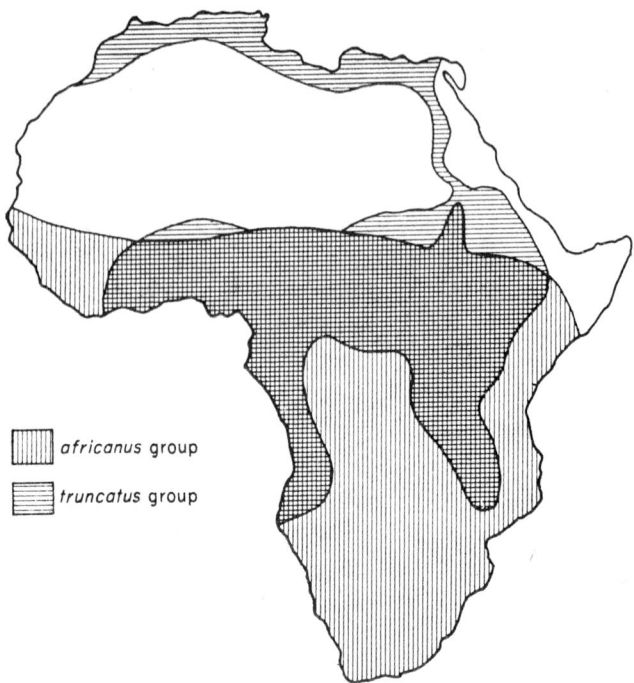

Fig. 2. The general distribution of the *africanus* and *truncatus* species groups of *Bulinus* in Africa.

the intensity of infection in the snail. Some snails survive the infection but most die while cercariae are being produced. The fecundity of the snails is also reduced, sometimes to the extent of sterility; and eggs from infected *Biomphalaria* have shown higher sterility, lower hatching success and greater post-hatching mortality than those laid by uninfected snails. Such effects on survival and fecundity must produce a selective pressure in favour of any form of innate resistance to infection which may develop among the snails. This pressure should be very strong if most snails are exposed to infection as they seem to be in many foci of transmission in Africa. Since the host–parasite complex is maintained there must be other selective factors which act in the opposite direction. One possible mechanism has been shown in recent field studies of trematode infections in the Australian marine snail *Velacumantus australis* (Quoy and Gaimard) (Ewers and Rose, 1966; Ewers, 1967). This snail occurs in two polymorphic forms producing a white-banded shell or an unbanded one. The infection rate in unbanded snails is four times higher than that in banded snails and infection has a severe effect on the gonads. In spite of this, banded snails are commoner among the young snails than among the older ones. Further investigations showed that the snails are eaten by fish and that a disproportionately high number of banded snails are taken. Such polymorphism with respect to resistance could be important in the snail hosts of schistosomes although no case has been reported yet.

THE *TRUNCATUS/TROPICUS* SPECIES COMPLEX

1. Taxonomic Difficulties

The conchological and anatomical similarities of the *truncatus* and *tropicus* groups have been mentioned above. Ecologically, the *truncatus* group occurs in northern Africa and the Near East and the *tropicus* group in southern Africa

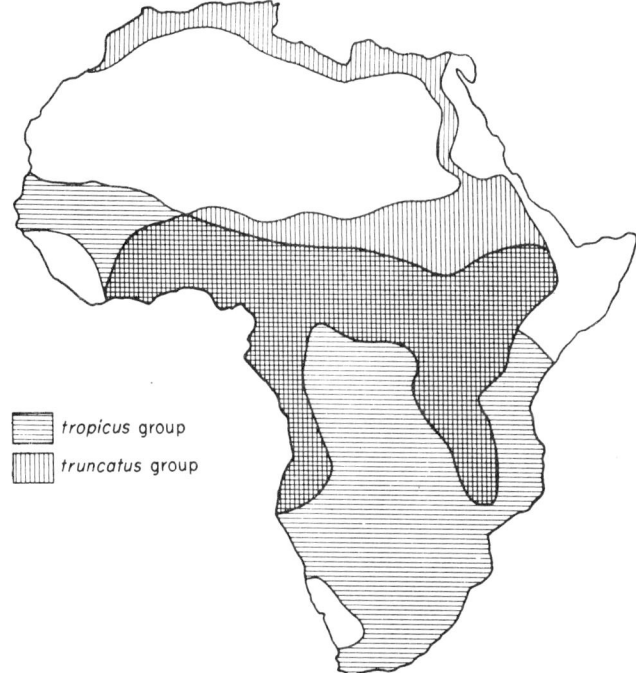

FIG. 3. The general distribution of the *truncatus* and *tropicus* species groups of *Bulinus* in Africa.

with a zone of overlap north of the equator which extends southwards around the East African rift and down the west coast to Angola (Fig. 3). Epidemiologically, *Bulinus truncatus* is the major intermediate host of *S. haematobium* north of the Sahara and in the Sudan and parts of West Africa. The *truncatus* group is less important in the south-eastern part of its range but has been considered potentially dangerous if suitable strains of *S. haematobium* become established in this area. It seems likely that such strains must have been introduced during the southward spread of Nilotic tribes but, if so, they have failed to become established. No member of the *tropicus* group has been shown to act as an intermediate host on the African mainland.

The affinities of *Bulinus natalensis* (Küster, 1841), which occurs in Central and South Africa, have been difficult to establish. It was placed formerly in the *truncatus* group because of the shape of the radula mesocones and the fact that a high proportion of aphallic specimens were found, although these criteria did not separate it clearly from either species group. Recent detailed studies in South Africa have shown no clear morphological differences between *B. tropicus* and *B. natalensis*, nor between *B. natalensis* and *B. truncatus* (Brown, Oberholzer and van Eeden, 1971a, b). Both shapes of mesocones may occur in different individuals from the same population of *B. natalensis* while other individuals may have intermediate shapes. Many populations have normal or predominantly normal copulatory organs. The haploid chromosome number of 18 and the electrophoretic pattern of the egg proteins both conform to the *tropicus* group. However, there is some evidence that *B. natalensis* may not be resistant to infection with *S. haematobium* (Pitchford, 1965; Lo et al., 1970).

2. Polyploidy in Ethiopia

Mandahl–Barth (1957) included all the named forms of *Bulinus* from Ethiopia in a single species, *B. sericinus* (Jickeli, 1874). This species was first placed in the *tropicus* group and then transferred to the *truncatus* group. Subsequently, variations in the chromosome numbers of different populations indicated the existence of a complex of sibling species forming a polyploid series with $2n = 36$, 72, 108 and 144 respectively (Brown and Burch, 1967; Burch 1967). Further investigations of this polyploid complex have been reported recently (Brown and Wright, 1972) and the distribution of the populations is shown in Fig. 4. Most of the diploid and tetraploid populations are found at comparatively low altitudes below 2130 m (7000 ft) in a variety of habitats including eutrophic lakes and temporary rain-pools. The hexaploid and octoploid populations are associated with high altitudes and are found only in streams or the pools left in the bed of streams during the dry season. The few snails examined cytologically from each locality always had the same chromosome number but the geographical ranges of the different forms overlap in certain areas. The diploid populations are very variable in their morphological characters and appear to be a northward extension of the *B. natalensis/tropicus* complex. Attempts to infect them with an Egyptian strain of *S. haematobium* failed. The tetraploid popu-

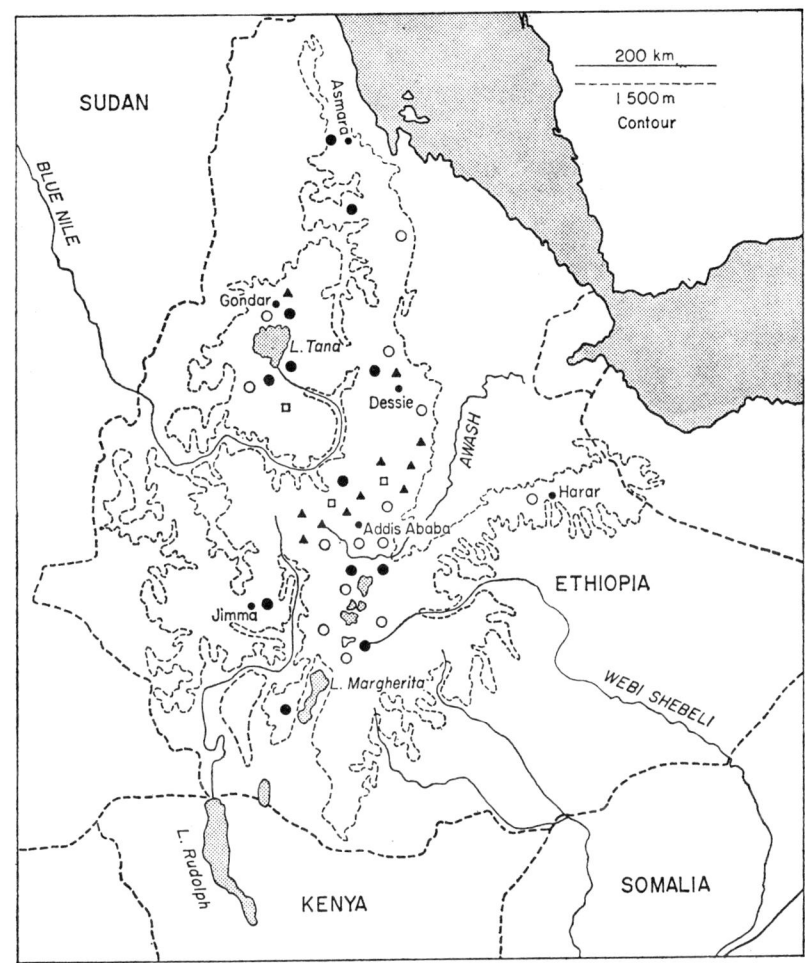

FIG. 4. The distribution of diploid and polyploid populations of *Bulinus* in Ethiopia Diploid (○), tetraploid (●), hexaploid (□), octoploid (▲). (From Brown and Wright, 1972. Reproduced with permission from *J. Zool., Lond.* **167**, 101. Academic Press, London.)

lations closely resemble Egyptian *B. truncatus truncatus* in morphological and biochemical characters. Since they proved susceptible to infection with the Egyptian strain of *S. haematobium* there seems little doubt that they belong to that species. The hexaploid and octoploid populations are fairly homogeneous in their morphological characters but are within the range of variation shown by the other groups. The spire of the shell is generally long and the eggs are large, with a distinctive protein pattern in the octoploids. Octoploid snails may have a very limited susceptibility to Egyptian *S. haematobium*.

Polyploids are often associated with habitats that have variable physical conditions. Brown and Wright (1972) consider that the tetraploidy of *B. truncatus* may have arisen in the Mediterranean region and the Near East during the marked climatic fluctuations which occurred there in the Pleistocene and may be favoured by the present annual climatic variation, which is much greater than that found through most of the range of continental Africa occupied by diploids. The high altitude streams of Ethiopia are also subject to considerable fluctuations and it is reasonable to suppose that the hexaploids and octoploids originated from the tetraploids in this area.

Two characteristics of tetraploids are their susceptibility to infection with *S. haematobium* and the aphallic condition. Since neither appears to be typical of octoploids, it is assumed that these have arisen by allopolyploidy.

In Ethiopia *S. haematobium* is only known to be transmitted at a few localities and the snail host is believed to be a member of the *africanus* group. Since snails thought to belong to the *truncatus* group are abundant it has seemed strange that they did not transmit the disease as they do farther north along the Nile valley. In fact, the tetraploid populations have a scattered distribution and are uncommon in some areas that are climatically suitable for transmission of *S. haematobium*. However, *B. truncatus* is highly successful in the irrigation canals and ditches of the Nile delta and there is a real danger that similar habitats will be created for it in Ethiopia as agricultural development progresses. Such developments also tend to aggregate a human population in close proximity with the snails and provide good conditions for the transmission of schistosomiasis. It has been suggested that such newly created habitats might be stocked with diploid snails in an effort to prevent their colonization by tetraploids (Brown and Burch, 1967).

ZOOGEOGRAPHIC PROBLEMS

1. The africanus group

Although the *africanus* group is reasonably distinct from the other groups of *Bulinus*, there are problems in distinguishing some of the species within the group. The approximate distributions of three of the species are shown in Fig. 5. *B. africanus* (Krauss, 1848) has two subspecies: *B. africanus ovoideus* (Bourguignat, 1879) in East Africa and *B. africanus africanus* to the south. *B. nasutus* (Martens, 1879) has two subspecies also: *B. nasutus nasutus* on the coastal plain of East Africa and its immediate hinterland, and *B. nasutus productus* Mandahl-Barth, 1960 on the inland plateau. *B. globosus* (Morelet, 1866) has several distinctive local forms within its wide range but no subspecies are recognized. There are three more species in the *africanus* group, each occupying a different and restricted area between the equator and the Sahara.

It is reasonably easy to identify B. *africanus*, B. *nasutus* and B. *globosus* over most of their ranges but considerable difficulties have arisen near the East African coast where B. *africanus ovoideus*, B. *nasutus nasutus* and B. *globosus* are all present. Mandahl–Barth (1957, 1965) refers to the situation in the Tanga

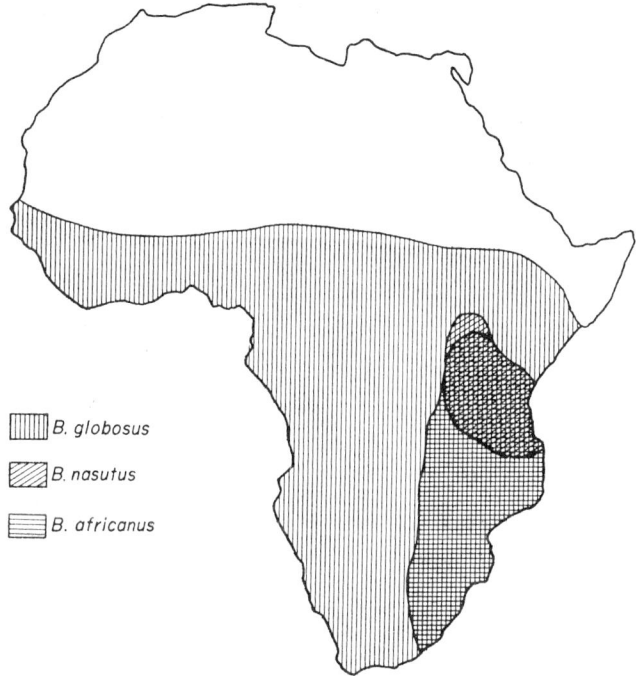

FIG. 5. The general distribution of three species of the *africanus* species group of *Bulinus* in Africa.

District of north-east Tanzania. Here short-spired B. *nasutus nasutus* seem to mingle with long-spired B. *globosus* and the characters of both merge with those of B. *africanus ovoideus*. Various intermediate forms are found and it is difficult to believe that firm species divisions exist. A similar situation involving B. *nasutus nasutus* and B. *globosus* exists in south-east Tanzania (Berrie, unpublished). Specimens showing the characters of each species can be found together with a range of intermediate forms in a single small pool. A possible explanation of this situation is that Tanzania may be an evolutionary centre from which these species have radiated (Wright, 1961).

The ability to distinguish these species is important to ecologists and epidemiologists. Although all three species are good intermediate hosts of *S. haematobium*, their importance varies in different areas. Figure 6 shows the relationship between snail size and infection rate in some populations of *B. globosus* and some that were at least predominantly *B. nasutus nasutus*. The

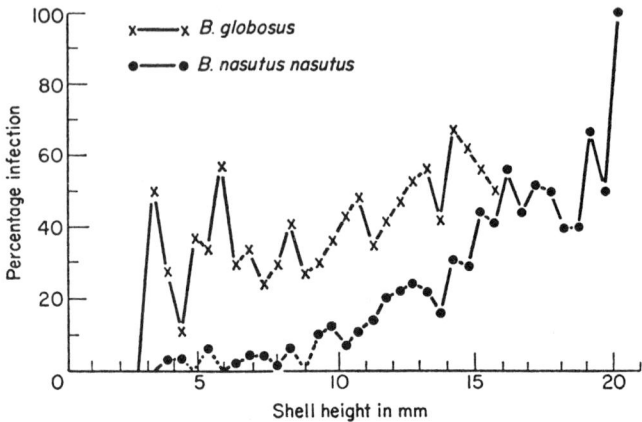

FIG. 6. The relationship between snail size and infection with *S. haematobium* in two populations of snails in southern Tanzania. (From Berrie, 1970. Reproduced with permission from "Advances in Parasitology", Vol. **8**, p. 70. Academic Press, London.)

patterns are quite different and small *B. globosus* are clearly more dangerous than small *B. nasutus nasutus*. Studies of the population dynamics of these snails are important to determine seasonal changes in snail numbers, sizes, infection rates and their relationship to environmental factors. These data can be used to assess the transmission potential of different habitats and to plan control measures. It is difficult to compare data from different areas and formulate general principles if the species under investigation cannot be identified reliably.

2. Bulinus nasutus

The two subspecies of *B. nasutus* present an interesting situation. The main physical difference between them is that *B. nasutus productus* has a higher shell with a longer spire. The spire constitutes about 43% of the total height compared with about 33% in the nominate form. There is no obvious ecological reason why this difference should be associated with their respective ranges. Nor is there any clear distinction in the types of habitat that they occupy, both being found commonly in small dams and in temporary pools. Data are available for the daily output of *S. haematobium* cercariae from 205 naturally infected *B. nasutus productus* collected near Mwanza in Tanzania (McClelland, 1967) and from 32 naturally infected *B. nasutus nasutus* collected near Nachingwea in Tanzania (Berrie, 1973). The two sets of data are in reasonable agreement (Table I). Most snails produce less than 1000 cercariae per day but outputs up to 2500 are fairly common. The highest production in the small group of *B. nasutus nasutus* was 3361 but six of the *B. nasutus productus* yielded over 3500 with the highest figure over 6000. This may indicate that the subspecies with the

TABLE I. Numbers of *S. haematobium* cercariae produced by two groups of *B. nasutus* from different areas of Tanzania

No. of cercariae produced	*B. nasutus nasutus* from Nachingwea		*B. nasutus productus* from Mwanza	
	number	%	number	%
1–500	11	34	69	34
501–1000	8	25	42	20
1001–1500	2	6	42	20
1501–2000	4	13	30	15
2001–2500	5	16	8	4
2501–3000	1	3	6	3
3001–3500	1	3	2	1
3501–4000	0	0	3	1
4001–4500	0	0	2	1
4501–5000	0	0	0	0
5001–5500	0	0	0	0
5501–6000	0	0	0	0
6001–6500	0	0	1	1

higher spire provides more space for the development of cercariae. The number and infectivity of the cercariae present in a body of water are an important aspect of its potential as a site for human infection, and a difference in maximum cercarial production could mean that *B. nasutus productus* is a more dangerous intermediate host.

3. *The Nile passage*

There is a remarkable discontinuity between the species of *Bulinus* found to the north and to the south of the Sahara. The two areas lie in different zoogeographic zones but, for freshwater animals, they are linked by a major highway—the River Nile. However, this route has several inhospitable features. It is some 5000 km from Uganda to the Mediterranean coast. The climate of the two areas is very different and, if snails adapted to East African conditions reach Egypt they may be unable to survive the winter conditions in competition with the Mediterranean species. Large rivers do not seem favourable habitats for *Bulinus*, although they may inhabit suitable marginal areas, and for much of its length the Nile passes through desert where there are no marginal swamps or inlets to offer shelter.

The only species of *Bulinus* which seems to have migrated successfully along the Nile is the ubiquitous *B. forskali* (Ehrenberg, 1831). The area around Khartoum is the meeting point of the southern limit of *B. truncatus truncatus* and the northern limit of *B. ugandae* (Mandahl-Barth, 1957); a member of the *africanus* group and the only other southern species of *Bulinus* to penetrate down the Nile.

The situation is similar with the species of *Biomphalaria* which are intermediate hosts of *S. mansoni*. *B. alexandrina alexandrina* (Ehrenberg, 1831) is the intermediate host in Egypt and it was endemic there until recently. *B. sudanica* (Martens, 1870) and *B. pfeifferi* (Krauss, 1848), the main transmitters south of the Sahara, extend down the Nile beyond Khartoum. Recent information indicates that *B. alexandrina* may be extending southwards and *B. sudanica* northwards so that the ranges of the three species now overlap in the northern Sudan (Williams and Hunter, 1967).

The history of human activity may have some relevance to the situation. Egyptian civilization established contact with the Sudan and Ethiopia but there seems to have been little trade with areas farther south. Even today communications along the Nile valley are not well developed. Ancient trade routes were established between the Nile valley and West Africa and it seems probable that the northern form of *S. haematobium*, and possibly the appropriate snail hosts, reached West Africa in this way.

CONCLUSION

The taxonomy of *Bulinus* is fraught with difficulties and this problem is fundamental to all aspects of their biology. We are still at the stage of seeking answers to apparently simple questions and have not reached a stage where many clear relationships can be demonstrated between systematics and ecology in this genus. What is clear is that the systematics, taxonomy, physiology, and ecology of the snails are inextricably linked and that it is very difficult to make progress in any of these fields without involving studies in the others. These studies are all important in understanding the epidemiology of schistosomiasis and, thus, have practical value in terms of human welfare. Ecologists and systematists ignore each other's problems at their peril.

REFERENCES

Amberson, J. M. and Schwarz, E. (1953). On African schistosomiasis. *Trans. R. Soc. trop. Med. Hyg.* **47**, 451–502.

Berrie, A. D. (1970). Snail problems in African schistosomiasis. *In* "Advances in Parasitology" (Ben Dawes, ed.), Vol. **8**, pp. 43–96. Academic Press, London and New York.

Berrie, A. D. (1973). The production of *Schistosoma haematobium* cercariae by *Bulinus globosus* and *Bulinus nasutus nasutus* in south-east Tanzania. In press.

Brown, D. S., Oberholzer, G. and van Eeden, J. A. (1971a). The *Bulinus natalensis/tropicus* complex (Basommatophora : Planorbidae) in south-eastern Africa: I. Shell, mantle, copulatory organ and chromosome number. *Malacologia* **11**, 141–170.

Brown, D. S., Oberholzer, G. and van Eeden, J. A. (1971b). The *Bulinus natalensis/tropicus* complex (Basommatophora : Planorbidae) in south-eastern Africa: II. Some biological observations, taxonomy and general discussion. *Malacologia* **11**, 171–198.

Brown, D. S. and Burch, J. B. (1967). Distribution of cytologically different populations of the genus *Bulinus* (Basommatophora : Planorbidae) in Ethiopia. *Malacologia* **6**, 189–198.

Brown, D. S., Schutte, C. H. J., Burch, J. B. and Natarajan, R. (1967). Chromosome numbers in relation to other morphological characters of some southern African *Bulinus* (Basommatophora : Planorbidae). *Malacologia* **6**, 175–188.

Brown, D. S. and Wright, C. A. (1972). On a polyploid complex of freshwater snails (Planorbidae: *Bulinus*) in Ethiopia. *J. Zool. Lond.* **167**, 97–132.

Burch, J. B. (1960). Chromosome numbers of schistosome vector snails. *Z. Tropenmed. Parasit.* **11**, 449–452.

Burch, J. B. (1964). Cytological studies of Planorbidae (Gastropoda : Basommatophora). I. The African subgenus *Bulinus* s.s. *Malacologia* **1**, 387–400.

Burch, J. B. (1967). Some species of the genus *Bulinus* in Ethiopia, possible intermediate hosts of schistosomiasis haematobia. *Ethiopian med. J.* **5**, 245–257.

Ewers, W. H. (1967). Selection pressure and polymorphism in *Velacumantus australis* (Quoy & Gaimard). *Aust. J. Zool.* **15**, 727–738.

Ewers, W. H. and Rose, C. R. (1966). Polymorphism in *Velacumantus australis* (Gastropoda : Potamididae) and its relationship to parasitism. *Aust. J. Zool.* **14**, 49–64.

LeRoux, P. L. (1958). The validity of *Schistosoma capense* (Harley, 1864) amended as a species. *Trans. R. Soc. trop. Med. Hyg.* **52**, 12.

Lo, C. T., Burch, J. B. and Schutte, C. H. J. (1970). Infection of diploid *Bulinus* s.s. with *Schistosoma haematobium*. *Malacol. Rev.* **3**, 121–126.

McClelland, W. F. J. (1967). Production of *Schistosoma haematobium* and *Schistosoma mansoni* cercariae in Tanzania. *Expl Parasit.* **20**, 205–218.

McCullough, F. S. (1959). The susceptibility and resistance of *Bulinus* (*Physopsis*) *globosus* and *Bulinus* (*Bulinus*) *truncatus rohlfsi* to two strains of *Schistosoma haematobium* in Ghana. *Bull. Wld Hlth Org.* **20**, 75–85.

Mandahl-Barth, G. (1957). Intermediate hosts of *Schistosoma* : African *Biomphalaria* and *Bulinus*: II. *Bull. Wld Hlth Org.* **17**, 1–65.

Mandahl-Barth, G. (1965). The species of the genus *Bulinus*, intermediate hosts of *Schistosoma*. *Bull. Wld Hlth Org.* **33**, 33–44.

NELSON, G. S., TEESDALE, C. and HIGHTON, R. B. (1962). The rôle of animals as reservoirs of bilharziasis in Africa. In "Bilharziasis: Ciba Foundation Symposium" (G. E. W. Wolstenholme and M. O'Connor, eds), 127–149. Churchill, London.

PILSBRY, H. A. and BEQUAERT, J. (1927). The aquatic mollusks of the Belgian Congo, with a geographical and ecological account of Congo malacology. *Bull. Am. Mus. nat. Hist.* **53,** 69–602.

PITCHFORD, R. J. (1965). Differences in egg morphology and certain biological characteristics of some African and Middle Eastern schistosomes, genus *Schistosoma*, with terminal-spined eggs. *Bull. Wld Hlth Org.* **32,** 105–120.

WILLIAMS, S. N. and HUNTER, P. J. (1967). Distribution of *Biomphalaria* species in Sudan. *Nature, Lond.* **215,** 1408.

WRIGHT, C. A. (1961). Taxonomic problems in the molluscan genus *Bulinus*. *Trans. R. Soc. trop. Med. Hyg.* **55,** 225–231.

WRIGHT, C. A. (1962). The significance of infra-specific taxonomy in bilharziasis. In "Bilharziasis: Ciba Foundation Symposium" (G. E. W. Wolstenholme and M. O'Connor, eds), 103–120. Churchill, London.

WRIGHT, C. A. (1972). Biochemical and immunological methods in the taxonomy of the Mollusca. In "Biochemical and Immunological Taxonomy of Animals" (C. A. Wright, ed.). Academic Press, London.

9 | Anthecology, Floral Morphology and Angiosperm Evolution

A. D. J. MEEUSE

Hugo de Vries-Laboratorium Voor Bijzondere Plantkunde,
University of Amsterdam, Netherlands

Abstract: Different aspects of floral evolution—pollination ecology, floral morphology and the origin of the flowering plants—are discussed on the basis of two alternative hypotheses concerning early flower types. After weighing the evidence, an aphananthous (non-showy, "apetalous") archetype, which was unisexual and originally wind-pollinated, is postulated. In the progenitors of the nowadays predominantly phaneranthous, monoclinous and zoophilous taxa, floral evolution passed from this very early phase of dicliny, anemophily and aphananthy of the prefloral reproductive region (anthocorm) to a phase of incipient entomophily, soon concomitant with a partial sex reversal within the floral region, leading to an incipient monocliny of the early flower type. This second phase culminated in the strictly monoclinous and primarily zoophilous flower types after the latter had acquired semaphylls, but some flower types are exceptional in that some of the elements of the characteristic zoophilous syndrome—bisexuality of the flower and the presence of optical lures or semaphylls—did not develop, whereas in the primarily diclinous and typically anemophilous groups the original aphananthy and dicliny persisted.

The second phase of floral evolution gave rise to different morphological flower types, in each of which the primary zoophily is almost invariably associated with the development of optical lures or semaphylls. There were two sources of these semaphylls, viz. bracteoid elements already present in the anthocorm, and androecial members (stamens) which became petaloid.

Early zoophilous flower types were aphananthous and clearly anthocormoid. Approximations are still extant in the reproductive region (the so-called inflorescence) of the Saururaceae and of some other taxa. Various morphogenetic evolutionary processes—contractions, condensations (brachyblasty), readjustments, oligomerizations, reductions, fusions, adnations, and modifications (semaphylls!)—resulted in a great diversity of morphological patterns, thus providing the background of the many adaptive flower types. Initially anemophilous flower types underwent a different semophylesis whilst retaining the anthecological syndrome of aphananthy ("apetaly" and dicliny), except in cases of secondarily developed zoophily with a special anthecological syndrome.

INTRODUCTION

Several aspects of the interconnected problems of the evolution of the Flowering Plants, the phylogenetic origin of the flower, and the development of ecological associations between flowers and pollinators have been understood for a considerable length of time (cf. Meeuse, 1972c for references; see also Leppik, 1960–1969; Baker and Hurd, 1968; Ehrendorfer, 1971). The interactions between floral morphology and pollination mechanisms form such an essential part of Angiosperm evolution that only a comprehensive treatment of all aspects can possibly result in well-founded conclusions. It is not an exaggeration if one states that the origin of the Angiosperms is largely synonymous with the problem of the evolution of the flower, the morphological evaluation of their fruits and seeds either emanating from a floral theory (such as the concept of the carpel) or following from the conditions already prevailing among cycadopsid gymnosperms. For this reason we need only concern ourselves here with the phylogenetic history of the fertile region from a primitive pre-floral stage towards a fully-fledged flower. This semophyletic process involved three more or less clearly interrelated aspects of floral evolution, namely, the changing morphological structure of the fertile region and its constituent elements, the sex distribution (i.e. monocliny versus dicliny), and the mode of pollen transfer (anemophily versus zoophily).

Another aspect of Angiosperm phylogeny has been dubbed "protection of the ovules" by Grant (1950) and "angiovuly" by Van der Pijl (1960). This is nothing but an anthecological interpretation of the conventional morphological difference between Gymnosperms (plants with "naked" ovules) and Angiosperms (plants with ovules contained in a pistil): the driving, selective force in floral semophylesis which causes the need for (additional) protection is said to be the advent of early zoophily, namely visits to the floral region by, primarily, dystrophic insects such as the larger anthophilous beetles. This implies that phytophagous animals acting as pollinators are thus prevented from attacking the vital ovules of angiospermous forms; but are, by inference, the ovules of gymnospermous taxa exposed to destruction by dystrophic animals? In my opinion it is rather doubtful whether the evolution of the flowering plants from advanced members of the cycadopsid line of evolution—from chlamydospermoid-cycadeoid forms to all intents and purposes—in fact involved an abrupt change from "unprotected" ovules to a condition of "angiovuly". In gymnospermous taxa the vital female reproductive tissue is located in the nucellus (the homologue of the megasporangium), and all its accessory organs (integuments, cupule, chlamys, sarcotesta, sclerotesta, aril, arilloids, etc.) are ultimately expendable. In point of fact, the sarcotesta, aril, arilloids, chlamys, etc., are often fleshy and intended to be sacrificed to phytophagous animals (in cases of endozoochory) while the nucellus and all its derivatives (perisperm, endosperm, embryo sac, embryo) remain protected by, for example, a sclerotesta. The ovules of gymnospermous taxa are often effectively protected within cone-shaped strobili by hard cone-scales (in many coniferopsid groups and in *Cycadales*) or by interovular scales (in *Cycadeoidales*), so that the necessity of protection apparently did not arise with the advent of Angiosperms with zoophilous flowers.

The morphological interpretation of the pistils of at least the more primitive flowering plants as modified derivatives of pteridospermous cupules (cf. Meeuse, 1971, 1972b, 1973) denies the traditional difference (the "gap") between the ovule-bearing structures of advanced cycadopsid forms and those of Angiosperms. In other words, the "protection of the ovules" is not different or better or more necessary than it is in various gymnospermous taxa: it may even be less effective! Protection of ovules (or of whole pistils) *in a flower* by the advent of a hollow receptacle and by such devices as hypogyny manifestly came about *after* the early evolution of the flower and we have again arrived at our starting point: the crucial question of *the phylogenetic origin of the flower* as a clue to the origin of the flowering plants.

ANTHECOLOGICAL CONSIDERATIONS OF FLORAL EVOLUTION

A rational ecological approach to the problem of floral evolution must commence with the presupposition of some basic morphological pattern adapted to a particular form of pollination, and subsequently deduce the various flower types from the prototype by semophyletic changes associated with a switch-over to a different pollination mechanism, the more conservative forms retaining some, or all, of the characteristic features of the original anthecological syndrome. This principle was recognized a long time ago, but the phytomorphological tenets relating to interpretative floral morphology prescribed the prototype of the angiospermous flower as a special morphological and anthecological floral unit called the *phaneranthous* reproductive region in Leppik's papers. This is accepted in nearly all relevant anthecological papers from about 1904 onward (and cited in van der Pijl, 1960, 1961; Leppik, 1960 *et seq.*; Faegri and van der Pijl, 1966; Baker and Hurd, 1968; Ehrendorfer, 1971: pp. 593–595; Meeuse, 1972c), although some workers (such as Kugler, 1970: pp. 303–304) express some doubt as to the conclusion that early Angiosperms were consistently phaneranthous, because phaneranthy implies monocliny and zoophily, and zoophily in turn requires the presence of coloured optical lures (petaloid elements or semaphylls; another feature of phaneranthy). The question of sex distribution—

dicliny or monocliny—was discussed by Heslop-Harrison (1957, 1958, 1964) and in spite of strong repudiations (e.g. by Parkin 1952, 1957) his arguments have not been refuted; the descent of Angiosperms from a group of the almost consistently diclinous Gymnosperms (there is no better alternative!) requires a change in the sex distribution at one time or another to account for the frequent occurrence of monocliny among the recent Angiosperms. The choice of the prototype in question is, therefore, very important because it provides the starting point for phylogenetic speculations and thus prescribes the semophyletic sequence (which is a directional process), at the same time pinpointing the more primitive and the more advanced conditions in floral morphology, and, hence, in the anthecological syndromes.

A study of recent taxa gives us the necessary data concerning the correlated occurrence of a morphological pattern, a certain type of sex distribution and a special form of pollination, because we may assume that one such syndrome characterized the floral region of the earliest Angiosperms and their immediate precursors. Given the conditions prevailing among the fossil and the recent Gymnosperms, the selection of the most likely archetype must be compatible with the still "gymnospermous" nature of the ancestral forms under discussion and, therefore, exhibit advanced "gymnospermous" rather than "angiospermous" characters. Early Angiosperms or even pro-angiospermous, advanced Gymnosperms may already have exhibited highly evolved characters in the structure of their ovules, integuments, and seed coat (cf. Sharma, 1970, 1970b; Crepet and Delevoryas, 1972, on cycadeoid ovules and seeds; Bouman, 1971a, 1971b; De Boer and Bouman, 1972 on the ovular coats of recent ranalean Angiosperms), in their wood anatomy, in the mode of embryo sac formation, in phytochemical features, etc., whereas floral evolution may have lagged behind. Gymnosperms, even if they are reputed to be animal-pollinated, never possess semaphylls as far as can be ascertained (the strobilus of Cycadeoidea is still adduced as an example of early zoophily and incipient phaneranthy by Leppik in his most recent papers, although Delevoryas, 1963, 1968, had already refuted the earlier reconstructions as incorrect; moreover, Cycadeoidea appears too late in the stratigraphical sequence to be a *precursor* of Angiosperms with phaneranthous flowers). Zoophily is apparently not always associated with semaphylls, and not necessarily with monocliny, so that there is no reason to maintain that the typical anthecological syndrome of the recent phaneranthous Angiosperms represents the initial phases of morphological flower evolution and pollination mechanisms among the earliest Angiosperms.

Students of floral ecology have neglected the evidence provided by small pollinating insects belonging to major groups other than the now predominant, larger flower-visiting beetles, and advanced Hymenoptera, Diptera, and Lepidoptera with their often adaptive, sucking mouth parts. The sensory perception of the more primitive Coleoptera, Diptera, and Hymenoptera, particularly colour vision and image formation, is less developed and in fact hardly comparable with that of the more advanced groups of insects. They are (and were) most probably guided by their sense of smell rather than by optical signals, as we see in the cases of coprophily and saprophily of non-anthophilous groups of insects (certain fly groups, Psychodidae, dung and carrion beetles, etc.). It is feasible that early zoophily existed in aphananthous, pro-angiospermous forms with dicliny (as in *Ficus*), or at most with incipient dicliny (as in Chlamydosperms), and that from this phase onward both the anthecological syndrome of the reproductive region and the evolution of the more advanced types of anthophilous insects proceeded as a form of co-evolution. The argument that flowers with an "excessively" elongated corolla-tube (e.g. *Pavetta, Ixora, Ruspolia, Ipomoea alba*) or with a very long spur (e.g. *Angraecum sesquipedale*) could only originate as the product of co-evolution with the macroglossid (long-tongued) hawk-moths and papilionid butterflies can be extrapolated into the past; gamepetalous and spurred flowers are derived from choripetalous and ecalcarate archetypes adapted to microglossid pollinators, petaliferous flowers from apetalous ones (the latter visited by insects with a poor colour perception or anemophilous), and bisexual flowers, ultimately, from unisexual precursory structures. One thus arrives logically at an aphananthous and unisexual archetype of a flower, but this original, morphological entity need not have already evolved into a flower: the definition or circumscription of a "flower" is also at stake.

The mutual comparison of the different ecological flower types of recent angiospermous taxa leads to the conclusion that showy (or "phaneranthous") flowers, which are normally monoclinous and almost invariably zoophilous, owe their phaneranthy to the presence of optical lures or intrafloral semaphylls (i.e. of white or coloured, laminar floral parts, usually called tepals or petals), whereas, barring a few exceptions, aphananthous (i.e. monochlamydeous, apetalous or even completely naked) flowers are not only devoid of such semaphylls, but also unisexual and anemophilous. Either of these conditions may be considered to reflect the ancient (protoangiospermous) condition. There are special cases, it is true, such as autogamy, cleistogamy, and zoophily of diclinous taxa with *extrafloral* semaphylls (e.g. Euphorbiaceae–Euphorbieae such as poinsettias, the cyperaceous *Dichromena ciliata* and the dioecious pandanaceous genus *Freycinetia*). Conceivably, these cases are either exceptional or specializations of a secondary nature, so that we can safely assume that the oldest Angiosperms reproduced sexually in the normal way, and usually by heterogamy rather than through selfing. In other words, the prototype of the angiospermous floral region exhibited, like the flowers of the recent Angiosperms, either the syndrome of (incipient) phaneranthy, monocliny and entomophily (the earliest form of zoophily), or the alternative correlated complex of aphananthy, dicliny and anemophily. This is roughly equivalent to the morphological and taxonomic controversy concerning the "basic" angiospermous group and its floral morphology, with the majority of the contemporary phanerogamists still favouring the ranalean hypothesis (postulating some phaneranthous form as the archetype), and only a few workers proposing some apetalous and diclinous forms as progenitors of the flowering plants.

MORPHOLOGICAL ASPECTS OF FLORAL EVOLUTION

The ecological aspects of floral evolution can be related to the two principal morphological categories of the phaneranthous (and bisexual) versus the aphananthous (and unisexual) flower types as was done by Leppik in a series of papers (1960-1969). As stated before, the selection of the most plausible, ancient condition cannot be decided solely on the basis of ecological relations between recent plants and their pollinators, but must also be compatible with evidence from other sources. If the immediate precursors of the flowering plants were advanced members of some cycadopsid line of evolution, which is almost a foregone conclusion, the conditions prevailing among cycadopsid forms must provide the clues rather than an allegedly primitive assembly of angiospermous types closely resembling certain recent members of the ranalean alliance. Gymnosperms are, generally speaking, manifestly aphananthous and strictly diclinous, a condition usually associated with wind pollination. Hypotheses presupposing the occurrence of a small, now extinct, group of monoclinous and phaneranthous preangiosperms are not very convincing, as we have seen: certain forms such as Cycadeoidea occurred stratigraphically in the "wrong" era, others are not sufficiently phaneranthous to qualify, and some of the reconstructions of fossil bennettitalean forms are based on spurious interpretations of the morphology of their reproductive region.

These doubts certainly merit the alternative approach, and I am convinced that the development of a relation between the reproductive regions of gymnospermous, and more particularly of proangiospermous, plants and certain groups of insects must of necessity have originated in an *aphananthous* phase, but the anthecological relation could not become general and efficient until a partial sex reversal had taken place within the reproductive region which eventually culminated in consistent monocliny in most of the recent groups with phaner-

9. Anthecology, Floral Morphology and Angiosperm Evolution

anthous and zoophilous flowers. The sex reversal became, subsequently or perhaps simultaneously, associated with the appearance of semaphylls in nearly all zoophilous taxa. It is especially in this last respect that my views differ from the still current ideas regarding early flower types which presuppose the erstwhile presence of hemi-angiospermous forms with more or less complete phaneranthy (cf. Leppik, 1964, 1969).

The fossil record of such (by inference also already monoclinous!) advanced gymnosperms is anything but convincing, whereas the tangible evidence provided by the recent Chlamydosperms, namely aphananthy coupled with a clear tendency towards a partial sex reversal (sometimes already culminating in *functional* monocliny in the predominantly dioecious group) and with incipient or perhaps already well-established entomophily, points to the alternative theory.

One cannot avoid a morphological evaluation at this stage, because the advent of the supposedly characteristic, early flower of the phaneranthous type must have roughly coincided with that particular phase of co-evolution in which the relation between flower and pollinator had attained a certain level of consistent interdependence. Other lineages, in which a tendency towards ambisexuality of the reproductive region did not become apparent and in which anemophily persisted, gave rise to as many recent groups with still aphananthous flowers. The still most current explanation of the origin of such taxa with diclinous, anemophilous and also otherwise aphananthous flowers by phylogenists and anthecologists is a progressive morphological reduction of a phaneranthous (and monoclinous) archetype associated with a supposedly secondary change to anemophily (and dicliny). Secondary anemophily is rare, in my opinion, and restricted to truly advanced forms such as the ragweeds among the Compositae and the plantains among the Gamopetalae-Tubiflorae, so that of necessity one must accept primary anemophily in such forms as palms, pandans, and many monochlamydeous Dicotyledons (cf. Meeuse, 1972a); in other words, the majority of the diclinous Angiosperms did not descend from monoclinous precursors of the favoured ranalean type but rather from aphananthous, diclinous and wind-pollinated prototypes.

One of the most crucial changes, apart from an altered sex distribution, in the floral region of the early zoophilous forms, was the development of intrafloral semaphylls, presumably by the selective pressure associated with pollination by younger groups of insects with a well-developed form of colour perception, as an expression of the co-evolution of flower type and pollen vector. This implies that an association of flowers with pollen carriers devoid of the sense of selective colour perception is conceivable. In nearly all speculations concerning ancient entomophilous flower types colour perception and phaneranthy are accepted as a foregone conclusion (Leppik, also Takhtajan, 1959, 1969; Baker and Hurd, 1968; and many others), but I believe (and have explained why) a precursory phase must have existed in which an association between primitive insects and aphananthous reproductive regions became established preceding the advent of the colourful optical lures.

Nowadays the entomophilous genus *Ficus* is associated with specialized Hymenoptera presumably descended from a primitive (early Braconid) ancestral taxon; *Gnetum* and *Ephedra* are reported to be visited by small insects; and the association of the Araceae-Areae with pollinating insects primarily guided towards the inflorescence by odour perception may well have started without an optical lure (if so, the extra-floral semaphyll in the form of the spathe was a secondary refinement).

The whole idea of co-evolution is based on the assumption that two groups of organisms with an ecological relationship become subject to correlated evolutionary changes which conceivably include the continual, mutual improvement by selection of morphological, physiological, and phytochemical features of adaptive significance. In the case under discussion the pollen vector presumably achieved a progressively improved capacity of colour perception, and the flower became enriched with optical lures. If this is the case, there are two likely precursory sources of the semaphylls, namely, bracteoid elements (already present within the archetype of the true flower or anthocorm *sensu mihi*) which changed from an originally greenish, brownish or at least not strikingly coloured condition into a showy tepaloid organ, and stamens which became staminoidal and subsequently petaloid in shape and in pigmentation. Such changes are compatible with the postulated evolution of the euanthous flowers from aphananthous archetypes, and, as far as the second kind of semaphyll is concerned, with the "transitions" between functional androecial members and petals present in, for example, the flowers of several Nymphaeaceae, whereas the advent of petals by the modification of androecial elements is decidedly at variance with the idea of early phaneranthy in floral evolution.

PHYLOGENETIC EVALUATION OF THE ANTHECOLOGICAL AND MORPHOLOGICAL EVIDENCE

If one defines the angiospermous flower as a condensed and variously modified anthocorm (Meeuse, 1965 onwards), one can only logically link up advanced Gymnosperms and Angiosperms by assuming that initially an aphananthous and unisexual anthocormoid floral region was of general occurrence. In evolutionary lines with incipient zoophily this became partially to completely ambisexual, ultimately to assume the characteristics of a phaneranthous flower. In other lineages anemophily and monocliny persisted, and phaneranthy (with zoophily) was only indirectly (and secondarily) achieved by aggregation of contracted anthocorms (flowers) into pseudanthia which acquired extra-floral semaphylls. There are some special cases and exceptions which complicate matters somewhat, but they are not necessarily at variance with the phylogenetic sequences sketched above. If incipient entomophily exists in a predominantly or completely diclinous group without semaphylls, as in *Ficus* and in the Chlamydospermae, some taxa, starting from this ancient condition, may have attained the level of complete zoophily without a change-over to monocliny during their evolution and subsequently acquired semaphylls (cf. Schizandraceae, Menispermaceae, Myristicaceae, several groups of the Laurales such as Monimiaceae, and some Lardizabalaceae); on the other hand an occasional group may have become strictly monoclinous and zoophilous but without ever developing semaphylls (*Trochodendron*). Intermediate cases exist in which certain elements of the syndrome are not yet fixed one way or the other. *Euptelea*, very primitive, but functionally monoclinous, is apetalous and both wind- and insect-pollinated (Endress, 1970); some apetalous Rosaceae–Poterieae (=Sanguisorbeae) are entomophilous and a number of them are diclinous to polygamous (monoclinous phaneranthous Roseae may be derived from diclinous aphananthous forms with beginning entomophily!); and similar transitional stages between dicliny and ambisexuality, apetaly and phaneranthy, and anemophily and entomophily are found among the Laurales, especially in the Monimiaceae *sensu lato*.

The occurrence of various intermediate and transitional forms of floral organization and of pollen transfer is not only compatible with the basic assumption of a primitive unisexual, anemophilous and apetalous floral region, but also explains why a primitive Angiosperm such as *Trochodendron* may be apetalous yet entomophilous. In the classical floral concept, an apetalous flower is supposed to be invariably derived from a phaneranthous floral region, so that the apetaly of the homoxylous (and, hence, without doubt primitive) genus *Trochodendron* would be highly incongruous and requires an undue degree of heterobathmy. Similar inconsistencies are encountered in the cases of *Euptelea*, Piperales, *Cercidiphyllum*, and Monimiaceae *sensu lato*.

The anthecological considerations become more plausible if a parallel morphological evolution of the floral region can be demonstrated. Starting from a preangiospermous archetype with primitive, unisexual anthocorms, one may expect to find primitive anthocormoid reproductive regions in a still prefloral stage of semophylesis among primitive Angiosperms. The occurrence of a prefloral organization of the floral region would be even more convincing if transitions towards ambisexuality occurred in the anthocormoids of such primitive forms. A number of such forms indeed exist and can be arranged in a "phylogenetic" sequence. *Euptelea* and *Saururus* have ambisexual anthocormoids in the pre-floral stage without semaphylls (unisexual types, reduced by, mainly, oligomerizations, are found in other Piperales); *Anemopsis* and *Houttuynia* have the lowermost bracts developed as bracteoid semaphylls and have a very strong tendency towards entomophily (otherwise the organization of the floral region is very much the same as it is in *Saururus*); *Trochodendron* has flowers representing an already much compacted monoclinous anthocorm but devoid of semaphylls; *Schizandra* and *Akebia* have attained phaneranthy and entomophily, but they remained diclinous; and *Drimys*, *Ranunculus* and *Magnolia* have attained the ultimate stage of phaneranthy with monocliny, entomophily and even androecial semaphylly. A parallel series with a divergent floral organization is formed by *Amborella*, Monimiaceae and Lauraceae. From a condition similar to that found in *Euptelea*, but unisexual or bisexual, a contraction of the anthocorms would result in flower types found in Hamamelidales, Amentiflorae and presumably Rosales. For details the reader is referred to some recent publications by the present author (Meeuse, 1971, 1972b, 1973).

CONCLUSIONS

The considerable diversity in the morphology and the ecology of the floral region of primitive Angiosperms is much more compatible with the postulation of a diclinous, aphananthous (and originally anemophilous) archetype of the flower (or anthocorm) from which several, partly divergent, morphological and ecological patterns arose, than with the alternative hypothesis of a phaneranthous, ambisexual precursor of the flower with established insect pollination (see, for example, Leppik, 1969; Takhtajan, 1959, 1969; Ehrendorfer, 1971). Ecological and phylogenetic considerations require the change-over from the original anemophily present at the early gymnospermous level of Angiosperm evolution to zoophily at some phase of the semophyletic history of the (generally phaneranthous and zoophilous) flower of the fundamentally monoclinous groups among the recent Angiosperms. This could have happened at an early (still gymnospermous) level, as suggested by the conditions in chlamydospermous forms, or considerably later, or possibly more than once, at unequal evolutionary levels in different lineages.

The special case of *Ficus* and the fig-wasps is instructive by showing entomophily in an aphananthous and diclinous taxon. Small, ancient Hymenoptera, Diptera, and Coleoptera may have become frequent visitors of certain diclinous forms without semaphylls, which relation must primarily have come about by the perception of olfactory signals but became furthered by the partial sex reversal within the reproductive region causing a different spatial arrangement of polliniferous and ovuliferous organs. The adaptive alteration of bracteoid elements of the condensed anthocorm and/or stamens (conceivably, in an occasional diclinous entomophilous genus such as *Schizandra*, also of gynoecial members?) into optical attractants constitutes the finishing touch in the evolution of the phaneranthous flower. In other phylogenetic lineages the floral region remained aphananthous, anemophilous and unisexual, only to become adapted to zoophily by a different pathway involving pseudanthy and the development of extra-floral semaphylls.

Among primitive Angiosperms the stages of the progressive evolution from a monosexual, primitive anthocorm can be followed from an anthocormoid floral region in such forms as *Saururus* and *Euptelea* to a phaneranthous ranunculoid, magnolioid, laureoid and roseoid flower. Various parallel, terminal semiphyleses, such as phaneranthy coupled with dicliny (*Schizandra*) and apetaly associated with monocliny and entomophily (e.g. *Trochodendron*) illustrate the many pathways of specialization. In other evolutionary lines apetaly, monocliny and anemophily persisted up to the present day.

REFERENCES

BOER, R. DE and BOUMAN, F. (1972). Integumentary studies in the Polycarpicae. II. Magnoliaceae. *Magnolia stellata* and *Magnolia virginiana*. *Acta Bot. Neerl.* **21**, 611–623.

BOUMAN, F. (1971a) The application of tegumentary studies to taxonomic and phylogenetic problems. *Ber. Deut. Bot. Ges.* **84**, 169–177.

BOUMAN, F. (1971b). Integumentary Studies in the Polycarpicae. I. Lactoridaceae. *Acta Bot. Neerl.* **20**, 565–569.

CREPET, W. L. and DELEVORYAS, T. (1972). Investigations of North American cycadeoids —early ovule ontogeny. *Am. J. Bot.* **59**, 209–215.

DELEVORYAS, T. (1963). Investigations of North American Cycadeoids—cones of *Cycadeoidea*. *Am. J. Bot.* **50**, 45–52.

DELEVORYAS, T. (1968). Investigations of North American cycadeoids—structure, ontogeny, and phylogenetic considerations of cones of *Cycadeoidea*. *Palaeontographica* **121** B, 122–133.

EHRENDORFER, F. (1971). Systematik und Evolution. *In* "Lehrbuch der Botanik für Hochschulen" (E. Strasburger *et al.*, eds). pp. 379–741. Gustav Fischer, Stuttgart.

ENDRESS, P. (1970). Gesichtspunkte zur systematischen Stellung der Eupteleaceae (Magnoliales). Untersuchungen über Bau und Entwicklung der generativen Region bei *Euptelea polyandra* (Sieb. et Zucc.). *Ber. Schweiz. bot. Ges.* **79**, 229–278.

FAEGRI, K. and VAN DER PIJL, L. (1966). "The Principles of Pollination Ecology." Pergamon Press, London.

GRANT, V. E. (1950). The protection of the ovules in flowering plants. *Evolution* **3**, 179–201.

Heslop-Harrison, J. (1957). The experimental modification of sex expression in flowering plants. *Biol. Rev.* **32**, 38–90.

Heslop-Harrison, J. (1958). The unisexual flower—a reply to criticism. *Phytomorphology* **8**, 177–184.

Heslop-Harrison, J. (1964). Sex expression in flowering plants. *Brookhaven Symp. Biol.* (1963) **16**, 109–125.

Kugler, H. (1970). "Blütenökologie" (2nd edition). Gustav Fischer, Stuttgart.

Leppik, E. E. (1960). Early evolution of flower types. *Lloydia* **23**, 72–92.

Leppik, E. E. (1963). Fossil evidence of floral evolution. *Lloydia* **26**, 91–115.

Leppik, E. D. (1964). Floral evolution in the Ranunculaceae. *Iowa St. J. Sci.* **39**, 1–101.

Leppik, E. E. (1968). Directional trend of floral evolution. *Acta Biotheor. (Leyden)* **18**, 87–102.

Leppik, E. E. (1969). Morphogenetic classification of flower types. *Phytomorphology* **18**, 451–466.

Meeuse, A. D. J. (1965). Angiosperms—past and present. *Advanc. Front. Pl. Sci.* (Spec. Vol.) **11**, 1–228.

Meeuse, A. D. J. (1966). "Fundamentals of Phytomorphology." Ronald Press, New York.

Meeuse, A. D. J. (1971). Interpretative gynoecial morphology of Lactoridaceae and the Winteraceae: A re-assessment. *Acta Bot. Neerl.* **20**, 221–238.

Meeuse, A. D. J. (1972a). Palm and Pandan Pollination—Primary Anemophily or Primary Entomophyly? *Botanique (Nagpur)* **3**, 1–6.

Meeuse, A. D. J. (1972b). Facts and fiction in floral morphology, with special reference to the Polycarpicae. I. *Acta Bot. Neerl.* **21**(2), 113–127, II. Ibid. (3): 247–264, III. Ibid. (4): 351–365.

Meeuse, A. D. J. (1972c). Angiosperm Phylogeny, Floral Morphology and Pollination Ecology. *Acta Biotheor. (Leyden)* **21**, 145–166.

Meeuse, A. D. J. (1973c). Taxonomic affinities between Piperales and Polycarpicae, and their implications in interpretative floral morphology, *In* T. M. Varghese (ed.), *Puri Comm. Vol.*

Meeuse, A. D. J. (*In press*): Some fundamental principles in interpretative gynoecial morphology.

Parkin, J. (1952). The unisexual flower—a criticism. *Phytomorphology* **2**, 75–79.

Parkin, J. (1957). The unisexual flower again—a criticism. *Phytomorphology* **7**, 7–9.

van der Pijl, L. (1960). Ecological aspects of flower evolution. I. Phyletic evolution. *Evolution* **14**, 403–416.

van der Pijl, L. (1961). Ecological aspects of flower evolution. II. Zoophilous flower classes. *Evolution* **15**, 44–59.

Sharma, B. D. (1970a). On the structure of a *Williamsonia* cf. *W. scotica* from the Middle Jurassic rocks of the Rajmahal Hills, India. *Ann. Bot.* N.S. **34**, 289–296.

Sharma, B. D. (1970b). On the structure of the seeds of *Williamsonia* collected from the Middle Jurassic rocks of Amarjola in the Rajmahal Hills, India. *Ann. Bot.* N.S. **34**, 1071–1078.

Takhtajan, A. L. (1959). "Die Evolution der Angiospermen." Gustav Fischer, Jena.

Takhtajan, A. L. (1959). "Flowering Plants—Origin and Dispersal." Oliver and Boyd, Edinburgh.

10 | Comments on Host-Specificity of Tropical Herbivores and its Relevance to Species Richness

DANIEL H. JANZEN

Department of Zoology, University of Michigan, Ann Arbor, Michigan, U.S.A.

Abstract: We have the working hypothesis that the high species richness of lowland tropical forests is maintained in major part by the herbivore community. The ecological process is that these consumers prevent the best competitors within a given life form from becoming common enough to eliminate competitively the other species of that life form from the community. In short, the herbivores are making space for the poorer competitors. The effectiveness of such a process depends in great part on the degree of host-specificity displayed by the members of the herbivore community. In general, the highly host-specific fraction of the herbivore community should be responsible for the density-dependent responses that result in heavier damage per plant as the plant or plant part becomes more abundant.

Even at this embryonic stage, we can identify three major aspects of the way host-specificity relates to plant species diversity. (1) How does host-specificity differ from the viewpoint of the animal as contrasted with that of the plant? The animal is concerned with what proportion it can eat of the plant species and plant parts in the habitat. It is further concerned with its relative fitness on each. The plant is concerned with what proportion of the herbivore community will respond in a density-dependent manner to the plant's frequency in time and space. (2) Why are tropical herbivores more host-specific than temperate ones (if they are)? The selective pressures in the coevolution of feeding efficiency and plant defenses are such that the larger the plant resource base, and the more constantly it is present (in both cases, from the perspective of the animal), the more host-specific will be the animals and the more species of animals can coexist in a given plant community. (3) Who should win the coevolutionary race between a particular plant population and the herbivores in its habitat? The more effectively the herbivore community counters the plant's chemical and behavioral defenses, the shorter should be the half-life of a plant population in a given habitat. We may conclude from this that extinction rates should be higher in the tropics than in temperate zones. Further, the number of plant species in a habitat should be only indirectly related to rates of speciation.

INTRODUCTION

We have the working hypothesis that the high species-richness of lowland tropical forests is maintained in great part by the herbivore community. The ecological process is as follows. The herbivores present the best competitors within a given life form by becoming common enough to eliminate competitively the other species with that same form from the habitat; the herbivores are making space for the poorer competitors. The effectiveness of such a process depends in great part on the degree and kind of host-specificity displayed by the members of the herbivore guild. In general, it is the highly host-specific fraction of the herbivore guild that should be responsible for the density-dependent responses which result in heavier damage per plant as the plant becomes more abundant (Janzen, 1970, 1971a, b, c, d, 1972a, b, c; May *et al.*, 1970; Connell, 1971).

I shall discuss three aspects of the relationship of herbivore host-specificity to tropical tree species richness. The first is methodological: How does host-specificity differ from the viewpoints of the animal and the plant? The second deals with the species richness of the herbivore guild: Why are tropical herbivores more host-specific than temperate ones (if they are)? The third links herbivory and extinction rates: What is the structure of the coevolutionary race between a particular plant population and the herbivores in its habitat?

HOST-SPECIFICITY FROM THE ANIMAL AND PLANT VIEWPOINT

We must first clarify an important aspect of terminology. There are two pairs of words which are often confused in the literature: generalist versus specialist, and host-specific versus not host-specific (and various Latin and Greek transliterations). The first pair usually refers to the length of the list of plant parts or species on which the individuals of the herbivore species may be found feeding. The second pair often refers to this as well, but may also incorporate some measure of the relative abundance on each host species. Statements on host-specificity may even incorporate some measure of how reliably the population displays this relative abundance in different habitats, at different times, etc. Furthermore, the issue is almost never discussed in respect of the obvious problem that there is variance in host-specificity by the individual and by the population as a whole; it is almost impossible to discriminate between these two types of variance with a sample taken at just one time. If one quarter of the herbivore population is found on one species of plant and three-quarters on another, this may be due to all the herbivores feeding indiscriminately on both and the plants being in a 1 : 3 ratio, or due to the herbivore population being heterozygous at a 1 : 3 ratio with respect to its genetic programming in host-specificity. Finally, we must accept the possibility that the host population is heterozygous with respect to susceptibility (Jones, this volume).

From the herbivore's viewpoint, we want to know what proportion of the total nuts, fruits, leaves, etc. in the habitat are acceptable food. Furthermore, we want to know its relative fitness when feeding on each species. In such a calculation we cannot afford to forget that herbivores do not eat Latin binomials. They feed on specific parts of the plant, or if they feed on several parts of one species (e.g. old and new foliage) they have quite different relative fitnesses on each. They may even have different fitnesses on the same plant part when that part is in different microhabitats or when taken from plants growing under different nutrient–competition–herbivory regimes.

Let us consider a *Curculio* weevil that matures in the nuts of an oak and two hickories in a temperate zone forest composed of maple, tulip, poplar, beech, oak and two species of hickory. The weevil may be regarded as a generalist in that 50% of the tree species serve as hosts. On the other hand, if we add in a measure of nut abundance, the weevil may be labeled a specialist if the oaks and hickories are rare. Furthermore, if only a few of the trees produce large nut crops at intervals of several years, then the trees may be much rarer for the weevil than for the forester. To complicate the picture further, we must remember that the acorns may have a high tannin concentration and thus female weevils reared from them might have much lower fecundity than those from the hickory nuts.

There is, however, another set of confounding factors in host-specificity from the animal's viewpoint. If we look at all the weevils in the genus *Curculio* (which feed on the large nuts of Fagaceae and Juglandaceae in North America (Gibson, 1969)), we find that most have 3–5 host species and it appears that if these hosts are growing in the same site, a single species of weevil may be breeding in all of them. In this context the weevil mentioned in the previous paragraph is neither an extreme specialist nor extreme generalist. On the other hand, out of well over 100 species of seed-eating weevils and bruchids reared to date in a host-specificity study in Costa Rica (Janzen, 1972c), only 8 occur on more than one host species. Here, a weevil that matures in seeds of three tree species could be labeled an extreme generalist even though it feeds on less than 1% of the plant species in the habitat. This actually makes sense biochemically. The marked specialization cited above is probably based on differences in secondary compounds in the seeds, and thus an insect that feeds on more than one or two may have to have an extremely generalized gut biochemistry.

From the plant's viewpoint, the critical variable is what proportion of the plant is consumed by the herbivore each time the plant produces a new set of whatever parts the herbivore eats. This is related in a very complex manner to the fate of the herbivore population when the food item is absent (e.g. during a year between seed crops). For convenience, at this point we may recognize a dichotomy between the physical environment and the biotic environment as represented by alternate hosts. Forgetting alternate hosts for the moment, the herbivore population will be variably decimated when crossing the same metric distances in space and time between hosts. For example, to a herbivore moving between two host trees, 300 m along a dry ridge crest may be ten times as far as 300 m across a humid swamp. A two-month wait between seed crops may be twice as lethal to a bruchid population during the rainy season as during the dry season.

With respect to alternate hosts, the situation is equally complex. When the primary host is absent, we may profitably distinguish between alternate hosts that simply slow the rate of population decline, and those that lead to population increase. Slowing of the rate of population decline may be accomplished in two ways: (1) Commonly an alternate host only provides water, sugar or other compounds which help keep the animal alive but do not sustain reproduction (e.g. "non-hosts as well as leaves and stems of host plants serve for water supply" —Eggerman and Bongers, 1971). This phenomenon is manifested in tropical deciduous forest by the fact that many species of insects pass the dry season as active adults in reproductive diapause; during the dry season their host-specificity decreases greatly as they feed on a number of plants and plant parts they ignore in the rainy season (Janzen, 1973a, b). (2) Small amounts of reproduction on a sub-optimal host may also suppress the rate of population decline when the primary host is absent. A striking example is provided by temperate zone conifers. When the host population has a three-year reproductive (mast) cycle, some species of cone-infesting insects have a three-year diapause. However, a small percentage of their populations does not go into diapause, and reproduces on the cone crops of the few trees that are out of phase with the remainder of the tree population. They do poorly on these "alternate" hosts, yet it is this sub-population that keeps the herbivore in the game when the tree occasionally waits four years between mast crops (Janzen, 1971b).

Alternate hosts, on which the herbivore does not reproduce, may also strongly influence the herbivore's fitness when it finally locates the primary host. For example, the seed-eating bruchids mentioned earlier lay their eggs on the host fruits or seeds, and the larvae mature within. The percentage of the seed crop eaten is directly related to the number of eggs that the female beetle can lay. Her fecundity may be nearly doubled by having had the opportunity to feed on flower nectar (and pollen?) during the time between emerging from last year's seed crop and finding this year's seed crop (Janzen, 1971b). Here we may expect female bruchids, whose larvae are so specialized, to be generalists owing to the chemical similarity of flower nectar and pollen.

The abundance of the primary host should be especially sensitive to the presence of alternate hosts which provide a food supply on which the herbivore can multiply. On a contemporary time scale, the plant should be indifferent to whether the non-host plants are of many or few species; these plants may almost be regarded as inert "stuffing" between the primary host plants and the alternate hosts. We are then concerned with how many species of alternate hosts are present, their timing, their relative abundance, and their spacing with respect to the primary host. Cotton stainer bugs (*Dysdercus* spp.) and Malvales provide a good tropical example (Janzen, 1972a). These bugs feed and reproduce on the seeds of tropical Malvaceae and Sterculiaceae, and have received special attention because of their importance as cotton pests. They build up large densities on the seed crops of wild Malvales; then, as that food supply is exhausted, they move *en masse* to nearby cotton fields. The reverse also occurs. It is easy to see how there would be selection favoring either strong interspecific synchrony (to satiate the bugs) or maximal asynchrony among the wild species of Malvales in a tropical forest. Every time a new malvaceous species immigrated to a given habitat, there was a strong possibility of its either raising the amount of damage done to the seed crops of other Malvales, or incurring unusually heavy destruction of its own seed crop.

We should also turn the question upside down and ask what fraction of the total herbivore guild can feed on any given species of plant. Since ecologists are not yet in the habit of characterizing entire herbivore guilds, I have to let the reader's imagination wander on the subject. For a start, however, we can be assured that in a tropical habitat it will be only a tiny fraction, while in a temperate habitat it may be a very large fraction. Ironically, this suggests that a new chemical defense in a temperate tree may be subject to a much more severe herbivore test than it would in a tropical tree. We may also note that the answer depends to a large extent on what portion of the herbivore guild is vertebrate and what part arthropod, as vertebrates appear to have much greater generalist abilities as based on the detoxification ability of their complex gut microflora.

WHY SHOULD TROPICAL HERBIVORES BE MORE SPECIALIZED?

There are two causally related short answers to this question. First, the food items of tropical herbivores display more spatial, chemical and behavioral heterogeneity than do those of temperate herbivores. Second, there is a theoretical answer that has been around a long time: the more predictably available a specific food item (e.g. new leaves on the lower branches of heavily shaded crowns of tree species x), the more a herbivore can afford to specialize in the face of interspecific competition for that general class of food items (e.g. all new leaves). We can then see that the number of herbivore species in the habitat should be a function of the size of each kind of food relative to the size of the herbivore, the distinctness and predictability of the food items in the eyes of the herbivore, and the total amount of food that can be harvested without generating effective defense mechanisms through natural selection (see also Southwood, 1961; Janzen, 1968). Let us examine several applications of these ideas.

I have recently found that there are substantially more species and individuals of herbivorous insects (and associated parasites) at intermediate elevations (900–1300 m) on tropical mountains than at sea level. The provisional explanation is that the increased insect community is living off the photosynthate that is not metabolized by the plants during the cool nights at higher elevations (Janzen, 1973a, b). However, there is no dramatic increase in plant species at this elevation. In accordance with the ideas expressed in the previous paragraph, I interpret these data in the following manner. In their biochemical–behavioral "eye", the herbivores divide each plant in the community into a number of parts (shoot tips, upper crown-leaves, shade-leaves, flowers, anthers, etc.). In the lowlands, a number of these plant fractions will not be an adequate food supply to support a completely specialized herbivore. The herbivore that might have been supported by it will be more of a generalist (and competitively displace some other herbivore) or be itself absent. As the size of the plant's total energy budget increases with elevation, the replacement rate for specific fractions eaten off the plant should increase, allowing any given species of plant to support more species or herbivores. This assumes that the plant has some absolute percentage of its total budget that must be expended on maintenance; if the percentage needed for maintenance declines with an increasing overall budget, there should be even more which the herbivore can remove without generating selective pressures sufficient to produce a chemical defense to exclude or debilitate the herbivore.

Throughout this and the following discussion, I have been treating the habitat as ecologically "full". I assume that the number of herbivore species in the habitat is set by energy and competitive relationships, and has little to do with rates of speciation; these rates are assumed to be high enough such that far more species arrive at any given habitat than that habitat can absorb. This assumption is explicitly contradictory to the hypothesis that the tropics have so many species because they are a benevolent repository for species from harsher environments.

What happens when the resource base becomes more finely divided? In a temperate zone deciduous forest–field mixture, there is a large guild of seed-eating insects. Many anecdotal reports suggest that the majority of these species of seed-eaters are parasitized by one or more species of parasitic Hymenoptera. Further, the proportion of the individuals that are parasitized appears to be high enough to have at least the potential for influencing the dynamics of the density of their prey. In a tropical forest–field mixture, the story is quite different. Of more than 100 species of weevils and bruchids reared from seeds in Costa Rica, not more than 35% had any parasites at all. Furthermore, most of those with parasites experienced parasitization well below 10%. There is no reason to postulate that the overall biomass of seed-eating insects is any less in the tropical forest than in the temperate zone forest. If anything, it might be higher in the tropics owing to the steadier input of seeds.

However, at the tropical site, the parasites' food resource is much more finely divided than in the temperate zones; there are more species of bruchids at the tropical site (about 50×100 miles) than in all the United States (which has about 100 species (Johnson, 1970)). One of the major problems facing a hymenopterous parasite is to have the morphology and behavior to penetrate the fruit and seed wall with its ovipositor so as to reach the bruchid or weevil larva within. It may also have to deal with toxic secondary compounds that the larva has derived from the seed. We may expect a parasite that can deal with any one bruchid, but the tropical parasite will have to have an attack repertoire that allows effective parasitization of several species of bruchids or weevils, if it is to have a host population large enough for survival. It is significant in this connection that most of the Costa Rican bruchid parasites recorded to date parasitize several species of bruchids, and the hosts of these bruchids usually have very similar fruit morphology.

A third elaboration of tropical herbivore specificity is brought to mind by the mixed dipterocarp forest growing on white sand hillside soils in Bako National Park (sea level to 100 m elevation), Sarawak. At first glance, this rain forest appears unexceptional. It has a high tree species richness, a generally closed canopy at 30–40 m, and a shrub- and sapling-filled understory. However, two important things strike the observer on closer examination. First, the rate of regeneration of cleared forest is extraordinarily slow. A 40 acre field on a broad and flat ridge top had scattered shrubs and small trees 2–4 m tall among a sedge and herb ground layer; it had never been grazed by cattle and was reputed to have been last cleared in the early 1940s. For a site with about 4 m of relatively evenly distributed rainfall per year (similar to Kuching, cf. Fogden, 1971), this suggests exceptionally low primary productivity, which is probably due to the white sand soil. Second, the vertebrate community was almost non-existent, despite this being a national park that was rarely visited. During two weeks of field work in the rainy season, I saw only two insectivorous lizards; I rarely saw more than three small to medium sized birds during a clear morning. There were almost no mammal tracks on the clean sand and mud paths, and rodent runways were only very rarely observed.

In this strange community, foliage-inhabiting herbivorous insects (and their arthropod predators) were almost completely absent. Unfortunately, standardized samples were not taken, but my experience with taking tropical sweep samples (Janzen, 1973a, b) leads me to feel certain that the quantity of insects taken would have been similar to that on the top of a 3300 m Costa Rican mountain. This would be about one tenth what one would get in a Costa Rican sample from an environmental regime like the Sarawak site, but on better soil. Yet the Sarawak site has a very high diversity of plants in the forest (the old field, however, was very impoverished). If tropical herbivorous insects are not forced to be highly specialized, as the overall productivity declines in the habitat they should increase the list of acceptable host plants. The number of species of insects is expected to decline, but some should remain common. The only other option would be for each species to become very rare; however, the small number of species of insects observed on the foliage does not support this idea.

WHAT IS THE STRUCTURE OF THE COEVOLUTIONARY RACE?

Let us construct a hypothetical evolutionary history for a tree species in a particular tropical habitat in a particular region (e.g. the site represented by well drained north-facing lateritic slopes of valleys with permanent rivers). Shortly after the species arrived in the habitat the ensuing population probably increased after a period of physiological evolutionary adjustment to as high a density as would ever be experienced in that habitat. This is because in the initial immigration it left all its herbivores behind and was at first faced only by those herbivores that could shift over from the resident plants. It should thus have been at its competitive best. This phenomenon is currently manifested in the practice of establishing tropical tree plantations on foreign continents. As the tree progresses through evolutionary time, it should gradually acquire an array of host-specific herbivores, as ways evolve to breach the tree's chemical and behavioral defenses. Since the herbivores have many more generations per year than the tree, it seems likely that were they to be super-proficient at locating their food, the tree would have a short history indeed. However, as the herbivore load builds up, the density of the tree should decline, causing some of its herbivores to switch to other plant species or even become locally extinct. The rate of decline of the tree's density may also be slowed by the local extinction of other species of trees that were major competitors with our exemplar tree. We may expect the tree eventually to become so rare that it becomes locally extinct through a perturbation of the habitat or through failure to outcross. It is noteworthy that in the latter case the failure to outcross may not only be directly lethal for an obligatory outcrosser, but also may result in new resistant genes or recombinants being lost from a facultatively outcrossing population before they can spread through the susceptible tree population.

We now have a plot of tree density (survivorship) against evolutionary time, that looks roughly like a jagged negative exponential that eventually drops abruptly to the x-axis. At any one time, we would expect most of the tree species in the habitat to be fairly low on the long right-hand tail of their evolutionary survivorship curve (i.e. most of the trees should be moderately rare). There should be very few common species. The abundance of the common species should be directly proportional to how recently they immigrated and how different a habitat each immigrated from. There should be no excessively rare species (e.g. one or two adults per habitat).

How would the evolutionary survivorship curves differ for the tree species in a temperate zone habitat? They should start off in the same manner, but once into their downward decline, they should level off well above the x-axis. This sustained high density and low rate of local extinction is postulated because long before a temperate zone tree species gets very rare, it can escape from potential herbivores through greater than annual synchronized timing of its nut, flower, leaf, etc. production on a population- or habitat-wide basis (Janzen, 1971b). A newly invading temperate zone tree should also have a more difficult time in becoming established than its tropical counterpart, because the temperate zone tree's competitors will be a few very common species, each highly adapted to the peculiar weather and edaphic circumstances of that habitat. In a tropical forest, almost every time an invasion is attempted, the new seedling or sapling will be competing with a new array of species, each of which has similar abiotic needs to the others and to the immigrant. On the average, we may expect that at least 50% of the time the immigrant should be at a competitive advantage when compared with the tree with which it is most directly competing.

I expect to find that once established in a temperate zone habitat, a tree species persists almost indefinitely. On the other hand, there should be a comparatively high local extinction rate from any given tropical habitat. That is, extinction rates should be higher in the tropics, rather than lower, as is often assumed to be the case.

I should emphasize that owing to the highly heterogeneous physical environment of the tropics as a whole, there should be intra-tropical gradients in extinction rates as well. We should expect the highest extinction rates in those sites where conditions favor the rapid development of an effective and highly host-specific herbivore load on each tree species, and where the tree species lack environmental cues adequate to allow population-wide behavioral synchrony on a greater than annual basis.

In closing, I might add that the more skewed to the right the average evolutionary survivorship curve and the closer to the x-axis it is, the more species there should be in the habitat at any one time. In short, the more the conditions favor survivorship of the average tree population at a low density, the greater should be the species richness of the habitat.

ACKNOWLEDGEMENTS

I am indebted to students and faculty of the Organization for Tropical Studies for providing much of the discussion leading to the generation of these ideas. C. M. Pond constructively criticized the manuscript. The research reported here was supported by NSF grants GB-7805, GB-7819, and GB-25189, and by a travel grant from my mother. The Zoology Department at Oxford kindly provided working space.

REFERENCES

CONNELL, J. H. (1971). On the role of natural enemies in preventing competitive exclusion in some marine animals and in rainforest trees. *In* "Dynamics of Populations" (P. J. den Boer and G. R. Gradwell, eds), pp. 298–312. Centre for Agricultural Publishing and Documentation, Wageningen.

EGGERMAN, W. and BONGERS, J. (1971). Wasser- und Nahrungsaufnahmean Pflanzen unter besonderer Berücksichtigung der Wirtsspezifität von *Oncopeltus fasciatus* Dallas. *Oecologia* **6**, 303–317.

FOGDEN, M. P. L. (1972). The seasonality and population dynamics of equatorial forest birds in Sarawak. *Ibis* **114**, 307–343.

GIBSON, L. P. (1969). Monograph of the genus *Curculio* in the New World (Coleoptera: Curculionidae). Part I. United States and Canada. *Ent. Soc. Am. Misc. Publ.* **6**, 239–285.

JANZEN, D. H. (1968). Host plants as islands in evolutionary and contemporary time. *Am. Nat.* **102**, 592–595.

JANZEN, D. H. (1970). Herbivores and the number of species in tropical forests. *Am. Nat.* **104**, 501–528.

JANZEN, D. H. (1971a). Escape of juvenile *Dioclea megacarpa* (Leguminosae) vines from predators in a deciduous tropical forest. *Am. Nat.* **105**, 97–112.

JANZEN, D. H. (1971b). Seed predation by animals. *A. Rev. Syst. Ecol.* **2**, 465–492.

JANZEN, D. H. (1971c). Escape of *Cassia grandis* L. beans from predators in time and space. *Ecology* **52**, 964–979.

JANZEN, D. H. (1971d). The fate of *Scheelea rostrata* fruits beneath the parent tree: pre-dispersal attack by bruchids. *Principes* **15**, 89–101.

JANZEN, D. H. (1972a). Escape in space by *Sterculia apetala* seeds from the bug *Dysdercus fasciatus* in a Costa Rican deciduous forest. *Ecology* **53**, 350–361.

JANZEN, D. H. (1972b). Association of a rainforest palm and seed-eating beetles in Puerto Rico. *Ecology* **53**, 258–261.

JANZEN, D. H. (1972c). The role of the seed predator guild in a tropical deciduous forest with some reflections on tropical biological control. *Proceedings of a BEC Symposium* (in press).

JANZEN, D. H. (1973a). Sweep samples of tropical foliage insects: Description of study sites, with data on species abundances and size distributions. *Ecology* **54**, 659–686.

JANZEN, D. H. (1973b). Sweep samples of tropical foliage insects: Effects of seasons, vegetation types, elevation, time of day, and insularity. *Ecology* **54**, 687–708.

JOHNSON, C. D. (1970). Biosystematics of the Arizona, California, and Oregon species of the seed beetle genus *Acanthoscelides* Schilsky (Coleoptera: Bruchidae). *Univ. Calif. Publ. Ent.* **59**, 1–113.

MAY, V., BENNETT, I. and THOMSON, T. E. (1970). Herbivore–algal relationships on a coastal rock platform (Cape Banks, N.S.W.). *Oecologia* **6**, 1–14.

SOUTHWOOD, T. R. E. (1961). The number of species of insect associated with various trees. *J. Anim. Ecol.* **30**, 1–8.

11 | Co-evolution and Cyanogenesis

DAVID A. JONES

*Department of Genetics, University of Birmingham, England**

"Ut quod ali cibus est aliis fuat acre venenum."
Lucretius

Abstract: The interaction between the polymorphism of cyanogenesis in *Lotus corniculatus* and *Trifolium repens* and climatic, edaphic and biotic selective agents is discussed. There is good evidence that cyanogenesis is a defensive mechanism in some plants and that a variety of physiological processes occur in other plants and in animals, any one of which enables a particular organism to graze or parasitize cyanogenic plants. Some general principles relating to chemical interactions between species are outlined and possible modes of evolution of cyanogenic substances are considered briefly.

INTRODUCTION

It is now being argued that secondary substances play a major role as passive defence mechanisms in plants and that both passive and aggressive chemical defence can be used by animals (Whittaker and Feeny, 1971). The evidence is largely circumstantial, albeit eminently reasonable, because there are few species in which it is possible to show that individuals which do not possess the defensive substance are at a selective disadvantage with respect to their fellows (Jones, 1971).

On the other hand, there is evidence that many of these secondary plant substances are used by parasitic and grazing animals as the labels by which they recognize their food (Fraenkel, 1969). Clearly these behavioural patterns can only evolve alongside (or after) the development of the putative defensive substance by the plant and must involve tolerance of, or some form of resistance to, this substance. There will, therefore, be a stepwise co-evolutionary process of resistance and virulence of the type well known with bacteria–bacteriophage (e.g. Stent, 1963) and *Linum usitatissimum* L.–*Melampsora lini* (Pers) Lev. interactions (Flor, 1956).

In another system, viz. the polymorphism of cyanogenesis, there are also complex interactions between species which demonstrate elegantly the pro-

* Present address: Unit of Genetics, Department of Plant Biology, University of Hull, England.

cesses of co-evolution (Jones, 1972a). Upwards of 20 species of animal and plant have now been shown to interact with the polymorphism in *Lotus corniculatus* L. and *Trifolium repens* L. Added to this are the effects of the non-biotic environment, temperature and water being particularly important for these two species.

Cyanogenic compounds have been used extensively for chemotaxonomy (Hegnauer, 1971) and in conjunction with cytotaxonomy (Grant and Sidhu, 1967), but there is the embarrassment, however, that some cyanogenic glucosides are synthesized by essentially identical pathways in three species from different families (Abrol and Conn, 1966) and yet closely related species do not possess cyanogenic compounds in measurable quantities. The evolutionary aspects of this paradox are intriguing.

SECONDARY SUBSTANCES AND EVIDENCE FOR CO-EVOLUTION

For many years there have been arguments over the function in plants of the so-called secondary substances. These substances appear to play no major part in metabolism and have frequently been described as waste products. In 1918, however, Combes warned that it was becoming a general habit to regard as waste products substances whose function was unknown. A splendid comment by Cain (1951) expresses the sentiment in a different way: "It seems only reasonable, therefore, to suggest that those characters or variation patterns that have been described as non-adaptive or random should properly be described as uninvestigated. One must not assume randomness or selection without proof!"

There is an enormous literature on the formal chemistry of secondary substances, but it has been left to biologists to produce a synthesis of the abundant, apparently unrelated information accumulated over the years by biochemists. The lone voice of Fraenkel (e.g. 1959) has insisted that these secondary substances are the means by which insects recognize food plants. Ehrlich and Raven (1965) in discussing the precise relationships between butterfly larvae and their food plants concluded that the two groups of organism must have evolved in an interdependent or co-evolutionary way. That such chemical interactions between organisms were widespread was revealed in the allelochemic article of Whittaker and Feeny (1971), although part of the crucial argument was missing.

The notion of co-evolution is a synthesis of interactions between organisms which is inferred in retrospect. Fundamental to the idea developed by Ehrlich and Raven (1965) is the principle that secondary plant substances, and secondary animal substances, are primarily for defence against parasites, grazers or predators. Yet surprisingly little direct evidence in favour of a defensive role for these substances has been offered. De Varigny (1892, pp. 120 and 121) was acutely aware of this problem, but gave no satisfactory examples. The concept

of interactions between species is not new; Lucretius (55 B.C.) was aware of it—
"To us hellebore is rank poison; but goats and quails grow fat on it". Similarly
Darwin (1862) discussed at length the relations between plants of the orchid
family and their insect pollinators. It appears that Stahl (1888) was the first to
suggest that a particular substance, in his example a cyanogenic glucoside against
snails, was being used for defence, but Treub (1907) and Robinson (1930) were
not convinced. Treub noticed that the leaves of *Prunus javanica* Miq. and
Pangium edule Reinw. containing cyanogenic substances, seemed even to attract
leaf-eating parasites and concluded that protection was not the role of HCN in
plants. In 1962, Jones was able to show experimentally that the acyanogenic
form of *Lotus corniculatus* L. was more susceptible to grazing by molluscs and
voles than the cyanogenic form and since that time the principle of selective and
differential eating has become well established (Jones, 1966; Crawford-
Sidebotham, 1972b).

At its simplest, co-evolution is a leapfrog process. The host or prey gains the
ability to synthesize or enhance the synthesis of a chemical which conveys an
advantage as a defensive mechanism. Sequentially some parasites or predators
develop a mechanism which will overcome this defence. In some animals
devices appear which are sensitive to the secondary substance and this enables the
animal to recognize its host or food plant.

Almost no direct evidence for such a scheme has been offered in the articles
on co-evolution published to date although there is some available (Jones, 1971).
To the microbial geneticist, however, the process is commonplace and a simple
experiment with bacteria and bacteriophage makes the point very forcibly. If a
clone of bacteria derived from a single individual is exposed to infection by a
clone of a suitable phage, most of these bacteria are destroyed. A few survive
and on analysis it can usually be shown that these are either resistant to or
tolerant of the virus. Resistant forms generally give rise to forms showing the
same mechanism of resistance; the new character is inherited. Because we
started with a single individual these new forms must have arisen during the
course of the experiment. This should not be surprising for we should expect to
obtain some mutations to resistance when we are dealing with very large
numbers of individuals. If we now multiply up the resistant bacteria and expose
these to the same strain of phage it is possible to isolate progeny phages which
are able to infect the resistant bacteria (Luria, 1945). Second stage resistance by
bacteria and second stage virulence of phages can be obtained by repeating this
cyclical process *ad nauseam*. Genetical analysis of the various mutants obtained
often reveals that the resistance/virulence is associated with the chemistry of the
outer surfaces of the bacteria on the one hand, and of the phage on the other. The

phage recognizes its host chemically. By changing the chemical nature of its outer surface the bacterium is no longer recognizable as a host. The changes in the structural chemistry of the substances involved are usually trivial, but they are sufficient.

Experiments like these constitute proof of co-evolution because the whole process can be studied and analysed as it occurs. There is no retrospective inference on events which happened many years previously.

CYANOGENESIS AND SELECTION

Levin (1971) reminds us that plants are resistant to the majority of microorganisms and other parasites in their environment. But this in no way invalidates the argument (Jones, 1971, 1972a) that where a species is monomorphic for a secondary substance we have no means of telling whether a particular non-parasite could and would attack that species if the putative defensive substance were not present. Ideally we should examine species polymorphic for secondary substances and look for differential survival, infection or tolerance, but quantitative variation in secondary substances can also reveal the defensive nature of these chemicals (Schneider, 1952; Clauss, 1961; Jones, 1962; Nayer and Fraenkel, 1963).

We are fortunate with cyanogenesis in having a system in which we can combine and compare experiments in the laboratory with experiments and observation in the field. We can test whether the selection we observe in the laboratory occurs in the wild and vice versa. The problem comes when, as is often the case, the two types of experiment are not mutually corroborative. The polymorphism of cyanogenesis, particularly in *Lotus corniculatus* and *Trifolium repens*, will be used to demonstrate the roles of known selective agents and the co-evolutionary processes associated with cyanide resistance. At the same time it will become clear that generalized conclusions from empirical observations and experiments frequently break down in practice.

I have recently discussed at length cyanogenic glucosides and their function (Jones, 1972a) and so the details of the genetics and biosynthesis of these compounds will not be repeated here. Work is continuing on the pathway of biosynthesis of linamarin (Tapper and Butler, 1972; Tapper *et al.*, 1972) and more information is available about the associated β-glucosidase (Hughes and Maher, 1973; Maher and Hughes, 1973). Better techniques for the estimation of cyanide in plant tissues are being developed (e.g. Easty *et al.*, 1971; Zitnak, 1973). We have now confirmed the observations of Guérin (1929) that flowers of *Lotus corniculatus* are cyanogenic. T. J. Crawford-Sidebotham has shown that the four phenotypes relating to cyanogenesis in leaves (Table I) can be cross-classified with the same array of phenotypes in flowers. In a large group of plants growing near Birmingham all 16 phenotypic combinations occur, but not at the frequencies expected from the product of the frequencies of the respective leaf and flower phenotypes.

TABLE I. The phenotypes of cyanogenic and acyanogenic *Lotus corniculatus*

Plant contains	Shorthand notation	Gross phenotype	Varietal type
Cyanogenic glucosides and enzyme	G + E	cyanogenic	amara
Cyanogenic glucosides but no enzyme	G + no E	acyanogenic	dulcis
Enzyme but no cyanogenic glucosides	no G + E	acyanogenic	dulcis
Neither cyanogenic glucosides nor enzyme	no G + no E	acyanogenic	dulcis

1. Selection by Temperature

Figure 1 shows the classic map of the distribution of cyanogenic *Trifolium repens* in Europe (Daday, 1954) modified by substituting phenotype frequencies for allele frequencies. Even though the genetic system involved in cyanogenesis is better understood than is the case for the majority of characters used in genecological studies, there are good reasons for using phenotype frequencies in preference to the allele frequencies for this work. We usually collect samples (occasionally it is possible to exercise complete enumeration) but the heterozygote cannot be detected except by test crossing and two generations of test crosses are required for autotetraploid *Lotus corniculatus*. Too many people assume the Hardy–Weinberg theorem when estimating allele frequencies without realizing that these estimates have enormous errors. Indeed, the estimates are often meaningless. For example, when 100 *L. corniculatus* plants are sampled it is quite likely that the aglucosidic allele will not be detected even if it is present at a frequency of 34%. And with diploids, a sample of 100 can overlook a recessive allele frequency of 10%. Thus it is clearly safer to use phenotype frequencies for this type of study. In addition, when a plant clones allele frequencies are nonsensical because we do not know what constitutes an individual. We are interested in what proportion of the cover of a species is of a particular phenotype and hence the use of phenotypes for comparison is reasonable.

Daday argued (1954) that temperature was a major factor in maintaining the polymorphism in *Trifolium repens*, warm winter temperatures favouring cyanogenic plants, cold winter temperatures favouring acyanogenic plants. The selection exerted by low temperature against cyanogenic plants presents little

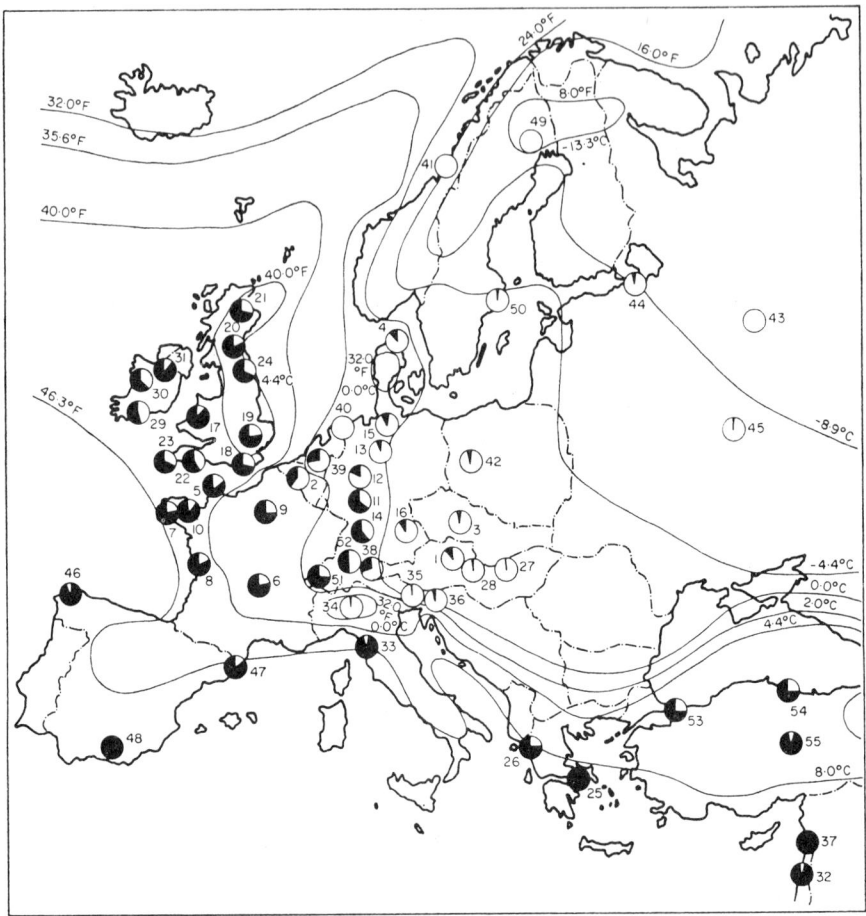

Fig. 1. Distribution and frequency of the cyanogenic form (see Table I) in European and Near Eastern wild populations of *Trifolium repens* L. Black section: frequency of the cyanogenic form. White section: frequency of the acyanogenic forms. —January isotherm. (Modified after Daday (1954a) by permission of the editor and publishers of *Heredity*.)

difficulty (Jones, 1972a) because of the poisoning effect of HCN released after a frost. It is the high temperatures which are the embarrassment. Daday (1965) found that G+E and G+no E were equally fit in a reproductive sense and had higher fitness than no G+E and no G+no E in warm climates. Clearly, therefore, he was not measuring the fitness of cyanogenic plants, but of glucosidic plants. Hence the problem becomes one of determining, firstly, why the enzymatic form occurs in samples at such a high frequency (because it is not, in

Daday's view, the cyanogenic form itself which is at an advantage at high temperature) and secondly, why the aglucosidic allele remains in climates with a warm winter. Invoking effects of *mean* temperature are unsatisfactory. It is much more likely that cyclical effects of temperature (Jones, 1970) are of consequence in *T. repens* but we cannot study this in the wild at the moment.

I have investigated whether temperature could explain differences in the frequency of cyanogenic forms over short distances in what appear to be linear habitats, and have concluded that the level of resolution we have in the techniques available at present is too coarse. Analysis of variance revealed no differences between testing locations even though replicate probes (thermistors) were used at all of half a dozen different locations within neighbouring groups of plants.

2. Biotic Selection

The first experiments in the laboratory on selective eating of the acyanogenic form of *L. corniculatus* by slugs and snails were quite convincing (Jones, 1962, 1966). We have yet to demonstrate this as convincingly in the wild, although work by Whitman (1973) and Angseesing (1973) is promising. In wild situations, alternative food becomes important and, of course, the available food varies according to the plant and animal community and in any one place will be cyclical and oscillating. Experiments by Jones (1966) and Crawford-Sidebotham (1972b) have shown the effect of alternative food on selective eating of cyanogenic plants. The vole *Microtus agrestis* L. ate all the plant material (*L. corniculatus*) when both cyanogenic and acyanogenic plants were offered together as sole food. Selective eating of the acyanogenic form occurred when an alternative food (oats and carrot) was offered in addition (Fig. 2). There was also a quantitative effect, because the higher the glucoside score the lower was the chance that the plant would be eaten. Crawford–Sidebotham (1972b) recognized differential eating of *L. corniculatus* and *T. repens* by several species of slug and snail when there was no alternative food. Sliced carrot offered as an alternative food markedly reduced differential eating, but it was still significant for some species.

Observations by Nayer and Fraenkel (1963) with the Mexican bean beetle (*Epilachna varivestis* Muls.) show beautifully how the concentration of linamarin (phaseolunatin) in cell-free extracts of *Phaseolus vulgaris* L. (syn. *P. nanus* L.) and *P. lunatus* L. (syn. *P. limensis* Macfad.) determines the biting response of the beetle larvae. Table II gives a resumé of the results from which it can also be concluded that there is an interaction with glucose. Unfortunately, the absolute concentration of the linamarin was not estimated, but parallel tests with a

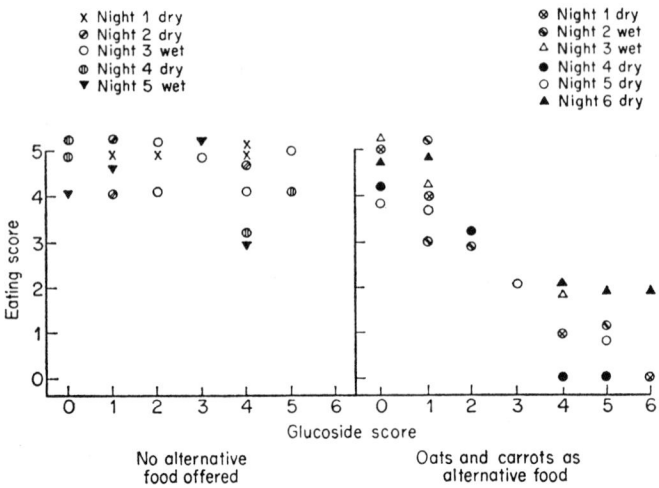

FIG. 2. Selective eating of *Lotus corniculatus* by *Microtus agrestis* L. (two voles). The eating of leaves and the glucoside content of plants were scored on arbitrary scales, 0–5 and 0–6 respectively. (Reproduced from Jones (1966) by permission of the editor and publishers of the *Canadian Journal of Genetics and Cytology*.)

TABLE II. Biting response of the Mexican bean beetle to different concentrations of crude extracts of linamarin from leaves of *Phaseolus limensis* applied to filter paper semicircles. (From Nayer and Fraenkel, 1963)

Relative concentration of crude extracts:	in 1 ml of 0·15M glucose	in 1 ml with no glucose
Nil	weak	not recorded
0·5 mg	strong	weak
1·0 mg	strong	weak
2·0 mg	medium	weak
5·0 mg	weak	nil
10·0 mg	nil	nil

preparation of linamarin and lotaustralin (2 mg ml^{-1}) from *Trifolium repens* gave a response similar to the 1·0 mg ml^{-1} concentration of the lima bean extract.

It is clear, therefore, that low concentrations of linamarin in the presence of glucose are a positive attractant to the beetle larvae, whereas at higher concentrations the substance acts as a deterrent. These results also show that work that does not report the concentration of the putative protective substance can only be regarded as unreliable, essentially because there is no means of making comparisons between conflicting results. There is, unfortunately, an enormous literature in this category.

3. Selection by Soil Moisture

By experimental means, Foulds and Grime (1972b) have shown that the glucosidic form of *T. repens* is killed more easily by soil water stress than is the aglucosidic form. In natural populations they found that the difference in the frequencies of the various phenotypes in *T. repens* was related to the glucosidic form—as predicted—and not to the cyanogenic form. With *L. corniculatus*, however, the effect of soil moisture does appear to influence the frequency of the cyanogenic phenotype, but it does so by reducing the frequency of the enzyme-producing allele (see Table III).

This appears most elegant until one looks at the size of the droughted areas in an essentially damp area. We have few data at present, but the results are promising. We have taken advantage of the local knowledge recently gained by a research student (J. W. Snape) working in a valley in Derbyshire on *Arabidopsis thaliana* (L) Heynh, an ephemeral which with other species can be used as an indicator of dryish soils. Crawford-Sidebotham and I have deliberately sought out adjacent groups of plants, one group on dry soil and the other on moist. We have also distinguished between dry areas of less than 4 m^2 from larger areas of over 20 m^2. The results of testing at one location near Ballidon are presented in Table IV. These groups of plants were tested originally in 1971 and further visits in 1972 confirmed the frequency differences.

The results suggest that the size of the individual components of the mosaic pattern are important in this species. Small areas appear to have ineffective selective effects when selection is not of the extreme type as that exerted, for example, by heavy metals.

4. Population Studies

The high sand dunes on the west coast of the Netherlands are used extensively as filter beds for domestic water, but the general impression is of aridity. In spite of this the cyanogenic form of *L. corniculatus* occurs at higher frequency in

TABLE III. Phenotype frequencies of populations of *T. repens* and *L. corniculatus* from moist and droughted habitats. (Reproduced from Foulds and Grime (1972a) by permission of the editors and publishers of *Heredity*)

	Phenotypes				Phenotypic frequencies	
	AcLi	*Acli*	*acLi*	*acli*	Cyanogenic glucoside	β-glucosidase
T. repens						
Moist						
Monsaldale	6	30	0	14	0·7200	0·1200
Littondale	5	32	0	13	0·7400	0·1000
Droughted						
Monsaldale	3	13	0	28	0·3636	0·0681
Dovedale	5	21	3	21	0·5200	0·1600
L. corniculatus						
Moist						
Littondale	42	7	0	1	0·9800	0·8400
Dudley	12	4	1	0	0·9412	0·7647
Droughted						
Monsaldale	27	23	0	0	1·0000	0·5400
Dovedale	28	20	0	2	0·8160	0·5600

AcLi, glucoside and enzyme; *Acli*, glucoside no enzyme; *acLi*, enzyme, no glucoside; *acli*, neither glucoside nor enzyme.

TABLE IV. Frequency of the cyanogenic form of *Lotus corniculatus* in large dry areas (left) and small dry areas (right) in relation to the frequency in adjacent damp areas near Ballidon, Derbyshire

State of soil	Phenotype			Phenotype		
	G+E	Not G+E	N	G+E	Not G+E	N
Damp	29	7	36	22	13	35
Dry	17	23	40	39	16	55
Totals	46	30	76	61	29	90

χ_1^2 (Yates correction) = 9·94 χ_1^2 (Yates correction) = 0·31
    $0.01 > P > 0.001$ $P > 0.5$

the south than in the north (Jones, 1972b). Clearly there is a relationship between frequency and latitude (Fig. 3), but is it meaningful? Unfortunately, *L. corniculatus* is difficult to find on the polder (the low-lying reclaimed land between the dunes and Isselmeer) other than in cemeteries and even there the plants are too infrequent to sample sensibly. Three localities have been found where the plants are relatively common, but two are on recent roadside verges (near Hoorn) and the other is on a derelict dyke. The frequencies at these locations are not significantly different from corresponding dune samples at the same latitude.

FIG. 3. The frequency distribution of the cyanogenic form of *Lotus corniculatus* in the Netherlands. Data of Jones (1972b). For explanation of the symbols see Fig. 1.

The Dutch meteorological records have been studied and there are no trends corresponding with the cline. There is, however, circumstantial evidence of another kind. Van Leeuwen and van der Maarel (1971) have described the effects of rabbit grazing on the vegetation of the dunes of the southern islands of the Netherlands. Photographs I have taken at localities where acyanogenic plants are common (Egmond Binnen) and where they are rare (Wassenaar) show that rabbit grazing is a major difference between the two habitats (Jones, 1972b), rabbits being common in the latter habitat. It is now necessary to check on this circumstantial evidence by direct experimentation and an appropriate experiment is in progress in Staffordshire. Corkill's comment (1952) that rabbits preferentially ate the acyanogenic *T. repens* of his field trials is encouraging.

Another aberrant situation can be dismissed quickly. There is evidence that alien cyanogenic plants have been introduced to roadside habitats in Denmark,

England, West Germany, the Netherlands and Wales (Jones, 1973). Apart from cyanogenesis these aliens are morphologically distinct from the native plants, and yet under natural conditions there seem to be no barriers to crossing with the natives. Attempts to locate the source of these plants have been frustrating, but plants of the same phenotype are being grown agronomically in Austria. A Danish seedsman quaintly said that he had supplied second-hand grass seed to the road constructor and that he believed that the seed came from Italy. There is little doubt that cyanogenic *L. corniculatus* occurs as a contaminant of the low grade grass seed used all over western Europe for sowing on the verges of new roads.

Near Büsum in Schleswig Holstein there is a different situation. There is a pocket of low-lying land west of Heide in which *L. corniculatus* is almost exclusively cyanogenic. Elsewhere on the peninsular the plants are as exclusively acyanogenic (Fig. 4). *L. corniculatus* was used agronomically in that part of Germany until 1939 and it seems likely that these plants represent the relics and descendants of alien cultivars. Apart from cyanogenesis, the general morphology

FIG. 4. The frequency distribution of the cyanogenic form of *Lotus corniculatus* in the area near Büsum, West Germany. Original data. For explanation of the symbols see Fig. 1.

of these plants is similar to their acyanogenic neighbours. Thus, if these were agronomic strains selection has been stronger for morphological characters than against cyanogenesis during the past 30 years.

CYANOGENESIS AND CO-EVOLUTION

1. Cyanide Poisoning and Cyanide Resistance

The primary poisoning effect of cyanide is associated with the formation of —CN compounds with the iron atom of cytochrome oxidase (cytochrome a_3), so inhibiting the normal functioning of this enzyme. The inhibition of respiratory enzymes by cyanide is very similar in plants (Lundegårdh, 1966), insects (Gilmour, 1961; Keister and Buck, 1964) and mammals (Keilin, 1966), although only part of the respiratory system in higher plants is cyanide sensitive (Marsh and Goddard, 1939). In some plants in the presence of cyanide, it is possible to detect what Lundegårdh (1966) terms "basal respiration" which he considers to be associated with a shunt via cytochrome b; this enzyme is partly autoxidizable and so oxidation is not entirely inhibited by cyanide. Bendall and Bonner (1971), on the other hand, concluded that cytochrome b plays no part in the cyanide resistant pathways in the cyanogenic plants *Symplocarpus foetidus* Nutt. (skunk cabbage) and *Phaseolus mungo* L. (Syn. *P. aureus* Roxb., mung bean), even though they obtained evidence that cytochrome b undergoes oxidation and reduction in the presence of cyanide.

The discussion by Hewitt and Nicholas (1963) shows that cyanide poisoning is not a straightforward process because it relates to the general metabolic state of the cell, particularly substrate availability. Resistance to cyanide is related to an increase in Fe^{+++} concentration, to detoxification mechanisms, or to modifications of the cytochrome system. Naturally man is concerned about the cyanide content of plant foods both for himself and for his animals and the effects on man have been discussed recently by Montgomery (1969).

Of the animals' associated with *L. corniculatus*, detoxification occurs in *Polyommatus icarus* Rott. (Common Blue butterfly) and in the weevil, *Hypera plantaginis* Degeer. The larvae of these insects feed on the leaves and the flowers (Parsons and Rothschild, 1964) and it is known that the larvae of the butterfly do not show selective grazing (Lane, 1962). The moths *Zygaena lonicera* von Schev. and *Z. filipendula* L. are themselves cyanogenic, but it seems unlikely that the cyanogenic substance is obtained directly from the cyanogenic food plant (Jones et al., 1962). The moths of this genus are notoriously difficult to kill in the cyanide bottle yet there is a threshold concentration of HCN above which Zygaenidae are poisoned (Rocci, 1914), the symptoms of intoxication being similar to those of other Lepidoptera.

Kurland and Schneiderman (1959) have re-examined the effect of cyanide on diapausing pupae of *Hyalophora cecropia* (L.) and other silk worms and proposed that the resistance is due to an excess of cytochrome oxidase relative to cytochrome *c*. Such an excess of cytochrome oxidase could occur when an animal has a low metabolic rate, a state manifestly reached by these pupae. The adult Zygaenidae are normally lethargic insects, warningly coloured, but it has not been shown whether the resistance is in any way comparable to that in silk moths.

Some species of millipede are also cyanogenic (Eisner *et al.*, 1963a, b) and recent work has shown that *Euryurus leachii* (Gray) and *Pleuraloma flavipes butleri* (McNeill) are cyanide tolerant (Hall *et al.*, 1969), the mechanism apparently being a resistant terminal oxidase rather than an excess of cytochrome oxidase (Hall *et al.*, 1971). The mechanism of cyanide resistance in larvae of *Malacosoma neustria* (L.) (Lackey moth), which have *Prunus laurocerasus* L. as food plant, is not known (Parsons and Rothschild, 1964).

Detoxification of cyanide is usually associated with a class of enzymes called rhodanese E.C. 2.8.1.1. (Lang, 1933). In the presence of a sulphur donor, e.g. $Na_2S_2O_3$, cyanide is converted to thiocyanate. The enzyme is widespread (Lang, 1933; Rosenthal, 1948) and was shown by Jones *et al.* (1962) to be the means whereby the *Apanteles zygaenarum* (Marshall) parasite of *Zygaena* larvae was able to overcome the cyanide produced by the host insect. Clearly we have here an elaborate interaction between host plant, grazing insect and parasitic insect. Indeed, the *Apanteles* can itself be parasitized by the insect *Mesochorus temporalis* (Thompson). The hyperparasite contains neither cyanogenic substances nor detoxifying enzymes.

In his original description of the action of rhodanese, Lang (1933) reported that the enzyme occurs in cattle, sheep, frogs, guinea-pigs and dogs. Although it is easy to understand why grazing mammals should possess enzymes for detoxifying the cyanide often contained in their food, it is more difficult to appreciate why carnivorous animals like the dog should contain high concentrations in the liver. It seems that some of the products of digestion and other normal metabolic processes must be cyanogenic but I can find only passing reference to this in discussions of digestion (e.g. Lovatt Evans, 1952). More precise is the comment that smokers secrete more potassium thiocyanate in saliva than do non-smokers, presumably because traces of HCN are absorbed from the smoke (e.g. Wilson, 1973).

2. *The Biotic Environment of* Lotus corniculatus

Neunzig and Gyrisco (1955) have listed the insects which they found associated with *L. corniculatus* in New York State. Some of these animals are leaf or seed parasites, but others are concerned with pollination (bees) and with predation of the parasites (ladybeetles). Similar data have been collected in Poland (Ruszkowski, 1968).

A survey of the literature, by no means exhaustive, reveals a large number of species which have been found in, on, or grazing *L. corniculatus* plants. Table V

TABLE V. Organisms found associated with *Lotus corniculatus*

	Comments	References
Bacteria		
Rhizobium spp.	A complex of many strains	e.g. Bailey *et al.* (1971)
Fungi		
Stemphylium loti Graham	Copper spot disease of leaves. The fungus is tolerant of the HCN which it releases from host cyanogenic glucosides	Graham (1953) Millar and Higgins (1970)
Nematoda	(There are no cyanogenic substances in the roots)	Guérin (1929)
Ditylenchus dipsaci (Kühn)	Oat and clover races attack *L. corniculatus*	Jones and Jones (1964)
Paratylenchus penetrans (Cobb.)	Races?	Willis and Thompson (1969)
Mollusca		
Slugs		
Arion ater (L.)	Differential	Jones (1962)
Arion hortensis (Fer.)	eating of acyanogenic	Crawford-Sidebotham (1972b)
Arion subfuscus (Drap.)	form	Crawford-Sidebotham (1972b)
Agriolimax reticulatus (Müll.)		Jones (1962)
Snails		
Cepaea hortensis (Müll.)	Differential	Crawford-Sidebotham
Cepaea nemoralis (L.)	eating of acyanogenic form	(1972b)
Helix aspersa (Müll.)		
Theba pisana (Müll.)		

Table V (continued)

	Comments	References
Insecta		
Orthoptera	5 spp. in New York	Neunzig and Gyrisco (1955)
Thysanoptera	2 spp.	
Hemiptera	9 spp.	
Homoptera	9 spp.	
Lepidoptera	5 spp.	
	23 spp. in British isles	Allan (1949)
Sparaganothis xanthoides (Walker)	Leaf roller	Ridgway and Gyrisco (1959)
Zygaena spp.	Contain cyanogenic substances	Jones et al. (1962)
Polyommatus icarus (Rott.)	Detoxifies HCN	Lane (1962)
Coleoptera	2 spp. in U.S.A.	Neunzig and Gyrisco (1955)
Weevils		
Apion loti (Kirby)		e.g. Morris (1967)
Hypera variabilis (Herbst.)		Jones and Jones (1964)
Hypera plantaginis (Degeer)	Detoxifies HCN	Parsons and Rothschild (1964)
Brachyrhinus ligusti (L.)		Neunzig and Gyrisco (1955)
Sitona spp.		Jones and Jones (1964)
Hymenoptera	21 spp. in New York	Neunzig and Gyrisco (1955)
Tenthredo aceurima (Benson)		Waterhouse and Sanderson (1958)
Bruchophagus kolobovae (Fedoseeva)		Fedoseeva (1956)
	Larvae eat developing seeds	Batiste (1967)
Diptera	5 spp. in New York	Neunzig and Gyrisco (1955)
Bremiola spp.		Edwards and Heath (1964)
Contarinia loti (Degeer)	Gall midges in ovaries	e.g. Ruszkowski (1968)
Mammalia		
Cattle	Detoxify	
Sheep	HCN	Lang (1933)
Microtus agrestis L.	Selective eating of acyanogenic form in presence of alternative food.	Jones (1962)

gives some details, particularly where an interaction with the character of cyanogenesis has been found. For many of the organisms listed it is not known whether they are regularly associated with *L. corniculatus*, but certainly some are deterred from grazing or parasitizing cyanogenic plants, whereas others have solved the problem of cyanide poisoning by evolving detoxifying mechanisms and other forms of resistance. See also Seaney and Henson (1970) for the name of other parasites. Some of these grazers and parasites are themselves parasitized (Table VI).

TABLE VI. Parasites of the insects listed in Table V

	Host	References
Hymenoptera		
8 species in U.S.A.	*Bruchophagus kolobovae*	Hansen (1955)
		Neunzig and Gyrisco (1959)
		Batiste (1967)
Apanteles zygaenarium (Marshall)	*Zygaena* spp. Parasite can detoxify HCN	Jones et al. (1962)
Diptera		
Zenilla longicauda (?) (Wainwright)	*Zygaena* spp. Parasite can detoxify HCN	Jones et al. (1962)
Hyperparasites		
Mesochorus temporalis (Thompson)	*Apanteles zygaenarum* *M. temporalis* contains neither cyanogenic substances nor detoxifying enzymes	Jones et al. (1962)

Speculations on such a multitude of interactions are obviously very hazardous, but one wonders whether the species which have evolved detoxifying systems are those which were restricted in their food plant or host range before the host became cyanogenic.

3. Other Interactions

While investigating differences in the frequency of the cyanogenic forms of *L. corniculatus* and *T. repens* over very short distances, the most intangible interaction associated with cyanogenesis was discovered (Jones, 1968). *T. repens* is the more ubiquitous species, being found on both dry and damp soils, and consequently it is rarely possible to find groups of *L. corniculatus* plants free of *T. repens*. On the other hand, in eight localities studied near Birmingham adjacent groups of *T. repens* plants occur with and without *L. corniculatus*. In four of these localities it was found that the frequency of the cyanogenic form of *T. repens* was significantly lower when *L. corniculatus* was present than when it was not (Table VII).

TABLE VII. Populations of *Trifolium repens* and *Lotus corniculatus* tested for the cyanogenic phenotype

Location	Mixed populations				Pure populations of *Trifolium repens*	
	Lotus corniculatus		*Trifolium repens*			
	N	% cyanogenic plants	N	% cyanogenic plants	N	% cyanogenic plants
1. Clent	154	87·6	22	36·4	73	80·8
2. Wawensmoor	87	80·5	47	19·1	50	42·0
3. Malvern	120	70·8	44	34·0	50	92·0
4. North Oxford	34	100·0	29	41·4	24	87·5
5. Walsall	68	98·5	50	42·0	50	58·0
6. Rubery, Old Station	148	77·0	95	33·6	25	44·0
7. Dunton Wharf	52	98·1	45	31·1	50	26·0
8. Hampton in Arden	80	98·7	70	38·5	49	26·5

(Reproduced from Jones (1968) by permission of the editors and publishers of *Heredity*.)

I pointed out that unless the environmental factors which were limiting the distribution of *L. corniculatus* were the same as those which determined the high frequency of cyanogenic *T. repens* the evidence suggested an interaction between the species. Foulds and Grime (1972a) have shown that soil water stress plays a part in the distribution of cyanogenic *L. corniculatus* and *T. repens* and certainly *L. corniculatus* can be found in abundance on dryer soils, giving way to *L. uliginous* Schkur. on damp soils. There is, however, no evidence that it is soil moisture which is limiting the distribution of *L. corniculatus* in the eight Birmingham habitats.

Three of these localities have been the subject of the inconclusive investigations on temperature differences (p. 219). We have not yet examined the distribution of the animals known to graze *T. repens* and *L. corniculatus*, but we are studying the parasites of flowers and seeds.

DISCUSSION AND CONCLUSIONS

1. Origin of Cyanogenic Substances

Are secondary plant substances normal metabolites for which selection has favoured extravagant production?

When a particular substance is confined to a genus or even a species, e.g. some alkaloids (Price, 1963), it is reasonable to infer that the synthetic pathway has evolved recently. Other substances like nicotine and linamarin, on the other hand, are widely distributed in plants and it is obvious that we must be careful to distinguish between homologous and analogous metabolic pathways (Hegnauer, 1963). When the same synthetic pathway exists in several different families, even though the end product does not appear in all species of these families in measurable quantities, extravagant production of a normal metabolite is a likely explanation. It is extremely difficult to determine that a particular substance is completely absent because, as Flück (1963) has argued, it is impossible to be certain that the techniques available have a fine enough resolution to be able to distinguish minute quantities from non-existence. The microbial geneticist has a simple criterion. Does the individual *grow* on minimal medium? If it does not, then some essential substance is missing or is being synthesized below a level necessary for life.

Be that as it may, we need to know whether the apparently identical pathways of biosynthesis of linamarin in *Linum usitatissimum* L. (Linaceae), *Osteospermum jucundum* Nordlindh (Compositae), *Lotus* spp. and *Trifolium repens* (Abrol and Conn, 1966) are determined by exactly the same genetic systems. If they are, it is likely that cyanogenic substances do play a necessary but as yet unappreciated part in the normal metabolism of flowering plants. Certainly cyanogenic substances are not mere end products of metabolism; active turnover of linamarin has been observed in *Trifolium repens* (De Waal, 1942), in *Lotus* and in other species (Abrol *et al.*, 1966).

Sequence analysis of the amino acids in proteins with the same basic function derived from a wide variety of organisms has given extensive evidence of homology (e.g. Dayhoff, 1969). And clearly homology is the basic premise of the suggestions and arguments over whether amino acid substitutions are neutral in selective value (Clarke, 1970; Kimura and Ohta, 1971). As far as I know the enzymes concerned with the biosynthesis of linamarin have not been subjected to sequence analysis.

Extravagant production of a normal metabolite usually results from mutation in regulatory genes. If, therefore, plants which we score as aglucosidic possess a strict control over the synthetic pathway of linamarin, addition of precursors should not increase to any great extent the concentration of linamarin.

Conversely, were the aglucosidic state due to an auxotrophic mutant, addition of the appropriate precursor would markedly increase linamarin production. It is imperative that experiments of this type are carried out on the aglucosidic forms of polymorphic species (See, however, Conn, 1973). Whether or not the difference between glucosidic and aglucosidic plants is regulatory, there is no doubt that there is considerable variation in the production of linamarin and lotaustralin amongst glucosidic plants. The genetic mechanism involved is unknown, but does not appear to be gene dosage.

There is clearly a similar problem with the genetic control of β-glucosidase synthesis, but here there is evidence of an allelic dosage effect (Maher and Hughes, 1973). Even in plants which as far as the substrate linamarin is concerned would be scored as lacking β-glucosidase, activity against other β-glucosides can be detected. This suggests that extravagant production of the enzyme is necessary to raise the concentration of the β-glucosidase above the threshold level of activity with linamarin. Whichever of the explanations suggested is correct other evolutionary problems can be posed. When regulatory mutation conveys a selective advantage by allowing extravagant production of a normal metabolite, one wonders why more species have not taken advantage of the process, because it is likely that such regulatory mutations will have occurred in those other species as well. The auxotrophic explanation of the aglucosidic form presupposes that the species was monomorphic for cyanogenesis. In which case why have the aglucosidic species in a cyanogenic genus lost the character while the poly- and monomorphic ones have retained it?

Cyanogenesis is an obvious character for chemotaxonomy and it has been used successfully in several studies (e.g. Paris, 1963; Ruijgrok, 1966; Hanelt and Tschiersch, 1967, and Hegnauer and Ruijgrok, 1971). Recently, however, Hegnauer (1971) has discussed the problems involved, particularly in relation to genetic polymorphism and other inconsistencies which plague a clean taxonomy. Gibbs (1963), for example, suggested that all the New World species of *Lotus* L. should be referred to the genus *Hosackia* because at that time all the North American species were thought to be acyanogenic. Grant (1966) and Grant and Sidhu (1967), however, demonstrated that at least four North American species are cyanogenic and so clearly cyanogenesis cannot be used as a decisive character in this genus. Within the *Lotus corniculatus* aggregate many morphological varieties have been described in eastern European material (Larsen and Žertová, 1963; Borsos, 1966) while a start has been made on relating the form of varieties to the ecology of the habitats in which they were found (De Vries, 1968).

2. Directed Evolution

Following Darwin (1859) we normally argue that evolution is a passive process (selection being active) merely allowing individuals with suitable characteristics to live, grow and reproduce. In addition there have been strong arguments against and little evidence for Lamarckian-type evolution, whereby the environment actually induces the appearance of a suitable character—so-called directed mutation. There appear, on the other hand, to be clear cases of directed evolution. Occasionally, under catastrophic circumstances, a species is forced to change, not merely to exploit a new niche, but in order to survive (Lewis, 1962). It could be argued that this is a special case (bacteria, bacteriorhage and other micro-organisms may behave this way in the laboratory) but Miriam Rothschild (personal communication, 1960) has rightly pointed out that if the mutation to the melanic state in *Biston betularia* L. had not been a recurrent phenomenon, the peppered moth could well have become extinct in the industrial areas of England (see also Kettlewell, 1965).

I have developed elsewhere (Jones, 1972a) a scheme of co-evolution of cyanogenesis and cyanide resistance (Fig. 5). Suppose an insect species is "fixed", on,

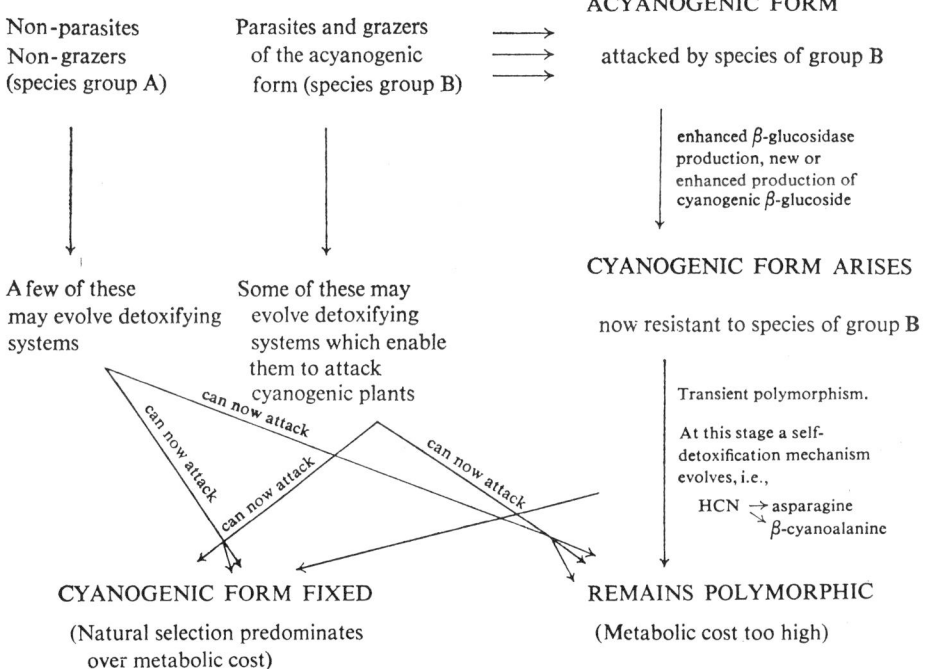

FIG. 5. A model to explain the evolution of cyanogenesis in plants (From Jones, 1972a).

say, *Prunus laurocerasus* L. as its larval food plant and the plant subsequently develops a cyanogenic system; then unless the insect changes food plant or is able to produce or already has produced forms which are tolerant of, resistant to or could metabolize CN^-, that species is doomed.

This is a population problem. Species that occur in large numbers are more likely to contain, by chance, the appropriate mutant which, normally disadvantageous, suddenly possesses an enormous selective advantage. Antonovics et al. (1971) have described several examples in *Agrostis tenuis* Sibth. where seed samples from parents with no history of exposure to heavy metals contain a small proportion of individuals that are tolerant of copper or of zinc. Thus given the environment of a heavy metal spoil heap, these mutant forms are the only ones that can colonize the contaminated soil.

3. Principles of Chemical Interaction

Implicit in the foregoing discussion are some general principles which can be applied to chemical interactions between species.

(a) Organisms are subject to attack by a very small minority of the other organisms in their environment.

(b) The organism possessing a particular protective substance must either be resistant to the harmful effects or sequester the substance in an inactive form until stimulated to release it.

(c) The substance is unlikely to be 100% efficient, because when the host is partly or wholly resistant a parasite could also evolve a similar system.

(d) Parasites can detoxify or sequester away harmful substances, or modify the metabolic pathways which are poisoned.

(e) Although some parasites are able to attack the host, it does not necessarily mean that the secondary substance is not protective.

(f) It is easy, however, to confound the protective substance with the species and it is only by showing that a parasite will attack a variety lacking the substance or demonstrate differential attack related to the concentration of the substance that a protective role can be proved.

(g) Like the theory of evolution itself, this field of research relies heavily on circumstantial evidence.

4. Cyanogenesis, Co-evolution and the Complexities of Natural Environments

In the early years of the study of genetical polymorphism it was noticeable how frequently one dominant selective agent (usually biotic) could be identified.

Because the genetical models which explain a stable polymorphism require a balance of selective forces it has often been necessary to invoke some counteracting physiological selection in order to accommodate these models. In few cases has the precise nature of the physiological selection been determined and consequently considerable scepticism has been expressed over the explaining away of apparent difficulties.

With the polymorphism of cyanogenesis in *Trifolium repens* and *Lotus corniculatus*, however, it appears that there are interactions between a physiological character and known physical and biotic components of the environment. Indeed, not only has a selective process of a physiological nature been demonstrated, but the chemistry involved is also well understood. But it is becoming clearer that none of the selective factors is in any way absolute; exceptions continue to be discovered with disturbing yet exciting regularity and it now becomes a question of piecing together the most likely explanation for each particular situation. Indeed the more we learn about cyanogenesis, the more certain I become that it is fortuitous that we have found three selective agents which have clear effects under certain circumstances. Jones and Wilkins (1971) have commented that we must not necessarily expect the responses of a plant to reflect the same categories of a habitat as those which we can measure with our instruments or, indeed, those which we can appreciate by observation. Thus the toxicity of a soil cannot be predicted from its heavy-metal context alone, as there is a strong interaction with the amount of calcium present. Again, some soils of low pH have a high soluble iron content and many plants which we regard as calcifuge actually show symptoms of iron deficiency when grown in chalk soils. Addition of iron in a suitable soluble form can remedy the reduction of chlorophyll without the need for any change in either the pH or the calcium content. It is, therefore, not the presence of calcium that restricts the edaphic range of these plants, but the availability of iron.

De Waal (1942) has studied the concentration of cyanogenic substances in leaves of *Trifolium repens* taking samples at several fixed times on successive days. In addition he recorded temperature, rainfall, light intensity and glucose content. Unfortunately the data presented are not adequate for multivariate analysis, but De Waal's lengthy discussion does show the complexity of the interactions between the various environmental components he measured. Three generalizations can be made from his work which are relevant to the protective role of cyanogenic glucosides. The available HCN, measured as a proportion of dry weight, is negatively correlated with glucose production. During the day, cyanogenic glucosides are often at their highest concentration shortly after a fall in light intensity, suggesting that free glucose is incorporated into linamarin and lotaustralin. Not infrequently, the concentration of cyanogenic glucosides rises after sunset and reaches a peak at approximately 02.00 hours the following morning.

Low light intensity during the day usually heralds rain and Crawford-Sidebotham (1972a) has shown that, of all the climate parameters he measured and estimated, the activity of slugs is related mainly to vapour pressure deficit (negative correlation). Although these animals feed mainly during the early part of the night, during the day they can also be active after a shower of rain, particularly following a period of dry weather. Whether it is significant or coincidental, it is remarkable that the feeding activity of the slugs, which have been shown to eat acyanogenic *Trifolium repens* differentially, slows down rapidly shortly before 02.00 hours (Barnes and Weil, 1945). If this timing is significant, then a most elegant picture can be built up. The plant responds to the fall in light intensity preceding rain or sunset by increasing the concentration of the cyanogenic glucoside so that by the time that the slugs become active after rain or during the early part of the night the plant is fully armoured against grazing.

Crawford-Sidebotham (1970) has reviewed work on the density of slug populations. Estimates from 105·5 per m^2 by trapping techniques to 276·7 per m^2 by soil sampling methods have been obtained so it is apparent that the grazing effects of slugs cannot be regarded as trivial even though the normal food of these animals is not known for certain.

In many cases (see Whittaker and Feeny, 1971) plants are monomorphic for secondary plant substances which can be considered as protective. Why therefore are plants such as *T. repens*, *L. corniculatus*, and *Vicia sativa* L. polymorphic for cyanogenesis? The metabolic cost may be too high (Fig. 5). Clearly there is an advantage in being cyanogenic when differential eating is important. The plant does, however, need to be damaged before any HCN can be detected. Apart from grazing, damage can also occur at low temperatures and so when the cytoplasm of the cell is likely to be frozen, cyanide poisoning must be regarded as a serious possibility (Loyd *et al.*, 1971; Jones, 1972a). Hence the metabolic cost of being cyanogenic in cold environments is likely to be too high unless an efficient self-detoxifying mechanism has evolved.

Thus Daday's linkage explanation (1965) using the *reproductive* fitness of glucosidic and aglucosidic plants could well be misleading. It is the *survival* fitness which matters basically and superimposed on this is the reproductive fitness which is concerned with the maintenance of disadvantageous phenotypes when selection is cyclical. Selection here is disruptive over seasons (time) and not necessarily in space.

Daday's results (1954) show that temperature appears to have an *overall* effect on the distribution of the cyanogenic form of *T. repens* and yet my own work (Jones, 1968) suggests that the temperature effect is not important over short distances. With seed production and winter survival, the four phenotypes of *L. corniculatus* show no difference in fitness (Jones, 1970). Foulds and Grime (1972a) revealed that soil water stress selects against cyanogenic *L. corniculatus* (and glucosidic *T. repens*) while we have evidence that the size of droughted areas must also be considered.

Selective and differential eating of the acyanogenic forms of both species has been demonstrated (Jones, 1962, 1966; Crawford–Sidebotham, 1972b). Nevertheless, there are several organisms that graze cyanogenic plants indiscriminately and some of these are known to be resistant to cyanide or even possess cyanogenic substances themselves (Jones *et al.*, 1962; Lane, 1962; Jones, 1963). The grazers and parasites of cyanogenic organisms often contain the detoxifying enzyme rhodanese and it is reasonable to argue that the development of the enzyme is a co-evolutionary response to cyanide production by the host. Bishop and Korn (1969) failed to detect selective eating in their experiments, but their work has been criticized in respect of scale (Crawford–Sidebotham, 1972b) and of the food (cabbage) on which the molluscs were fed before and during the experiment. Hegnauer (1971) suggested that because the Cruciferae contain sulphur glucosides closely related to cyanogenic glucosides, the animals may have become accustomed to β-glucosides. If Hegnauer is correct, then the plant food which an animal has just eaten may predetermine whether it shows selective or non-selective eating subsequently. The differential eating experiments of Crawford–Sidebotham (1972b) did not detect this effect even though, by the nature of their design, they were capable of doing so. Clearly the ecological genetics of cyanogenesis remains intriguing.

ACKNOWLEDGEMENTS

I am grateful to the Science Research Council (B/SR/6494) for financial support for some of the work reported here. I also thank Professor John Beardmore for translating the thesis of D. de Waal from the original Dutch, and Dr T. J. Crawford-Sidebotham for his constructive comments on the review.

REFERENCES

ABROL, Y. P. and CONN, E. E. (1966). Studies on cyanide metabolism in *Lotus arabicus* L. and *Lotus tenius* Waldst. et Kit. *Phytochemistry* **5**, 237–242.

ABROL, Y. P., CONN, E. E. and STOKER, J. R. (1966). Studies on the identification, biosynthesis and metabolism of a cyanogenic glucoside in *Nandina domestica* Thunb. *Phytochemistry* **5**, 1021–1027.

ALLAN, P. B. M. (1949). "Larval Foodplants." Watkins and Doncaster, London.
ANGSEESING, J. P. A. and ANGSEESING, W. J. (1973). Field observations on the cyanogenesis polymorphism in *Trifolium repens* L. *Heredity* (in press).
ANTONOVICS, J., BRADSHAW, A. D. and TURNER, R. G. (1971). Heavy Metal Tolerance in Plants. *In* "Advances in Ecological Research" (J. B. Cragg, ed.), Vol. **7**, pp. 1–85. Academic Press, London and New York.
BAILEY, R. W., GREENWOOD, R. M. and CRAIG, A. (1971). Extracellular polysaccharides of *Rhizobium* strains associated with *Lotus* species. *J. gen. Microbiol.* **65**, 315–324.
BARNES, H. F. and WEIL, J. W. (1945). Slugs in gardens: their numbers, activities and distribution. Part 2. *J. Anim. Ecol.* **14**, 71–105.
BATISTE, W. C. (1967). Biology of the trefoil seed chalcid *Bruchophagus kolobovae* Fedoseeva (Hymenoptera: Eurytomidae). *Hilgardia* **38**, 427–469.
BENDALL, D. A. and BONNER, W. D. (1971). Cyanide insensitive respiration in plant mitochondria. *Pl. Physiol.* **47**, 236–245.
BISHOP, J. A. and KORN, M. E. (1969). Natural selection and cyanogenesis in white clover, *Trifolium repens*. *Heredity* **24**, 423–430.
BORSOS, OLGA. (1966). Mikrotaxonomische bearbeitung der Artengruppe *Lotus corniculatus* L. agg. in der Pannonischen und Karpatischen Flora. *Acta. Bot. Acad. Sci. Hung.* **12**, 255–283.
CAIN, A. J. (1951). So-called non-adaptive or neutral characters in evolution. *Nature, Lond.* **168**, 424.
CLARKE, B. C. (1970). Selective constraints on amino-acid substitutions during the evolution of proteins. *Nature, Lond.* **228**, 159–160.
CLAUSS, E. (1961). Die phenolischen Inhaltstoffe der Samenschalen von *Pisum sativum* L. und ihre Bedeutung für die Resistenz gegen die Erreger der Fusskrankheit. *Naturwiss.* **48**, 106.
COMBES, R. (1918). Immunité des végétaux vis-à-vis des principes immédiate qu'ils élaborent. *C. r. hebd. Séanc. Acad. Sci. Paris* **167**, 275–278.
CONN, E. E. (1973). Cyanogenic glycosides: their occurrence, biosynthesis and function. *In* "Chronic Cassava Toxicity" (B. Nestel and R. McIntyre, eds), pp. 55–63, International Development Research Centre, Ottawa.
CORKILL, L. (1952). Cyanogenesis in white clover (*Trifolium repens* L.). VI. Experiments with high-glucoside and glucoside-free strains. *N.Z. J. Sci. Tech. A.* **34**, 1–16.
CRAWFORD-SIDEBOTHAM, T. J. (1970). Differential susceptibility of species of slugs to metaldehyde/bran and to methiocarb. baits. *Oecologia* **5**, 303–324.
CRAWFORD-SIDEBOTHAM, T. J. (1972a). The influence of weather on the activity of slugs. *Oecologia* **9**, 141–154.
CRAWFORD-SIDEBOTHAM, T. J. (1972b). The role of slugs and snails in the maintenance of the cyanogenesis polymorphisms of *Lotus corniculatus* and *Trifolium repens*. *Heredity* **28**, 405–411.
DADAY, H. (1954). Gene frequencies in wild populations of *Trifolium repens* L. I. Distribution by latitude. *Heredity* **8**, 61–78.
DADAY, H. (1965). Gene frequencies in wild populations of *Trifolium repens* L. IV. Mechanism of natural selection. *Heredity* **20**, 355–366.
DARWIN, C. R. (1859). "On the Origin of Species by Means of Natural Selection." Murray, London.
DARWIN, C. R. (1862). "The Various Contrivances by which Orchids are Fertilized by Insects." Murray, London.

DAYHOFF, M. O. ed. (1969 et seq.) "Atlas of Protein Sequence and Structure." Nat. Biometrical Research Foundation, Silver Spring, Maryland 20901, U.S.A.

DE VARIGNY, H. (1892). "Experimental Evolution." Macmillan, London.

DE VRIES, W. (1968). Oecologische variabiliteit van Lotus corniculatus L. Thesis. Instituit voor Systematische Plantkunde te Utrecht.

DE WAAL, D. (1942). "Het Cyanophore Karakter van Witte Klaver, *Trifolium repens* L. Thesis, H. Veenman en Zonen N.V., Wageningen.

EASTY, D. B., BLAEDEL, W. J. and ANDERSON, L. (1971). Continuous electrochemical determination of cyanide. Application to cyanogenic glycosides in Sudan grass. *Anal. Chem.* **43**, 509–514.

EDWARDS, C. A. and HEATH, G. W. (1964). "The Principles of Agricultural Entomology." Chapman and Hall, London.

EHRLICH, P. R. and RAVEN, P. H. (1965). Butterflies and plants: a study in coevolution. *Evolution* **18**, 586–608.

EISNER, H. E., EISNER, T. and HURST, J. J. (1963a). Hydrogen cyanide and benzaldehyde produced by millipedes. *Chemy. Ind.* **1963**, 124–125.

EISNER, T., EISNER, H. E., HURST, J. J., KAFATOS, F. C. and MEINWALD, J. (1963b). Cyanogenic glandular apparatus of a millipede. *Science, N.Y.* **139**, 1218–1220.

FEDOSEEVA, L. I. (1956). New species of the genus *Bruchophagus* Ashm. (Hymenoptera: Chalcidoidea) developing upon leguminous plants (in Russian). *Dokl. Akad. Nauk. S.S.S.R.* **111**, 491–493.

FLOR, H. H. (1956). The complementary genic systems in flax and flax rust. *Adv. Genetics* **8**, 29–54.

FLÜCK, H. (1963). Intrinsic and extrinsic factors affecting the production of natural products. *In* "Chemical Plant Taxonomy" (T. Swain, ed.), pp. 167–186. Academic Press, London and New York.

FOULDS, W. and GRIME, J. P. (1972a). The influence of soil moisture on the frequency of cyanogenic plants in populations of *Trifolium repens* and *Lotus corniculatus*. *Heredity* **28**, 143–146.

FOULDS, W. and GRIME, J. P. (1972b). The response of cyanogenic and acyanogenic phenotypes of *Trifolium repens* to soil moisture supply. *Heredity* **28**, 181–187.

FRAENKEL, G. S. (1959). The raison d'être of secondary plant substances. *Science, N.Y.* **129**, 1466–1470.

FRAENKEL, G. S. (1969). Evaluation of our thoughts on secondary plant substances. *Ent. Exp. App.* **12**, 473–486.

GIBBS, R. D. (1963). History of Chemical Taxonomy. *In* "Chemical Plant Taxonomy" (T. Swain, ed.), pp. 60–88. Academic Press, London and New York.

GILMOUR, D. (1961). "The Biochemistry of Insects." Academic Press, London and New York.

GRAHAM, J. H. (1953). A disease of birdsfoot trefoil caused by a new species of *Stemphylium*. *Phytopathology* **43**, 577–579.

GRANT, W. F. (1966). Cyanogenic glucoside distribution in the speciation of the genus *Lotus* and its evolutionary significance. *Proc. XI Pacific Sci. Conf. (Tokyo)* **5**, 38.

GRANT, W. F. and SIDHU, B. S. (1967). Basic chromosome number, cyanogenic glucoside variation and geographic distribution of *Lotus* species. *Can. J. Bot.* **45**, 639–647.

GUÉRIN, P. (1929). La teneur en acide cyanhydrique des *Lotus*. *C. r. hebd. Séanc. Acad. Sci., Paris* **187**, 1011–1013.

HALL, F. R., HOLLINGWORTH, R. M. and SHANKLAND, D. L. (1969). Cyanide tolerance in millipedes: comparison of respiration in millipedes and insects. *Entom. News* **80**, 277–282.

HALL, F. R., HOLLINGWORTH, R. M. and SHANKLAND, D. L. (1971). Cyanide tolerance in millipedes: the biochemical basis. *Comp. Biochem. Physiol.* **38B**, 723–737.

HANELT, P. and TSCHIERSCH, B. (1967). Blausaureglykosid-Untersuchungen am Gaterslebener Wickensortiment. *Kulturpflanze* **15**, 85–96.

HANSEN, H. L. (1955). "The Host Relationships of the Seed Chalcid *Bruchophagus gibbus* (Baheman)." Thesis, University of California, Berkeley, pp. 1–96.

HEGNAUER, R. (1963). The taxonomic significance of alkaloids. *In* "Chemical Plant Taxonomy" (T. Swain, ed.), pp. 389–427. Academic Press, London and New York.

HEGNAUER, R. (1971). Probleme der Chemotaxonomie, erläutert am Beispiel der cyanogenen Pflanzenstoffe. *Pharm. Acta. Helv.* **46**, 585–601.

HEGNAUER, R. and RUIJGROK, H. W. L. (1971). Die Verbreitung der Blausäure bei den Cormophyten. 6. Mitteilung: Weitere Beobachtungen über die Cyanogenese bei den *Junci* Septati. *Pharm. Weekblad.* **106**, 263–270.

HEWITT, E. J. and NICHOLAS, D. J. D. (1963). Cations and Anions: inhibitors and interactions in metabolism and in enzyme activity. *In* "Metabolic Inhibitors" (R. M. Hochster and J. H. Quastel, eds), pp. 311–436. Academic Press, London and New York.

HUGHES, MONICA A. and MAHER, E. P. (1973). Studies on the nature of the *Li* locus in *Trifolium repens* L. I. Purification and properties of the enzyme components. *Biochemical Genetics* **8**, 1–12.

JONES, D. A. (1962). Selective eating of the acyanogenic form of the plant *Lotus corniculatus* L. by various animals. *Nature, Lond.* **193**, 1109–1110.

JONES, D. A. (1963). "Polymorphisms and Antibodies in Lower Organisms." D. Phil. Thesis, University of Oxford, England.

JONES, D. A. (1966). On the polymorphism of cyanogenesis in *Lotus corniculatus* L. I. Selection by animals. *Can. J. Genet. Cytol.* **8**, 556–567.

JONES, D. A. (1968). On the polymorphism of cyanogenesis in *Lotus corniculatus* L. II. The interaction with *Trifolium repens* L. *Heredity* **23**, 453–455.

JONES, D. A. (1970). On the polymorphism of cyanogenesis in *Lotus corniculatus* L. III. Some aspects of selection. *Heredity* **25**, 633–641.

JONES, D. A. (1971). Chemical defence mechanisms and genetic polymorphism. *Science, N.Y.* **173**, 945.

JONES, D. A. (1972a). Cyanogenic Glycosides and their Function. *In* "Phytochemical Ecology" (J. B. Harborne, ed.), pp. 103–124. Academic Press, London and New York.

JONES, D. A. (1972b). On the polymorphism of cyanogenesis in *Lotus corniculatus* L. IV. The Netherlands. *Genetica* **43**, 394–406.

JONES, D. A. (1973). On the polymorphism of cyanogenesis in *Lotus corniculatus* L. V. Denmark. *Heredity* **30**, 381–386.

JONES, D. A., PARSONS, J. and ROTHSCHILD, M. (1962). Release of hydrocyanic acid from crushed tissues of all stages in the life-cycle of species of the Zygaeninae (Lepidoptera). *Nature, Lond.* **193**, 52–53.

JONES, D. A. and WILKINS, D. A. (1971). "Variation and Adaptation in Plant Species." Heinemann Educational Books, London.

JONES, F. G. W. and JONES, M. G. (1964). "Pests of Field Crops." Edward Arnold, London.

KEILIN, D. (1966). "The History of Cell Respiration and Cytochrome." Cambridge University Press.
KEISTER, MARGARET and BUCK, J. (1964). Respiration: Some exogenous and endogenous effects on rate of respiration. In "The Physiology of Insecta" (M. Rockstein, ed.), Vol. III, pp. 617–658. Academic Press, London and New York.
KETTLEWELL, H. B. D. (1965). Insect survival and selection for pattern. Science, N.Y. **148,** 1290–1296.
KIMURA, MOTOO and OHTA, TOMOKO. (1971). "Theoretical Aspects of Population Genetics." Princeton University Press, Princeton, N.J.
KURLAND, C. G. and SCHNEIDERMAN, H. A. (1959). The respiratory enzymes of diapausing silkworm pupae: a new interpretation of carbon monoxide insensitive respiration. Biol. Bull. **116,** 136–161.
LANE, C. (1962). Notes on the Common Blue (Polyommatus icarus (Rott.)). Egg-laying and feeding on the cyanogenic strains of the birdsfoot trefoil (Lotus corniculatus L.). Ent. Gaz. **13,** 112–116.
LANG, K. (1933). Die Rhodanbildung im Tierkörper. Biochem. Z. **259,** 243–256.
LARSEN, K. and ŽERTOVÁ, ANNA (1963). On the variation pattern of Lotus corniculatus in Eastern Europe. Bot. Tidsskr. **59,** 177–194.
LEEUWEN, C. G. VAN and MAAREL, E. VAN DER (1971). Pattern and process in coastal dune vegetation. Acta. bot. Neerl. **20,** 191–198.
LEVIN, D. A. (1971). Plant Phenolics: An Ecological Perspective. Am. Nat. **105,** 157–181.
LEWIS, H. (1962). Catastrophic selection as a factor in speciation. Evolution **16,** 257–271.
LOVATT EVANS, C. (1952). "Principles of Human Physiology" (11th edition). Churchill, London.
LOYD, R. C., GRAY, E. and SHIPE, E. (1971). Effect of freezing on hydrocyanin release from sorghum plants (Sorghum bicolor (L) Moench.). Agron. J. **63,** 139–140.
LUCRETIUS CARUS, TITUS (ca. 55 B.C.) "De Rerum Natura" [On the Nature of the Universe], line 637, translated by R. Latham, Penguin Books, 1951.
LUNDEGÅRDH, H. (1966). "Plant Physiology." Oliver and Boyd, Edinburgh.
LURIA, S. E. (1945). Mutations of bacterial viruses affecting their host range. Genetics **30,** 84–99.
MAHER, E. P. and HUGHES, MONICA A. (1973). Studies on the nature of the Li locus in Trifolium repens L. II. The effect of genotype on enzyme activity and properties. Biochemical Genetics **8,** 13–26.
MARSH, P. B. and GODDARD, D. R. (1939). Respiration and fermentation in the carrot Daucus carota. I. Respiration. Am. J. Bot. **26,** 724–728.
MILLAR, R. L. and HIGGINS, V. J. (1970). Association of cyanide with infection of birdsfoot trefoil by Stemphylium loti. Phytopathology **60,** 104–110.
MONTGOMERY, R. D. (1969). Cyanogens. In "Toxic Constituents of Plant Foodstuffs" (I. E. Liener, ed.), pp. 143–157. Academic Press, London and New York.
MORRIS, M. J. (1967). Differences between the invertebrate faunas of grazed and ungrazed chalk grassland. I. Responses of some phytophagous insects to cessation of grazing. J. appl. Ecol. **4,** 459–474.
NAYER, J. K. and FRAENKEL, G. (1963). The chemical basis of the host selection in the Mexican bean beetle Epilachna varivestis (Coleoptera, Coccinellidae). Ann. ent. Soc. Am. **56,** 174–178.
NEUNZIG, H. H. and GYRISCO, G. G. (1955). Some insects injurious to birdsfoot trefoil in New York. J. econ. Ent. **48,** 447–450.

NEUNZIG, H. H. and GYRISCO, G. G. (1959). Parasites associated with seed chalcids infesting alfalfa, red clover and birdsfoot trefoil in New York. *J. econ. Ent.* **52,** 898–901.
PARIS, R. (1963). The distribution of plant glycosides. *In* "Chemical Plant Taxonomy" (T. Swain, ed.), pp. 337–358. Academic Press, London and New York.
PARSONS, J. and ROTHSCHILD, M. (1964). Rhodanese in the larva and pupa of the Common Blue Butterfly, *Polyommatus icarus* (Rott.) (Lepidoptera). *Ent. Gaz.* **15,** 58–59.
PRICE, J. R. (1963). The distribution of alkaloids in the Rutaceae. *In* "Chemical Plant Taxonomy" (T. Swain, ed.), pp. 429–452. Academic Press, London and New York.
RIDGWAY, R. L. and GYRISCO, G. G. (1959). Control of insects injurious to birdsfoot trefoil in New York. *J. econ. Ent.* **52,** 836–838.
ROBINSON, MURIEL E. (1930). Cyanogenesis in plants. *Biol. Rev.* **5,** 126–141.
ROCCI, V. (1914). Sulla resistenza degli *Zigenini* all'acido cianidrico. *Z. allg. Physiol.* **16,** 42.
ROSENTHAL, O. (1948). The distribution of rhodanese. *Fed. Proc.* **7,** 181–182.
RUIJGROK, H. W. L. (1966). The distribution of ranunculin and cyanogenetic compounds in the Ranunculaceae. *In* "Comparative Phytochemistry" (T. Swain, ed.), pp. 175–186. Academic Press, London and New York.
RUSZKOWSKI, A. (1968). Wstepne obserwacje nad szkodliwa fauna komonicy rozkowej— *Lotus corniculatus* L. *Pamiet. Pulawski* **35,** 297–298.
SCHNEIDER, A. (1952). Über das Vorkommen gerstoffartiger Kondensationsprodukte von Anthocyanidinen in den Samenschalen von *Pisum arvense*. *Naturwiss.* **39,** 452–453.
SEANEY, R. R. and HENSON, P. R. (1970). Birdsfoot trefoil. *Adv. Agronomy* **22,** 120–157.
STAHL, E. (1888). Pflanzen und Schnecken. Eine biologische Studie über die Schutzmittel der Pflanzen gegen Schneckenfraass. *Jena. Z. Naturw.* **22,** 557.
STENT, G. S. (1963). "Molecular Biology of Bacterial Viruses." Freeman, San Francisco.
TAPPER, B. A. and BUTLER, G. W. (1972). Intermediates in the biosynthesis of linamarin. *Phytochemistry* **11,** 1041–1046.
TAPPER, B. A., ZILG, H. and CONN, E. E. (1972). 2-Hydroxyaldoximes as possible precursors in the biosynthesis of cyanogenic glucosides. *Phytochemistry* **11,** 1047–1053.
TREUB, M. (1907). Note on the protective effect attributed to HCN in plants. *Ann. Jard. bot. Buitenzorg* **6,** 107–114.
WATERHOUSE, F. L. and SANDERSON, A. R. (1958). Geographical colour polymorphism and chromosome constitution in sympatric species of sawflies. *Nature, Lond.* **182,** 477.
WHITTAKER, R. H. and FEENY, P. P. (1971). Allelochemics: Chemical interactions between species. *Science, N.Y.* **171,** 757–770.
WILLIS, C. B. and THOMPSON, L. S. (1969). The influence of soil moisture and cutting management on *Paratylenchus penetrans* in birdsfoot trefoil and the relationship of inoculum levels to yields. *Phytopathology* **59,** 1872–1875.
WILSON, J. (1973). Cyanide and human disease. *In* "Chronic Cassava Toxicity" (B. Nestel and R. MacIntyre, eds), pp. 121–125. International Development Research Centre, Ottawa.
ZITNAK, A. (1973). Assay methods for hydrocyanic acid in plant tissues and their application in studies of cyanogenic glycosides in *Manihot esculenta*. *In* "Chronic Cassava Toxicity" (B. Nestel and R. MacIntyre, eds), pp. 89–96. International Development Research Centre, Ottawa.

12 | Some Anthecological Aspects of the Evolution of Nectar-Producing Flowers, Particularly Amino Acid Production in Nectar

HERBERT G. BAKER and IRENE BAKER

Department of Botany, University of California, Berkeley, California, U.S.A.

Abstract: The subject of anthecology, or pollination ecology, is developing rapidly at the present day. One very new revelation made on the basis of a survey of 266 species of flowering plants, is that nectar *usually* contains detectable quantities of amino acids as well as the familiar sugars. Although amino acids are present in the nectar of plants belonging to families usually considered "primitive", higher concentrations appear more consistently in more "advanced" families. A series of comparisons between plants showing a particular "primitive" character with those showing the "advanced" state for the same character reveals a higher concentration of amino acids and greater consistency of production in the latter in each comparison. Butterfly-pollinated flowers appear to produce nectar containing high concentrations of amino acids more consistently than bee flowers, a difference relatable to the utilization by bees of an important alternative source of amino acids in pollen. The amounts of amino acids ingested by many adult butterflies from nectar are likely to be nutritionally significant, but no one source of nectar can provide all of the amino acids that they need. Flowers pollinated by non-hovering moths resemble butterfly flowers in amino acid presentation but hawk-moth flowers show a wide range of concentrations with a rather low mean. Some "bird flowers" and the "generalized fly flowers" also show rather low means but "sapromyophilous" flowers may successfully lure their pollinators away from carrion and dung by presentation of nectar rich in amino acids as well as by their morphological features. Further studies are projected in California and in the tropics, and will be carried out on an ecosystem basis, with the expectation that they will reveal something of the significance of amino acid provision in flowering plant nectar for community evolution.

ANTHECOLOGY, OR POLLINATION ECOLOGY

Anthecology, a term already in use 45 years ago (Robertson, 1927), refers to the study of all aspects of the interactions between flower-visiting (anthophilous) animals and the flowers they visit, as well as to the pollination biology of those flowers that are pollinated by wind or water. Like all branches of ecology, the discipline is a synthetic one and its potential content is enormous. Progress towards the realization of this potential, from the middle of the nineteenth century onwards, has been discontinuous (Baker, 1961) but, at the present time, a strong surge is taking place.

The old natural history approach to anthecology has not been (and never should be) abandoned but it is being put on a firmer scientific basis by the utilization of novel, finer and more rigorous analytical techniques, by quantification of observations and by greater use of experiments. The morphological and behavioural aspects of anthecology have been reviewed quite recently by Baker and Hurd (1968), Faegri and van der Pijl (1971), Macior (1971), and others. As examples of quantitative treatments of pollen distribution, works such as those of Levin (1969, 1972) and Janzen (1971) may be cited. The energetics of flower visitation by animals has been studied by North American entomologists such as Hocking (1953, 1968) and Heinrich (Heinrich and Raven, 1972) as well as by ornithologists (e.g. Wolf and Hainsworth, 1971; Wolf et al., 1972).

Because the cited papers have been published in journals or books that are readily available, it is hardly necessary for us to make a general review of anthecology here. We prefer to devote this paper to a small part of the subject—in fact, to a subdivision that did not exist at the beginning of the summer of 1972—consideration of the provision of amino acids in the nectar of flowers and its ecological, evolutionary and phylogenetic implications.

1. Benefits to Animals of Flower Visitation

Animals can get at least three kinds of benefits from the flower visits that bring about pollination. First, there is the possibility of shelter inside the flower: shelter from predators, rain, wind and cold (cf. Hocking, 1953, 1968). Flowers also provide places in which to wait for prey or, alternatively, in the case of some insects, for laying eggs so that the larvae will be able to feast upon flower-parts. Sometimes the waiting is for a mate; flowers are excellent trysting places for small animals, especially for those that can eat whilst awaiting the arrival of their mates. For, most of all, flowers provide food materials for their visitors (or appear to do so in the not uncommon cases of deception). Consequently, an important question is, "What kinds of food are available to the animals that feed in flowers?"

The first visitors to flowers, when the angiosperms began to exert their attractive influences, are believed to have been beetles (review in Baker and Hurd, 1968). These insects paid attention to flowers that provided numerous fleshy floral parts and, no doubt, the beetles obtained carbohydrates, proteins and a range of other substances from these parts. It is interesting that a supposedly "primitive" flowering plant, *Calycanthus occidentalis* H. & A. (Calycanthaceae) produces flowers that contain protein-rich food-bodies that nourish their beetle visitors exceptionally well (F. Rickson, unpublished). More often, however, it seems that proteins (and the amino acids that are their building blocks) are obtained by these anthophilous insects from pollen grains.

Pollen not only provides proteins and amino acids in quantity but it is also remarkable for the range of essential and quasi-essential amino acids that it makes available to flower-visitors (Schuette and Baldwin, 1944; Auclair and Jamieson, 1948; Linskens and Schrauwen, 1969; Haydak, 1970).

Early in the history of flowering plants, nectar became another attractant for flower visitors and much attention has been given by anthecologists to the manner in which this liquid is made available to visitors. Less attention has been given to the determination of what chemical substances these visitors acquire when they imbibe nectar. Obviously, they get water and, dissolved in the water, sugars: predominantly sucrose, glucose and fructose. The conventional view is that this is all, and a consequence of this belief that nectar is simply sugar water is that chemical analyses of it have been confined almost entirely to investigations of the natures, amounts and proportions of the sugars present.

Vogel (1969) has shown that in some flowering plants belonging to the families Scrophulariaceae, Malpighiaceae and Orchidaceae, special glands (which he calls elaiophors) in the flowers secrete a lipid which takes the place of nectar and is collected by bees. Our own investigations have shown that the milky nectar of *Jacaranda acutifolia* H. & B. (Bignoniaceae) contains lipids, and other examples of lipid-containing nectar are coming to light.

Strikingly absent from the anthecological literature is the suggestion that nitrogenous nutrients might be present in amounts significant for the nutrition of nectar-drinkers. No doubt this general neglect of the possibility of the presence of proteins or amino acids in nectar stems from the belief that anthophilous insects which do not nurse their brood have need of nothing more than energy-giving sugars. Those that do nurse their brood, most notably the bees, can use pollen as a source of nitrogenous food material.

2. Utilization of Food Substances other than Sugars by Lepidoptera

Moths and butterflies are usually supposed to take all of their nitrogenous foodstuffs whilst in the larval stage. The caterpillars, feeding on leaves and flowers, store nitrogenous materials in the fat-body (Kilby, 1963) and this store is drawn upon by the adult, especially in egg-laying (Davey, 1965; Engelmann, 1970). In such Lepidoptera as the silk worm (*Bombyx mori* L.) or the Yucca moths (*Tegeticula* spp.), where the adult does not feed at all, this must be the case. But must it necessarily be so for other Lepidoptera?

According to Faegri and van der Pijl (1971, p. 133) some primitive Lepidoptera (family Micropterygidae) have chewing mouthparts even in the adult insects and they use them in feeding on the pollen of plants such as *Caltha* and *Ranunculus* (Ranunculaceae). But most adult Lepidoptera have a proboscis in the form of a long narrow tube, formed from the extremely elongated maxillae, and this restricts them to a liquid diet, usually of watery consistency. However, this is not always nectar!

Ford (1945, p. 94) points out that Purple Emperor butterflies (*Apatura iris* L.) are attracted to juicy carrion and "may also be found drinking at a foul puddle or at the edge of a stagnant pool". Others are attracted to the sap flowing from wounds in plants and to the "honey-dew" excretions of sap-sucking aphids and other insects. All of these liquids may be expected to contain nitrogenous substances.

Gilbert (1972) records the attraction of tropical butterflies to rotting fruit, fermenting sap, urine, dung and bird-droppings and, in Arctic Canada, Hocking (1971, p. 105) found butterflies attracted to his sweaty feet. Another entomologist, A. B. Klots (undated, pp. 86–91) has written, "One of the finest assemblies of butterflies that I have ever seen, composed of hundreds of individuals of many species, was gathered on the decomposing remains of a small crocodile on the bank of a tributary of the Amazon". He continues "... I have seen as many as two hundred of the large North American Tiger Swallowtail closely packed in less than half a square yard on a manure pile". Moths are known that variously drink fruit juices, sweat secretions from the eyes of animals, as well as blood passed by mosquitoes; in some cases they penetrate mammalian skin to suck blood (Bänziger, 1971). It is hard to believe that all of these Lepidoptera are interested only in the sugar or lipid contents of these liquids.

Very recently, Gilbert (1972) has shown that neotropical butterflies of the genus *Heliconius* (family Nymphalinae) collect pollen, steep it in nectar and subsequently ingest the amino acids that diffuse from the grains. By experiments he has demonstrated that amino acids taken in this way by female butterflies of this genus are incorporated quickly into the eggs they produce. Apparently, both the life span and the reproductive output of the adults are increased by such feeding.

Incidentally, 99 years ago, Hermann Müller (1873) published his observation that the long-tongued fly *Rhingia rostrata* L. (Syrphidae) behaves in a rather similar manner at the flowers of *Lythrum salicaria* L. (Lythraceae) in Europe, and he guessed correctly at the meaning of these actions: "This fly, standing on one or more of the petals, ... stretches its proboscis out to a length of 11-12 mm, and thrusts it down into the flower, letting it remain there from six to ten seconds. Immediately after withdrawing it from the tube, it usually manipulates one of the anthers with its labellae for a short time (one to two seconds) in order to add to the liquid non-nitrogenous food some solid nitrogenous matter in the shape of pollen-grains" (Müller, 1873; in English translation, 1883, p. 259).

The evidence that adult Lepidoptera, usually thought of as restricted to feeding on sugar-water, can be concerned with ingesting nitrogen-rich materials which presumably supply them with amino acids, suggests that nectar, itself, should be examined as a potential source of these nutrients.

Some flowering plants are extremely attractive to butterflies and some of these have received vernacular names like "butterfly bush" or "butterfly vine". Such species are *Buddleia davidii* Franch. (Loganiaceae), *Antigonon leptopus* H. & A. (Polygonaceae), *Clerodendron paniculatum* L. (Verbenaceae), *Caesalpinia pulcherrima* Sw. (Leguminosae–Caesalpinoideae) and many Compositae. Perhaps they are offering rewards to visitors comparable with those of the carrion and manure piles?

AMINO ACIDS IN NECTAR

There is no obvious reason why amino acids should not be present in nectar. Most nectar is derived from the phloem as well as the xylem of vascular bundles supplying the nectaries (Maurizio, 1962) and the relatively small molecules may be expected to be able to pass out into the nectar about as easily as sugars do.

However, few biochemists appear to have given attention to this possibility. Ziegler (1956) noted the presence of amino acids in floral nectar from three flowering plant species. Lüttge (1961, 1962) added five more species. Percival (1962) made a ninhydrin test on nectars of some unidentified species, obtaining a positive colour reaction, but apparently interpreted it as indicating the presence of enzymes. A number of comprehensive analyses of honey have been published (e.g. Pryce-Jones, 1950; White, 1957) but these are not very relevant because honey contains substances added by the honey bees.

Because of this apparent gap in our knowledge of nectar constitution, we have made a rapid survey of 266 species of flowering plant occurring in nature in California and in the University of California Botanical Garden. Our results are presented here as a working report.

1. Methods of Testing for Amino Acids and Proteins in Nectar

(a) *Amino acids.* Nectar was collected in clean micro-capillary tubes from fresh flowers at the time of day when they were judged to have the maximum amount present. For most species this was mid-morning, but for nocturnally-flowering species, night-time or early morning collections were made. Collections were not made at times of unusually high temperature and rain did not fall at any time during the studies.

Drops of nectar were placed on filter paper and dried so quickly that very little spreading took place. They were then treated with a drop of a 0·2% solution (in acetone) of ninhydrin (triketohydrindene hydrate). Up to 24 h at laboratory temperature were allowed for a colour to develop. Under these

TABLE I. Concentrations of aqueous histidine solution required to produce depths of colour scoring 0–10 in dried spots when ninhydrin solution is added

Score on "histidine scale"	Concentration of aqueous solution of histidine
0	<49 μM
1	49 μM (7·58 μg/ml)
2	98 μM
3	195 μM
4	391 μM
5	781 μM
6	1·56 mM
7	3·13 mM
8	6·25 mM
9	12·50 mM
10	25·00 mM (3·90 mg/ml)

circumstances, most α-amino acids give a violet coloration, or a colour approximating to it (Smith, 1969, p. 119). The notable exceptions to this are proline and hydroxyproline (yellow) and asparagine (orange–brown). Although some 4–500 compounds bearing an amino-group (both natural and synthetic) are known that can give a colour with ninhydrin under various treatments, when the test is carried out at laboratory temperatures in the manner described only α-amino acids are likely to do so (Smith, 1969, p. 119).

A comparison scale for visual use was created by making dilutions of an aqueous solution of the amino acid histidine and staining dried drops of these. The scale and the concentrations represented are shown in Table I. Each step in the scale represents a doubling of histidine concentration.

Each nectar sample was scored against this "histidine scale". Because of evidence from chromatographic analyses of some of the nectars (see below) that nectars from different species contain different groupings of amino acids and in differing proportions, the "histidine scale" does not provide a perfect measure of the amino acid concentration in each case, but it is felt that gross errors do not obtain in the results reported.

Wherever nectar was available in sufficient quantity a measurement of its sugar content was made by use of a Bellingham and Stanley pocket refractometer—the readings providing data in the form of "sucrose equivalents". The sugar content of the nectar was important because a concentrated nectar (up to 70% sugar w/v) would spread less on the filter paper than a dilute one (say 15–20% w/v). However, it was found by experiment with nectar dilutions that because of the rapidity of drying and the consequent restriction of spreading, between these extremes of sugar content along with a constant concentration of histidine, there would be no more than one unit change in the results obtained. Furthermore, nectars concentrated enough to rate above 40% w/v in sugar content were found only extremely rarely. Any difficulty in getting full penetration of the dried nectar spot by the ninhydrin when the sugar content was high was overcome by making dilutions before spotting, or, alternatively, by dissolving away the sugars with ethyl acetate.

For more than 40 of the species represented, two-way chromatograms (butanol–acetic acid–water; phenol–water) were run to give tentative identifications of the individual amino acids present in the nectar.

(b) *Proteins*. Zimmerman (1953, 1954) has shown that some enzymes (transglucosidases and transfructosidases) may be present in certain nectars. Other occurrences of protein in floral nectar have been noted (see Lüttge, 1961), usually with the implication that its presence results from bacterial activity. In addition, some kinds of honey are known for their relatively high protein contents, for example that derived from *Calluna vulgaris* Salisb. (Ericaceae) which reaches as high as 1·5% in samples (Percival, 1965, p. 68). The possibility that some of this protein comes directly from the nectar rather than admixed bacteria, pollen or secretions from the various glands of the honey bee meant that tests should be carried out for proteins in fresh nectar.

Starch gel electrophoresis was employed to test for the presence of proteins in the nectar from nine species (tris-citrate gel buffer, with lithium hydroxide-boric acid electrode buffer). Brom-phenol blue and amido-black were used as stains. In addition, tests were made on nectar drops from 21 other species (dried on chromatography paper) using brom-phenol blue stain.

2. Overall Results

The electrophoresis results for proteins were negative and, in the dried nectar spot tests, only three samples appeared to contain enough protein to give a detectable greenish-blue colour with the brom-phenol blue stain. These were nectars from *Calluna vulgaris* and *Erica mediterranea* L. (Ericaceae), and *Bergenia crassifolia* L. (Saxifragaceae).

Consequently, in all the results and conclusions that follow it should be borne in mind that proteins may be present in small quantities in some of the nectars concerned. These proteins could have some nutritional significance for bees, flies and birds which secrete proteolytic enzymes but would be much less likely to have any usefulness for adult Lepidoptera whose mouthparts and guts (by contrast with those of their larvae) secrete none of these protein-breaking enzymes (Wigglesworth, 1953, p. 351; House, 1965, pp. 842–843). The three plant species which gave positive protein stains in the present experiments are all pollinated by bees.

By contrast with the minority of samples giving positive protein tests, only six out of 266 of the species tested failed to give a positive ninhydrin reaction. *Thus, the vast majority of the species tested can be said to have amino acids in detectable quantity in their nectars.*

That these amino acids were not the metabolic products of microbial contaminants was demonstrated by three tests. First, the entire nectar-supply was removed from flowers of six species, spotted on slides, fixed and gram-stained. Bacteria present were counted in extremely small numbers, the largest being 73 in about 2 μl of nectar from *Cestrum parqui* L'Hérit. (Solanaceae). Second, nectars from flowers of *Kentranthus ruber* DC. (Valerianaceae), of a range of ages from freshly opened to four days old (and fading), were spotted and stained in the usual manner. No increase in staining reaction with age was to be seen. Finally, solutions of glucose and sucrose ranging in concentration from 10% to 80% were put into unsterilized test-tubes. Fungal growth took place in all tubes, but bacterial growth occurred only in the lower sugar concentrations. After one week of incubation at laboratory temperatures, these solutions were given the ninhydrin test but failed to develop distinguishable colour.

The conclusion seems inescapable, that amino acids are regular constituents of flowering plant nectar and that they are products of secretion by the nectaries.

3. Variations in Amino Acid Concentrations and Amounts

The amino acid concentrations in the 266 nectar samples studied were not uniform and there are some interesting patterns to be seen in relation to the pollination systems of the plants and to their taxonomic positions. There are also some suggestions of evolutionary trends.

Based on data from 23 species, there is a weak positive correlation ($r = +0.37$, assuming a straight line relationship) between amino acid concentration and

12. Evolution of Nectar-Producing Flowers

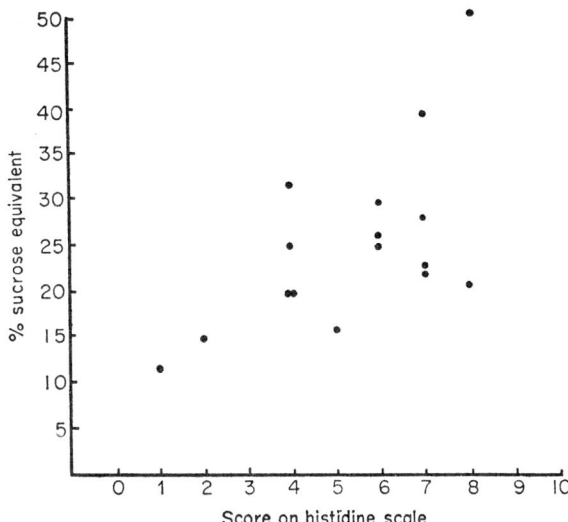

FIG. 1. Scatter diagram showing a positive correlation between sugar concentration (derived from refractive index measurements) and amino acid concentration (score on "histidine scale") for 16 species of flowering plants.

sugar concentration in the same nectar (Fig. 1). On the other hand, as Table II shows, there is no suggestion of a correlation between amount of nectar available in a flower and the concentration of amino acids in it, an important matter because the amounts of nectar produced by flowers of different species vary enormously (in the present study from a small fraction of a microlitre in such a species as *Solidago spathulata* DC. to 33 µl in *Hemerocallis fulva* L.).

When all the "histidine scale" scores for the various species are taken together, it can be seen that a frequency histogram (Fig. 2) approximates to a normal curve, although somewhat skewed to the left. The mode is at 6 on the "histidine scale", corresponding to a 1·56 mM solution of histidine.

TABLE II. Comparison of "histidine scale" scores of species judged to produce little nectar (<0·5 µl) and those producing moderate to abundant amounts

	Species	H.S. Mean	c.v.	P
Little nectar	107	5·52	40%	<0·90
Moderate–abundant nectar	159	5·57	37%	

H.S. = "histidine scale"
c.v. = coefficient of variation
P = probability that difference between means is due to chance

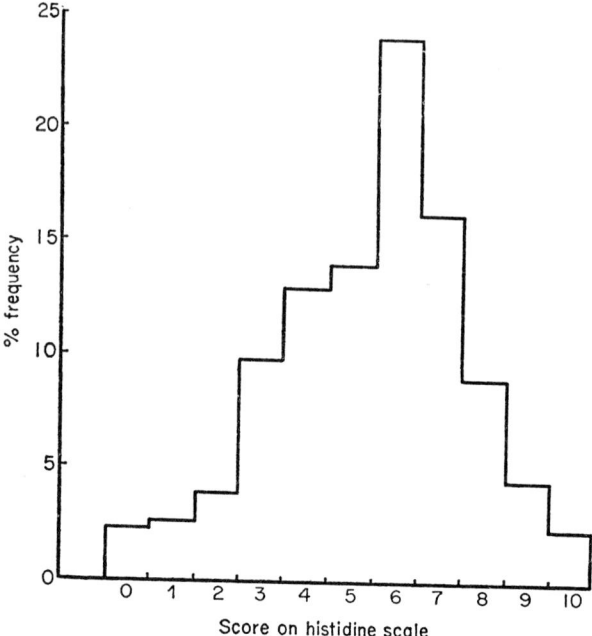

FIG. 2. Histogram of the percentage frequency of each score on the "histidine scale" in the sample of 266 flowering plants.

4. Amino Acid Production and Taxonomy

Although the sample of available species was chosen so as to avoid prejudicing the results by undue concentration upon only a few families, there are sufficient examples from certain families to allow tentative taxonomic conclusions to be drawn. Table III shows the families that gave the highest and the lowest average scores on the "histidine scale", as well as the range seen within each family.

It is notable that the families showing the highest scores tend to be the ones usually thought of as relatively "advanced" in any phylogenetic scheme, while the ones at the foot of the table are less obviously "advanced". Nevertheless, even *Drimys winteri* Forst. (Winteraceae) and *Liriodendron tulipifera* L. (Magnoliaceae) produce nectar containing amino acids (scores of 2 and 7, respectively) and those members of the Ranunculaceae and Palmae that were tested are well supplied. Consequently, it may be suggested that early angiosperm nectar probably contained amino acids but there may have been selection for an increase in their concentration or consistency of production as angiosperm evolution has proceeded.

It is not possible to arrange angiosperm families in any linear order that can be called a phylogenetic sequence (so that "histidine scale" scores could be cor-

TABLE III. Families showing the highest means and the lowest means on the "histidine scale"

	Spp.	H.S. Mean	Range
Asclepiadaceae	8	8·4	6–10
Liliaceae	8	7·4	6–10
Campanulaceae	6	7·0	5–9
Leguminosae	8	6·9	5–9
Amaryllidaceae	11	6·9	5–9
Compositae	25	6·3	4–8
Rosaceae	14	3·9	0–7
Myrtaceae	8	3·1	1–4
Saxifragaceae	6	2·7	1–4
Caprifoliaceae	6	2·2	1–3

related with positions in the sequence). However, there is general agreement that certain plant characters show "primitive" and "advanced" states (Davis and Heywood, 1963, etc.). It may be useful to take these characters one at a time and, in each case, compare the "histidine scale" scores of the species showing the "advanced" state with those of the ones in the "primitive" condition. As an example, Fig. 3 shows a percentage frequency histogram for the scores of woody plants (trees and shrubs) compared with a histogram for the scores of the herbaceous ones.

Although both histograms have bell forms, their shapes are clearly different. That for the woody plants is more clearly platykurtic and shows a higher proportion of values to the left of centre, while that for herbaceous plants, with a higher mode can be described as leptokurtic and has a higher proportion of values to the right of centre. These features suggest that the evolution of "more advanced" taxa has been accompanied by selection for the consistent production of higher concentrations of amino acids in nectar.

Similar pictures are presented by the other comparisons (which are more conveniently presented in tabular form—Table IV). In each case, the plants showing the more "advanced" character give a higher mean score on the "histidine scale". Furthermore, the coefficient of variability is lower for the "advanced" group in every case but one (where it is equal to that for the "primitive" group). In the last column of the table, "P" indicates the probability that the difference between the means in any pair is due to chance. It can be seen that with only two exceptions (both concerned with the gynoecium) the differences are very significant.

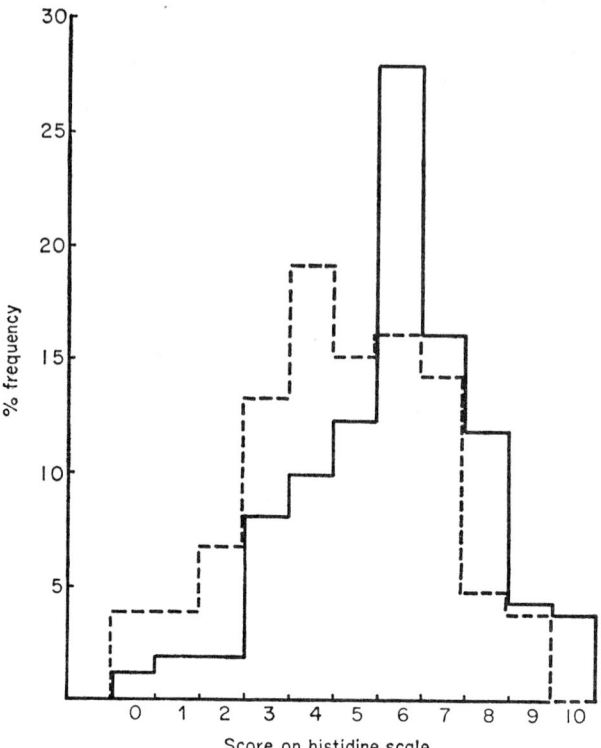

FIG. 3. Histograms of the percentage frequency of each score on the "histidine scale" for 92 species of woody plants (- - -) and 174 species of herbaceous plants (———).

These data are consistent with the thesis that, as the evolution of specialized pollination has proceeded, selection has taken place for the consistent production of amino acids in relatively high concentrations. An assumption, here, is that this has occurred because free amino acids in nectar are directly beneficial to the insects and other animals concerned (and the amino acids from pollen ingested by female *Heliconius* butterflies do appear to pass directly into eggs—Gilbert, 1972).

If we are allowed the assumption, the *implication* is that at least some of the animals that visit the flowers are able to detect the presence of amino acids that they need and that they take nectar selectively from the appropriate flowers. However, even if the visitors should not be sensitive to the concentration of amino acids present (let alone their separate chemical identities), they are known to be sensitive to the concentration of sugars and, as there is some correlation between sugar and amino acid concentrations, selection for one inevitably means some selection for the other.

12. Evolution of Nectar-Producing Flowers

TABLE IV. Comparisons of mean "histidine scale" scores for species grouped according to whether they show "primitive" or "advanced" states of a particular character

	No. of spp.	H.S. Mean	c.v.	P
Woody	92	4·45	48%	≪0·001
Herbaceous	174	6·00	32%	
Open nectar	73	4·88	48%	≃0·01
Concealed nectar	193	5·66	36%	
Polypetalous	105	4·75	45%	≪0·001
Sympetalous	161	5·90	34%	
Actinomorphic	158	5·15	42%	≃0·001
Bilateral	108	5·99	34%	
Many stamens	47	4·51	51%	<0·005
Few stamens	219	5·66	37%	
Polycarpous	24	4·88	39%	≃0·10
Syncarpous	292	5·54	39%	
Hypo- and peri-gynous	168	5·29	42%	<0·60
Epigynous	98	5·46	37%	

In any case, it is desirable to look separately at the various kinds of pollination system in relation to amino acid availability.

(a) *Pollination by bees (melittophily) and by butterflies (psychophily)*. Bees have a major source of amino acids in the pollen they collect and this also provides a wide range of individual amino acids. Consequently, amino acids in nectar may not be essential for the welfare of the bees that collect it nor to the brood that they nurse. As pointed out previously, proteins may also be digested by bees and these can be supplied by pollen as well as being detectable in the nectar of some flower species.

Therefore, the wide ranges (and large coefficients of variation) shown by short-tongue and long-tongue bee-flowers in Table V are not surprising, nor is the fact that the mean "histidine-scale" score for each of these kinds of flower is well below the mode (6) for the whole sample of flowering plants (Fig. 2).

Our chromatograms show that a wide variety of amino acids was available in the bee-flowers whose nectar was examined. All of the ten essential amino acids (Haydak, 1970), together with the three that are nearly essential, are available from one species or another, although tryptophan, methionine and threonine may be rather hard to come by.

Some plants, including many of the Compositae, are too freely visited by both bees and butterflies to be categorized as either "bee-flowers" or "butterfly-flowers". Therefore, they are grouped separately in our study as "bee and butterfly-flowers". These flowers give a significantly higher mean score on the "histidine scale" than the "bee-flowers" do (Table V). The mean score for "butterfly-flowers" is even higher (again, significantly so). In both cases the coefficient of variation is much less than that for the "bee-flowers".

TABLE V. Comparisons of mean "histidine scale" scores for species showing adaptations to bees and to butterflies as pollinators

	Fams.	Spp.	H.S. Mean	c.v.	P
Short-tongued bee	33	75	4·68	44%	<0·60
Long-tongued bee	9	20	5·05	54%	
Short- and long-tongued bee	38	95	4·76	47%	≪0·001
Bee + butterfly	19	44	6·02	23%	≪0·001
Butterfly	22	41	6·68	21%	<0·05

These results suggest that amino acids in nectar may be more important for the butterflies than they are for bees—which is not surprising in view of the fact that, with the exception of *Heliconius*, no butterfly is known to make use of pollen and none is known to produce a proteolytic enzyme (so that proteins in nectar would not be useful to them).

An important question is whether an amino acid concentration such as the "butterfly-flower" nectars show has any significance as far as a butterfly-visitor is concerned. It is possible to give a rough answer by making some calculations based upon Gilbert's (1972) work on *Heliconius* butterflies and the amino acids they get from pollen. One of these butterflies collects about 1·5 mg of pollen in a day. From the biochemical studies of Linskens and Schrauwen (1969) we know that 1·5 mg of *Petunia hybrida* Vilm. pollen will release about 783 nmol of amino acids in two hours into a 10% sucrose solution (which simulates nectar). Gilbert has assumed that these figures apply to other kinds of pollen, as well, and we can go along with the assumption for this calculation. He has also assembled evidence that this amount of amino acids each day in the diet of a *Heliconius* butterfly can have a profound effect on its life-span and reproductive output (Gilbert, 1972, pp. 1405–1406).

Based on a score of 6·68 on the "histidine scale" (the mean for our "butterfly-flowers", 783 nmol is contained in 298 μl of nectar. From our measurements, this amount of nectar can be produced by 12 flowers of *Asclepias curassavica* L. (Asclepiadaceae), 15 flowers of *Dianthus barbatus* L. (Caryophyllaceae) or 25 flowers of *Phlox paniculatus* L. (Polemoniaceae). These are all well-recognized "butterfly-flowers". It should be emphasized that extra visits are not necessary to collect the amino acids; the sugar-collecting flower visits perform this function and it should also be emphasized that the amount of amino acid in the calculation is one that Gilbert (1972) believes to have a profound effect on *Heliconius*; less than this amount could still be very useful to many insects.

The chromatograms of nectar from 27 species of flowers known to be visited by butterflies again show a wide range of amino acids to be available. However, once again tryptophan and methionine may be the hardest to come by. The "butterfly-flowers" also vary widely in the numbers of amino acids that they provide in their nectars. For example, *Kentranthus ruber* (Valerianaceae) provides only four; the nectar of *Dianthus barbatus* (Caryophyllaceae) contains at least twelve.

A butterfly visiting only one species would not pick up a balanced supply of the various amino acids necessary for protein-building; however, butterflies generally do not restrict their visits to a single plant species. If a butterfly's visits to several different flowers should be a random matter, they might still serve the function of assembling a range of amino acids although these random visits would not help *selection* for amino acid production in the flowers. However, there is increasing evidence that flower-visiting by butterflies is not on a random basis; rarely do they visit flowers with inappropriate structure or inadequate nectar consistency or amount. Butterflies of different species have different colour preferences (Dronamraju, 1960; Levin, 1972) and the outline of the flowers may greatly affect their attractiveness (Levin, 1969).

If bees should also be affected by amino acids in nectar, their considerable flower-fidelity could be useful in the selective process. Nearly 50 years ago, Robertson (1926) suggested that butterflies were not responsible for the selection of the basic characteristics of butterfly-flowers but that they have appropriated flowers that evolved under the influence of long-tongued bees. If such be the case, the butterflies would seem to have added extra selection for amino acid concentration *after* appropriating the flowers.

(b) *Pollination by moths (phalaenophily and sphingiphily)*. Relatively few "moth-flowers" were available for sampling, but those that were examined could be divided into two groups; those visited by non-hovering moths (such as the Noctuidae, Zygaenidae, Pyralidae, etc.) and those more obviously adapted to the swift-flying hawkmoths that usually hover when visiting flowers. Table VI shows that the former group score on the "histidine scale" very much like the "butterfly-flowers", a result that is not surprising in view of the similarities in life-style of their visitors. Amino acids would be likely to be as useful to these largely nocturnal moths as to the diurnal butterflies.

On the other hand, the situation appears to be different with the "hawkmoth-flowers". Some of these produce nectar that is rich in amino acids while in

TABLE VI. Comparisons of mean "histidine scale" scores for species showing adaptations to butterflies and to moths as pollinators

	Fams.	Spp.	H.S. Mean	c.v.	P
Butterfly	21	41	6·68	21%	
Hawkmoth	12	19	5·05	50%	<0·20
Non-hovering moth	6	11	6·55	23%	

others there is very little and this is reflected in the large coefficient of variation in the "histidine scale" scores (Table VI). Possibly the larvae of the hawkmoths are more voracious than their counterparts of the non-hovering moths, laying in a better store of nitrogenous materials early in life. In any case, the hawkmoth imagines need and get large quantities of nectar from the flowers they visit on their energy-expensive, darting flights (Heinrich, 1971). If they were to absorb amino acids at a rate in any way comparable with that of their sugar intake it is possible that amino acid toxicity could result (see discussion in House, 1965, pp. 776–777).

(c) *Pollination by birds (ornithophily)*. Among "bird-flowers" there are evidences of independent evolution of bird-flower relationships in the warmer regions of the New World and Old World (Baker, 1973), with adaptation primarily to hummingbirds in the former and to non-hovering birds in the latter.

Indeed, the results from our survey (Table VII) *appear* to indicate a difference in the "histidine scale" means for the two groups of plant species. Those

TABLE VII. Comparison of mean "histidine scale" scores for species showing adaptations to Old and New World birds as pollinators

	Fams.	Spp.	H.S. Mean	c.v.	P
New World bird	22	34	5·97	26%	≪0·001
Old World bird	10	15	3·53	47%	

characterized as "Old World bird-flowers" are adapted to visits by Nectarinidae, Meliphagidae and other birds that visit flowers in Africa, Asia and Australasia. The "New World bird-flowers" that were examined are exclusively "hummingbird-flowers" from *North* America. The difference between the low mean for the former and the relatively high mean for the latter appears to be very significant, but words of caution are necessary. Dr Paul A. Opler and Ms Kathleen Keeler have been making observations on tropical plants in Costa Rica. Examination of their results to date indicate that there, in the Neotropics, "hummingbird-flowers" often show little amino acid in the nectar. These results are understandable because hummingbirds as well as the "Old World flower-visiting birds" have a useful alternative source of protein-making materials in the insects that they catch. In Peru, hummingbirds of the species *Oreotrochilus estella* (Lafresnaye & d'Orbigny) appear to take pollen from nectarless flowers of *Chuquiraga spinosa* D. Don (Compositae—tribe Mutiseae) (Carpenter, 1972).

In the tropics of both hemispheres, birds and flowers have enjoyed a long relationship and it may be expected that they have become mutually adjusted. On the other hand, the North American "hummingbird-flowers" may have been improvised more recently (Grant and Grant, 1968). Few North American flowering plant genera contain more than an occasional species that has formed a pollinatory relationship with hummingbirds (Grant and Grant, 1968, p. 64), and it may be that in their nectar they are carrying the relics of their previous insect-pollination adaptations. An exception would be the genus *Zauschneria* (Onagraceae) in which all species have "hummingbird-flowers"; here the "histidine scale" scores are appropriately rather low (4 and 5 for two varieties of *Z. californica* Presl).

(d) *Pollination by flies (myophily) and by beetles (cantharophily)*. Once again, it may be desirable to make a distinction between two kinds of plant, a distinction that is usually made in handbooks on pollination (e.g. Faegri and van der Pijl, 1971). The first group contains those plants whose flowers or inflorescences are adapted to flies in a rather general way, with wide-open nectar display. Such plants are to be found in a number of families, for example in the Umbelliferae. The second group is more restricted, being represented in our survey only by some members of the Asclepiadaceae and Aristolochiaceae. This second group have developed what Faegri and van der Pijl call "sapromyophily" and, with flowers that look and smell like animal remains or excreta, have succeeded in luring carrion- and dung-flies away from their usual food-sources, so that they may even lay their eggs in the flowers.

The first group, the "generalized fly flowers" show widely variable "histidine scale" scores (Table VIII); apparently they have not been subjected to selection by discriminating visitors for any consistent amino acid production. Their visitors may or may not give them attention; selfing or geitonogamy may suffice as successful pollination. Their position may be much the same as that of "bee-flowers" in that their visitors have an alternative source of nitrogen nutrition at the same flowers in pollen, which such insects as hover-flies (Syrphidae) and other Diptera utilize readily, as well as animal sources of protein and amino acids.

In the second group, the "sapromyophilous" flowers, pollen-production is limited in amount and, in any case, the carrion- and dung-flies must be lured away from the animal materials that they usually utilize for protein and amino acids. This diversion of the flies might be expected to be most successfully accomplished if the flowers provide a source of amino acids for them in their nectar. Table VIII shows that these flowers are consistently high in their

TABLE VIII. Comparison of mean "histidine scale" scores for species showing generalized and specialized adaptations to flies as pollinators

	Fams.	Spp.	H.S. Mean	c.v.	P
Generalized fly	8	34	3·77	57%	$\ll 0.001$
Carrion/dung fly	2	8	9·25	12%	

"histidine scale" scores (when they do produce nectar at all, for deceit is rampant among these species, with shiny surfaces simulating active nectaries). Not only is the nectar strong in amino acids, the variety is also remarkable; at least twelve amino acids are present in the nectar of *Caralluma chrysostephana* Brgr. (Asclepiadaceae).

All "fly-flowers" appear to share at least one feature of their nectar. Produced in rather small quantity, it is extremely tacky in consistency. The mouthparts of flies appear to be capable of dealing with this viscous fluid (which could not be utilized by such insects as butterflies with long, narrow proboscises).

A particularly interesting group of "fly-flowers" are those that are visited by mosquitoes. They are now being investigated in our tropical studies (and they could also be looked at with advantage in the Arctic where they are important flower-visitors; cf. Hocking, 1953, 1968, 1971). For many female mosquitoes a blood-meal is a necessary preliminary to successful egg-laying; in other species, plant sap and nectar appear to supply whatever is needed (Spielman, 1971). Consequently, amino acids or proteins are being looked for in these plant sources. Some of the flower species listed by Hocking (1953, 1968), Sandholm and Price (1962) and Thien (1969) as being visited by mosquitoes in northern latitudes are included in our survey already and members of the same genus are represented in other cases. The amino acid concentrations of their nectars appear to be above average.

We have not yet made any specific attempt to study the amino acids in the nectar of "beetle-flowers". In part this has been due to the difficulty of distinguishing such a class from "bee-" and "fly-flowers", in part it is because "beetle-flowers" are often lacking in nectar, and mostly it is because such a study would better be made in the tropics—where an investigation of all aspects of the significance of amino acids in nectar is being carried out in Costa Rica as part of a study of the reproductive biology of wet- and dry-forest ecosystems.

(e) *General conclusions on pollinator relationships.* In general, we feel that an *a priori* case has been made that amino acids in the nectar from flowers play a significant role in the development and maintenance of flower-visiting behaviour of insects to whom such a supply could be advantageous. No longer can nectar be treated as simply sugar-water and future investigations in anthecology must take into account more than the energy requirements of the visitors.

FUTURE INVESTIGATIONS

As this work is continued, emphasis will be placed upon analyses made on an ecosystem basis. We believe that we shall have most success in this by giving attention to butterflies, where the complications introduced by the use of other food sources than nectar are minimized. All the species of flowering plant that are visited by one species of butterfly in a particular geographical location will be examined to see what they might contribute to a balanced amino acid supply.

This is already being attempted in Costa Rica, and there some observations upon which we can build have already been made. Thus, one of the anthecologists' favourite cases of mimicry leading to pollinator-sharing is to be seen. It concerns several species of butterfly (from at least three families: the Papilionidae, Nymphalidae and Pieridae). Together, these visit a number of flowering plants which, although only very distantly related taxonomically, share a yellow-red flower-colour pattern that is probably the result of convergent evolution. These butterflies visit *Lantana camara* L. (Verbenaceae), *Asclepias curassavica* (Asclepiadaceae), and *Senecio hoffmannii* Klatt (Compositae) as well as at least two other composites and, possibly, other plants as well. *Epidendrum radicans* Pav. (Orchidaceae) also receives visits from these butterflies, but this plant appears to be a "cheater", with little or no nectar in its flower-spur (van der Pijl and Dodson, 1967). Each one of the plant families involved in this pollinator-sharing has scored high on the "histidine scale" in our studies, and it should be particularly interesting to see what these species which share the butterflies' attention contribute to the total amino acid range.

Studies of butterfly pollination in an ecosystem context will also be carried out in California, where relatively undisturbed ecosystems (vitally necessary if the evolutionary aspects of the relationships are to be investigated) are still available in some montane areas.

One group of flower visitors that have not been considered at all in the present report but which are being included in the tropical studies are the nectar-lapping bats—an old concern of ours (reviewed by Baker, 1973). These animals need substantial quantities of protein-building materials; some take insects and other animals, some lap blood, some chew fruit and some take pollen. The flowers that they visit are being examined.

In all future studies, proteins will be looked for as well as free amino acids, so that a well-rounded picture may be obtained. Where pollen is utilized, as by bees, analyses will be made of both the free amino acids and those available from protein breakdown to see if there is any complementarity between what they supply and those which are available from the nectar.

Nectar, which, as mere sugar-water, seemed rather less than "the drink of the gods", may be on the way to recovering some of its lost status as a gourmet food item—for anthophilous animals.

REFERENCES

Auclair, J. L. and Jamieson, C. A. (1948). A qualitative analysis of amino acids in pollen collected by bees. *Science, N.Y.* **108**, 357–358.

Baker, H. G. (1961). The adaptations of flowering plants to nocturnal and crepuscular pollinators. *Q. Rev. Biol.* **36**, 64–73.

Baker, H. G. (1973). Evolutionary relationships between flowering plants and animals in American and African Tropical Forests. *In* "Tropical Forest Ecosystems in Africa and South America: a Comparative Review" (B. J. Meggers, E. S. Ayensu and W. D. Duckworth, eds). Smithsonian Institution Press, Washington, D.C.

Baker, H. G. and Hurd, P. D., Jr. (1968). Intrafloral ecology. *A. Rev. Ent.* **13**, 385–414.

Bänziger, H. (1971). Bloodsucking moths of Malaya. *Fauna* **1**, 4–16.

Carpenter, L. (1972). Ph.D. thesis (Zoology). University of California, Berkeley.

Davey, K. G. (1965). "Reproduction in the Insects." Freeman, San Francisco.

Davis, P. H. and Heywood, V. H. (1963). "Principles of Angiosperm Taxonomy." Van Nostrand, Princeton, N.J.

Dkonamraju, K. R. (1960). Selective visits of butterflies to flowers: a possible factor in sympatric speciation. *Nature, Lond.* **186**, 178.

Engelmann, F. (1970). "The Physiology of Insect Reproduction." Pergamon Press, Oxford.

Faegri, K. and van der Pijl, L. (1971). "The Principles of Pollination Ecology." Pergamon Press, Oxford.

Ford, E. B. (1945). "Butterflies." Collins, London.
Gilbert, L. E. (1972). Pollen feeding and reproductive biology of *Heliconius* butterflies. *Proc. natn. Acad. Sci. U.S.A.* **69**, 1403–1407.
Grant, K. A. and Grant, V. (1968). "Hummingbirds and their Flowers." Columbia University Press, New York.
Haydak, M. H. (1970). Honey bee nutrition. *A. Rev. Ent.* **15**, 143–156.
Heinrich, B. (1971). Thoracic temperature regulation in the sphinx moth, *Manduca sexta*. Parts I and II. *J. exp. Biol.* **54**, 141–166.
Heinrich, B. and Raven, P. E. (1972). Energetics and pollination ecology. *Science, N.Y.* **176**, 597–602.
Hocking, B. (1953). The intrinsic range and speed of flight of insects. *Trans. R. ent. Soc. Lond.* **104**, 223–345.
Hocking, B. (1968). Insect–flower associations in the high Arctic with special reference to nectar. *Oikos* **19**, 359–387.
Hocking, B. (1971). "Six-legged Science." Schenkman, Cambridge, Mass.
House, H. L. (1965). Insect nutrition. In "The Physiology of the Insecta" (M. Rockstein, ed.), Vol. 2, pp. 769–857. Academic Press, New York.
Janzen, D. H. (1971). Euglossine bees as long-distance pollinators of tropical plants. *Science, N.Y.* **171**, 203–205.
Kilby, B. A. (1963). The biochemistry of the insect fat body. In "Advances in Insect Physiology", Vol. 1, pp. 112–174. Academic Press, London.
Klots, A. B. (undated). "The World of Butterflies and Moths." McGraw-Hill, New York.
Levin, D. A. (1969). The effect of corolla color and outline on inter-specific pollen flow in *Phlox*. *Evolution* **23**, 444–455.
Levin, D. A. (1972). The adaptedness of corolla-color variants in experimental and natural populations of *Phlox drummondii*. *Am. Nat.* **106**, 57–70.
Linskens, H. F. and Schrauwen, J. (1969). The release of free amino-acids from germinating pollen. *Acta bot. neerl.* **18**, 605–614.
Lüttge, U. (1961). Über die Zusammensetzung des Nektars und den Mechanismus seiner Sekretion. I. *Planta* **56**, 189–212.
Lüttge, U. (1962). Über die Zusammensetzung des Nektars und den Mechanismus seiner Sekretion. II. *Planta* **59**, 108–114.
Macior, L. W. (1971). Co-evolution of plants and animals—systematic insights from plant–insect interactions. *Taxon* **20**, 17–28.
Maurizio, A. (1962). From the raw material to the finished product: honey. *Bee World* **43**, 66–81.
Müller, H. (1873). Die Befruchtung der Blumen durch Insekten. Engelmann, Leipzig.
Müller, H. (1883). "The Fertilisation of Flowers" (translated and edited by D'A. W. Thompson). Macmillan, London.
Percival, M. S. (1962). Types of nectar in angiosperms. *New Phytol.* **60**, 235–281.
Percival, M. S. (1965). "Floral Biology." Pergamon, Oxford.
van der Pijl, L. and Dodson, C. H. (1967). "Orchids and their Pollinators." University of Miami Press, Miami, Fla.
Pryce-Jones, L. (1950). The composition and properties of honey. *Bee World* **31**, 2–6.
Robertson, C. (1926). Quoted from Percival (1965). No reference given there.

ROBERTSON, C. (1927). Curiosities of anthecology. *Science, N.Y.* **65,** 472.
SANDHOLM, H. A. and PRICE, R. D. (1962). Field observations on the nectar feeding habits of some Minnesota mosquitoes. *Mosquito News* **22,** 346–349.
SCHUETTE, H. A. and BALDWIN, C. L., Jr. (1944). Amino-acids and related compounds in honey. *Food Res.* **9,** 244–249.
SMITH, I. (1969). "Chromatographic and Electrophoretic Techniques," Vol. 1.— "Chromatography" (3rd edition). Heinemann, London.
SPIELMAN, A. (1971). Bionomics of autogenous mosquitoes. *A. Rev. Ent.* **16,** 231–248.
THIEN, L. B. (1969). Mosquito pollination of *Habenaria obtusata* (Orchidaceae). *Am. J. Bot.* **56,** 232–237.
VOGEL, S. (1969). Flowers offering fatty oil instead of water. *Abstracts XI Int. Bot. Congr., Seattle,* p. 229.
WHITE, J. W. (1957). The composition of honey, *Bee World* **38,** 57–66.
WIGGLESWORTH, V. B. (1953). "The Principles of Insect Physiology." (4th edition.) Methuen, London.
WOLF, L. L. and HAINSWORTH, F. R. (1971). Time and energy budgets of territorial hummingbirds. *Ecology* **52,** 980–988.
WOLF, L. L., HAINSWORTH, F. R. and STILES, F. G. (1972). Energetics of foraging: rate and efficiency of nectar extraction by hummingbirds. *Science, N.Y.* **176,** 1351–1352.
ZIEGLER, H. (1956). Untersuchungen über die Leitung und Sekretion der Assimilate. *Planta* **47,** 447–500.
ZIMMERMAN, M. (1953). Papierchromatographische Untersuchungen über die pflanzliche Zuckersekretion. *Ber. schweiz. bot. Ges.* **63,** 402–429.
ZIMMERMAN, M. (1954). Über die Sekretion saccharoespaltenden Transglukosidasen im pflanzlichen Necktar. *Experientia* **10,** 3.

13 | Ecological Factors of Importance to *Columnea* Taxonomy

BRIAN MORLEY

Department of Agriculture and Fisheries, National Botanic Gardens, Dublin, Ireland

Abstract: Jamaican species in *Columnea* sect. *Columnea* are usually taxonomically untroublesome when epiphytic, but those found terrestrially, often in habitats disturbed by man, raise taxonomic problems. The factors responsible include breeding population size, niche availability, different amounts of gene flow and hybridization where two or more species cohabit.

In contrast, central American species in *Columnea* sect. *Collandra* give rise to taxonomic problems both when epiphytic and when terrestrial, partly due to a different breeding system from species in sect. *Columnea*.

Leaf anisophylly, a traditional taxonomic character in sectional classification of *Columnea* sensu lato, is discussed and suggested to have more relevance to ecology than taxonomy.

Corolla morphology, another traditional sectional character, is discussed in relation to pollination ecology and crossability data, and held to be a reliable taxonomic character on the basis of present knowledge.

INTRODUCTION

Ecological data can contribute to a better systematic understanding of phanerogams, but in the majority of cases, and *Columnea* (Gesneriaceae) is no exception, little detailed ecological work has been done. From a taxonomic viewpoint many groups like *Columnea* are so imperfectly known that it is necessary to devise simple schemes of classification, usually based on morphological characters, to differentiate taxa in preparation for more detailed systematic study. After such study the definition of various taxa may require change.

Perhaps the most important results can be expected from studies on the interrelationship between ecology and taxonomy at the population and individual level, because ecological pressures are directed primarily at the individual and adaptive variation first appears and becomes fixed within the individual. Whether it is profitable to propose schemes of classification below intraspecific rank, at microevolutionary level, is doubtful. It is more urgent to collect and integrate data, such is the state of our knowledge.

The ecological and physiological characteristics of epiphytic plants have been documented by a number of workers including Schimper (1903), Hosokawa (1943, 1949, 1953) and Richards (1952), but some systematic aspects of epiphytism in a particular genus, *Columnea*, is the subject of the first part of this contribution. The second and third parts concern the sectional taxonomic characters shoot morphology and corolla morphology as they may relate to habitat and pollination ecology respectively.

Relatively little autecological work has been done on vascular epiphytes and this is reflected in epiphyte terminology. For survey purposes the synecologist has distinguished between epiphytes and climbers, having subdivided the former into sun or photophytic, shade or skiophytic and extreme xerophilous epiphytes (Richards, 1952; Barkman, 1958). Climbers have been separated into "large climbers" and "small climbers" by Richards (1952), or photophytic and skiophytic climbers by Grubb *et al.* (1963). These categories break down as far as the autecology of *Columnea* is concerned, because species such as *C. hirsuta* Sw. may begin life as a skiophytic climber and later become an epiphyte, and others such as *C. rutilans* Sw. may be skiophytic epiphytes only to become terrestrial plants in certain conditions. Some species such as *C. urbanii* Stearn behave like photophytes and skiophytes in different ecological conditions. The small leaved and pendulous species *C. allenii* Mort. and *C. arguta* Mort., for example, most often appear as epiphytes, but it is not possible to generalize and say a species is always epiphytic.

TAXONOMIC ASPECTS OF EPIPHYTISM IN COLUMNEA

1. *Observations*

The elementary ecological work done in this group, and that by a taxonomist, concerns Jamaican columneas. They grow in areas experiencing an annual average temperature of about 27°C, areas receiving over 10 cm of rain in the driest month of March and over 38 cm in the wettest month of October, and areas experiencing an annual average relative humidity of about 75%. Given a relative humidity between 80 and 90%, temperatures up to 49°C were tolerated with only minor leaf scorch in cultivation. Leaf shedding was seen to begin at a relative humidity of about 50%.

(a) *Epiphytic populations.* When epiphytic, Jamaican columneas have a population density which is low and scattered in comparison with a terrestrial population. Epiphytic columneas are not found on young trees with trunks less than about 25 cm diameter at chest height, often because the bark is too smooth. The epiphytes are rooted in termitaria, pockets of humus, or amongst the mosses and ferns covering the bark of trees: columneas were not seen on palms but were occasionally found rooted on the "trunks" of tree ferns. Quantitative data were not collected, but in dense woodland it is unusual to see more than 7-8 individual columneas from any one observation point: the plants are mostly small. In terrestrial populations 50-60% more plants are visible.

Weston light meter readings 1 m above soil level were taken in undisturbed woodland with a closed canopy near Hardwar Gap in the Blue Mountains on 26 August 1965 at 11 a.m. The readings in Table I are related to the position of columneas above soil level. The readings were taken in full sunlight, and the species was *C. hirsuta*. Rainfall at Hardwar Gap is about 261 cm (altitude 1220 m) and *C. hirsuta* occurs in moist wooded ravines rather than on exposed wooded

TABLE I. Relationship between light intensity and position of columneas above soil level

Species	Light reading (lm ft^{-2})	Position of plant above soil (m)
C. hirsuta at Hardwar Gap	100	3
	100	4
	20	5
	12	7
C. urbanii at Top Hill	1600	0
	1600	0
	1600	0
	800	0

slopes. The position of the plants was clearly correlated with the amount of light filtering through the canopy, although the readings were intended only as a rough guide, radiation meteorology being a specialist subject.

In densely shaded epiphytic habitats there is evidence to show that flowering is reduced in some years because flower primordia seem to be initiated by a rainfall, temperature and light response in Jamaica. Initiation is sometime between November and January in most species and flowering lasts for 2-5 months depending on species (see Table II).

Hummingbirds are the pollen vector and being polytropic feeders, pollen mobility within the breeding population must be dependent on the number of flowers open at any one time. In Jamaica the birds were seen to fly at all levels in the forest canopy so that both terrestrial and epiphytic columneas were visited. There was some indication that *Trochilus polytmus* remembered food sources. Duration of flowering season only becomes important to the breeding system when more than one species cohabit, because Jamaican species differ slightly in flowering time and overlap in flowering can facilitate hybridization under certain circumstances.

In epiphytic habitats there is an observed paucity of niches for *Columnea* propagules of any sort: only two established seedlings were found on tree trunks after eight field trips to undisturbed woodland habitats. Vegetative

TABLE II. Flowering time in Jamaican columneas

Duration (months)	Species	A	S	O	N	D	J	F	M	A	M	J	J	
		+		++	++				+++	+++	+++	+		Precipitation maxima / Average temperature maxima
5	C. hirsuta		×	×	×	×	×	×						
4	C. fawcettii			×	×	×	×	×						
2?	C. rutilans				×	×	×							
4	C. urbanii	×	×		×	×	×							
3	C. argentea			×	×	×	×							
3	C. subcordata				×	×	×	×						
2?	C. hispida			×	×									
3	C. harrisii			×	×	×	×							
2	C. brevipila				×	×	×							
3?	C. jamaicensis	×	×	×						×				
5	C. proctorii		×	×	×	×	×	×						
2	C. urbanii × rutilans hybrid 4				×	×	×							
2	hybrid 7						×	×	×					
		A	S	O	N	D	J	F	M	A	M	J	J	

Where × = flowers seen in nature or on herbarium sheet; + = rainfall and temperature maxima; A S O etc. = months of the year.

propagation from fallen fragments of epiphyte appears to be an important means of spread when sufficient light reaches the woodland floor. When the floor was steeply sloping fragments of *Columnea* were seen to have rolled downhill and become established in significant numbers in *C. urbanii* and *C. fawcettii* (Urb.) Morton.

(b) *Terrestrial populations*. In contrast to epiphytic populations, terrestrial ones develop when the woodland canopy is thinned or destroyed, and if water supply permits, for there is considerable drying of a newly sunlit woodland floor. The relationship between light intensity and position of columneas above the soil in disturbed woodland at Top Hill, Manchester Parish, is shown in Table I. Light readings were made in full sunlight at 10 a.m. on 11 July 1965, and *C. urbanii* was the species involved. No precipitation figures are available for this area but surrounding stations receive 185 cm (Mandeville), 187 cm (Wait a bit) and 240 cm (Bull Head). Most columneas at Top Hill were terrestrial and those growing on bare rock, such as along the tops of drystone walling, had a stunted appearance with short internodes, almost leafless stems, thick bark and small leaves. As many as 20 individuals, some being large plants, could be seen from any one observation point.

The woodland at Top Hill, and in many parts of western Jamaica, grows on honeycomb limestone, so that water supply is of prime importance to succulent plants such as columneas. Removal and thinning of trees by a form of

TABLE III. Distribution of individual columneas and clump size in different niches at Top Hill

Niche	Number of individuals	Clump size
(a) On soil	1	8
(b) On tree	0	0
(c) On soil associated with bromeliad	20	214
(d) On tree associated with bromeliad	4	41

shifting cultivation has led to an altered ground flora where plants more tolerant of water stress begin to appear, such as *Stachytarpheta* sp. and *Lantana* sp. (Verbenaceae) and *Hohenbergia* sp. (Bromeliaceae). The hohenbergias are important as they make it possible for columneas to grow on almost soil-less scree when not growing amongst shrubs or along the tops of walls (Morley, 1970). Few other plants occupy the scree slopes and the roots of the columneas ramify the old moist leaf bases of the bromeliads. Not all bromeliads had columneas associated with them. Table III shows the numbers of individuals and their size in terms of numbers of leading shoots in different niches in parts of area A, Morley (1968) at Top Hill.

Because: (1) the distributions of *C. urbanii*, *C. argentea* Grisebach and *C. rutilans* overlap in the vicinity of Top Hill; (2) the woodland canopy has been disturbed; and (3) a diversity of unoccupied ecological niches continues to be created by agricultural activity, the behaviour of a common pollinator has brought about natural hybridization at least between *C. urbanii* and *rutilans*

(Morley, 1971). The dense terrestrial populations of parents and available niches have increased chances of backcrossing, and backcross progeny have been recognized at Top Hill. Without biosystematic data the columneas at Top Hill present a perplexing group of intergrading taxa.

2. *Discussion*

When an epiphytic population consists of a single taxonomic species the activity of the hummingbird pollen vector, with ensuing xenogamy, assures the genetic integrity of that population. However, poor flowering in a population may mean that gene flow can become weak and that a certain amount of inbreeding can take place from the observed occurrence of geitonogamy. The taxonomic products of such an inbred model population would be isolated taxa differing slightly from the norm, and it may be that the forms of *C. hirsuta* recognized from three different peaks in the Blue Mountains by Urban (1901) as varieties *hirsuta* (var. *genuina* Urb.), var. *concolor* Urb. and var. *pallescens* Urb. are the products of local limited inbreeding.

In Jamaica and central America at least, distinct but interfertile taxonomic species grow beside one another with an apparent absence of hybridization. This may indicate that epiphytism and consequent low population density, coupled with the behaviour of the pollen vector, provide some degree of reproductive isolation. Such isolation may be brought about not only by limiting the mobility of pollen amongst a restricted number of plants, but also by the observed paucity of niches available for seedlings where columneas are epiphytic. Even when a niche exists, intraspecific competition between germinating seedlings may reduce the chances of establishment. There is considerable seedling mortality in cultivation under controlled conditions. Ross and Harper (1972) have noted intraspecific competition between monocotyledon seedlings in high density sowings, and that individual performance was influenced by the density of previously emerged neighbours. *Columnea* seeds are dispersed in a gelatinous matrix which indicates that groups of seeds, not individuals, arrive at the few niches available. There is no observed dormancy so that competition between seedlings is likely.

The larger populations and more abundant flowering of terrestrial columneas at Top Hill suggests that conditions for gene flow are more favourable than in epiphytic populations. Had plants at Top Hill been epiphytic as on nearby Shooters Hill where *C. urbanii* and *C. argentea* occur, there would have been fewer niches to accommodate hybrid progeny, if formed, and a smaller epiphytic parent population on which to draw for backcrossing despite cohabitation of more than one species. No evidence for hybridization was found on

Shooters Hill. It is possible for terrestrial columneas to reclimb trees and become epiphytic in the event of the woodland canopy closing because of stems which in C. *rutilans*, for example, can elongate to 3 m or more and root adventitiously at the nodes.

In summary, small breeding populations of epiphytic columneas and low seedling establishment are seen as two reproductive isolation factors which help permit the coexistence of different interfertile taxonomic species in close proximity. Geographic and topographic isolation factors have played the pre-eminent role in *Columnea* speciation (Morley, 1972). The development of larger terrestrial populations favours hybridization when more than one species cohabit. The loss of the epiphytic habitat at Top Hill is directly correlated with increased taxonomic difficulty of the columneas which grow there.

Most species in sect. *Columnea* vary little in nature and are taxonomically simple. Many columneas are known to have restricted geographical distributions so it is not surprising they show little variation, but others such as *C. scandens* L. are more widespread and exhibit variation (Morley, 1972). *C. nicaraguensis* Oerst. has a more extensive distribution than many species and varies in leaf-shape, and on a smaller scale *C. urbanii* shows variation of leaf-vestiture in Jamaica. The variation patterns seen in species in sect. *Columnea* are unlike those found in species in sect. *Collandra*. Collandras exhibit continuous variation patterns which are rare in sect. *Columnea*, and study indicates that the breeding system of collandras may be responsible (see below). While the loss of the epiphytic habitat can provide taxonomic problems in sect. *Columnea*, the problems seen in sect. *Collandra* have a quite different origin.

LEAF ANISOPHYLLY AS A TAXONOMIC CHARACTER
1. Observations

Most gesneriads including *Columnea* sensu lato have opposite leaf insertion and depending on leaf-length and width it may or may not be possible to accommodate all laminas for light interception along a shoot without some overlap. Presumably for purposes of photosynthetic efficiency, especially in epiphytic niches, a leaf mosaic is developed in species possessing larger leaves where one of the leaves at each node becomes reduced in length. Reduction of leaf-size alternates from one side of the shoot to the other on successive nodes down the axis. In this way the long leaves find accommodation with minimal overlap and the short leaves fill the spaces near the stem between the petioles of the long leaves—a condition known as anisophylly. It is a striking character and early taxonomists used it at sectional level for defining sects. *Collandra* and *Cryptocolumnea*.

Isophylly has never been used in the same way as a sectional taxonomic character and it is confined to those chiefly epiphytic species with leaves less than about 3 cm long and long pendulous shoots. Isophyllous species occur in sects. *Columnea* and *Pterygoloma*. The pairs of leaves are equal in length whatever the shoot posture because laminas all find accommodation without significant overlap.

Many other species have neither short nor excessively long leaves, and occur in sects. *Columnea, Pentadenia, Pterygoloma, Ortholoma* and *Collandra*. Depending on leaf-length and shoot posture, anisophylly varies in these species. Terrestrial plants with vertical shoots have more isophyllous leaves than epiphytic with horizontal or pendent shoots (Morley, in press).

By understanding that anisophylly is an expression of adjustments of the leaf mosaic to light interception, and that the fundamental character determining these adjustments is leaf-size, it is possible to accept different amounts of anisophylly within the same taxon and to regard anisophylly as an ecological rather than taxonomic character. It is oversimplification to suggest that all small-leaved columneas are epiphytic and large-leaved terrestrial, but leaf-size may reflect adaptation to different sorts of niche. That the leaves have adapted to different ecological selection pressures is seen in the anatomical data of Morley (1972).

Species such as *C. arguta, C. allenii, C. oerstediana* Kl. ex Oerst. and *C. billbergiana* Beurl. with small succulent leaves, slender stems and a more or less pendulous habit, appear to be well fitted as sun epiphytes. They have a morphology which should give some tolerance to exposure to bright sunlight, wind and desiccation and they have lightweight stems which provide little wind resistance, and root at the nodes. These ecological conditions extrapolated from morphology are indicated on herbarium sheets in the unfortunate absence of field studies.

Species such as *C. sanguinea* (Pers.) Hanst., *C. aureonitens* Hook., *C. florida* Mort. and *C. purpurata* Hanst. with large and thin leaves in relation to surface area might be considered vulnerable to conditions favouring rapid transpiration, bright sunlight or high winds. Massive stems about 3 cm thick are required to support the foliage and the stems are brittle and succulent when young. These characters suggest ecological conditions of shelter, wetness, humidity, perhaps shade, and terrestrial niches where vegetative propagation can occur by basal suckering as observed in these taxa. Herbarium sheet data indicate that large-leaved taxa do occur terrestrially, but that they may also be found as epiphytes in suitable locations such as beside or overhanging forest streams. Large-leaved species could be regarded as shade epiphytes.

Species with intermediate-sized leaves are usually either vining or shrubby but not pendulous. Many show morphological plasticity of the shoot system in a way not found in small- or large-leaved species. Depending on habitat the plant may be a small addressed climber on trees, an erect scandent shrub, compact shrub, or a pendent epiphyte.

2. *Discussion*

The thesis that leaf-size, and therefore leaf-anisophylly is basically an ecological character leads to the acceptable existence of isophyllous and anisophyllous taxa in the same section, e.g. in sect. *Columnea C. allenii* and *C. wilsoni* Wiehler. The use of anisophylly as a sectional character has the result of splitting taxa into parallel groups as in the pairs sect. *Stygnanthe* and *Collandra*, or *Columnea* and *Cryptocolumnea*. Leaf anisophylly has been used as a critical character in the definition of sections such as *Collandra* and *Cryptocolumnea* but then ignored in other sections such as *Stenanthus* and *Ortholoma* in favour of leaf vestiture, itself of doubtful sectional significance. If a character is to be used in an acceptable intrageneric classification it must apply to each taxon defined. Leaf-size (anisophylly) does not fulfil these requirements but corolla-morphology does.

TAXONOMIC ASPECTS OF POLLINATION ECOLOGY IN *COLUMNEA* SENSU LATO.

1. *Observations*

A traditional character used in sectional classification, corolla-morphology, is perhaps better regarded as a suite of characters some of the components of which are: (1) corolla-vestiture; (2) tube-length; (3) tube-width; (4) curvature of tube walls; (5) length of unfused portion of corolla-dissection; (6) relative degree of lobe-fusion; (7) relative posture of lobes; and (8) stamen withdrawal mechanism. Figure 1 and Table IV summarize the main sections recognized on the basis of corolla morphology.

The pollen vector of only certain species in sect. *Columnea* has been studied (Morley, 1966, 1971) if the indirect evidence provided by Vogel for sect. *Collandra* is ignored (see below). But in the absence of field work it is reasonable to assume that the great diversity of corolla-morphology is, in part at least, a response to selection by pollen-vector visits. The question remains whether one kind of animal or many are involved. Hybridization may have contributed to the existing diversity of corolla-shapes.

Some workers already discount the reliability of corolla-morphology in sectional *Columnea* classification, claiming that its use simply serves to group species under their respective pollen-vector types. This is of course possible, but as the pollen-vectors are unknown there is no field observation to support the

FIG. 1. Flower shape in *Columnea* sensu lato. 1. *C. aureonitens* Hook. (sect. *Collandra*); 2. *C. mira* Morley (sec. *Ortholoma*); 3. *C. jamaicensis* Urb. (sect. *Pterygoloma*); 4. *C. kucyniakii* Raymond (sect. *Pentadenia*); 5. *C. aurantiaca* Decne. (sect. *Pentadenia*); 6. *C. incarnata* Mort. (sect. *Pentadenia*); 7. *C. urbanii* Stearn (sect. *Columnea*); 8. *C. tomentulosa* Mort. var. *tulae* (Urb.) Morley (sect. *Columnea*). The line scale by each flower represents 1 cm.

TABLE IV. Sectional classification of *Columnea* L.

Section	Synonymy
Collandra (Lem.) Benth. & Hook. (1876)	*Stygnanthe* (Hanst.) Benth. & Hook. (1876)
Ortholoma Benth. (1846)	*Stenanthus* (Oerst. ex Hanst.) Fritsch (1893)
Pterygoloma (Hanst.) Fritsch (1893)	*Trichantha* (Hook.) Kuntze (1904)
	Collandra sensu Morton, *pro parte* (1971)
	Stygnanthe sensu Morton, *pro parte* (1971)
Pentadenia (Planch.) Benth. & Hook. (1876)	
Columnea	*Cryptocolumnea* (Hanst.) Benth. & Hook. (1876)

view, and as Vogel may have shown, there is an equal possibility that sections other than *Columnea* are pollinated by hummingbirds. Species such as *C. incarnata* Mort. (sect. *Pentadenia*) may be one of the few non-ornithophilous taxa, perhaps pollinated by bats (see Fig. 1).

The floral mechanism of the two most commonly cultivated sections has been studied. Section *Collandra* differs from *Columnea* in having tubular flowers with an almost actinomorphic limb instead of a funnel-shaped tube and zygomorphic limb with an obliquely dissected throat. In nature, and the absence of cross pollination, collandras may be autogamous but columneas are not. This is seen in cultivation under controlled conditions. Both set fruit normally if artificially pollinated. Both *Collandra* and *Columnea* flowers are protandrous, but they differ in herkogamy (the spatial separation of anthers and stigma during anthesis). The withered stamens of collandras, still retaining pollen if the flower is unvisited, retract horizontally by coiling of the filaments to make way for the receptive stigma in the narrow confines of the corolla-tube. The pollen is not waxy but is adherent and forms soft lumps: the extine is slightly sculptured (see Plate I). During stamen retraction pollen is seen to lodge on the stigma of *Collandra* flowers. The pollen vector of collandras is not known but if observations made by Stefan Vogel (personal communication) on the related genus *Alloplectus* are any indication, hummingbirds could be involved. *Alloplectus* has tubular, almost actinomorphic corollas which closely resemble those of species in *Columnea* sect. *Collandra*, and Vogel notes that pollen is deposited at the base of the bill of feeding birds instead of on the crown of the head as is the case in species in sect. *Columnea*.

In contrast, columneas have a funnel-shaped tube and much dissected limb which allows the withered stamens still retaining pollen to retract, partly by coiling but also by swinging in a vertical arc away from the stigma. Almost without exception this vertical displacement prevents autogamy in sect. *Columnea*. The autogamy seen in collandras is a measure of either the breakdown of the allogamous system seen in columneas, or the inefficiency of the floral mechanism for outcrossing; I favour the latter view.

The predominantly red corolla-colour in columneas may help attract the hummingbird pollen vector in accord with classical bird pollination theory, but not all columneas have red flowers, and the birds are seen to feed from white, yellow and pink flowers in Jamaica: *T. polytmus* was seen to visit many nectariferous flowers including *Asclepias* (Asclepiadaceae), *Hedychium* (Zingiberaceae) and *Elleanthus* (Orchidaceae). Jamaican columneas have either yellow or red and yellow corollas. The yellow carotenoid pigment is in the form of plastids as illustrated in Kirk *et al.* (1967) and follow the corolla vascular traces. The red

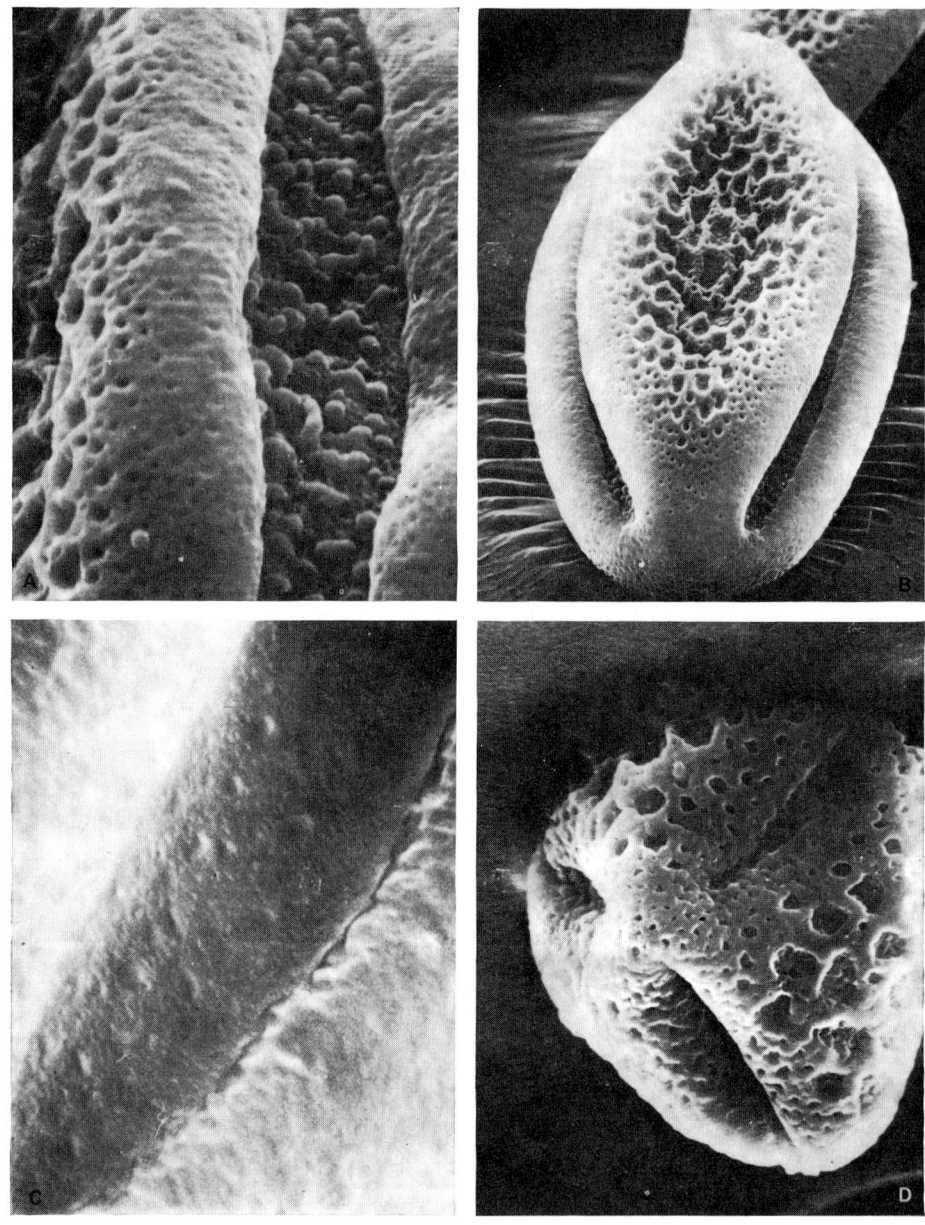

3-desoxyanthocyanin pigment, called columnin, is present in solution in at least *C. microphylla* Kl. & Hanst. ex Oerst. (sect. *Columnea*) and *C. teuscheri* (Mort.) Moore (sect. *Pterygoloma*), Harborne (1966), and occurs in the interstitial tissue between corolla-traces and at the edges of the lobes. Corolla-colour is sometimes variable in columneas with red and yellow flowers depending on the relative expression of component colours. Multicellular hairs may contain red pigment and in species with yellow corollas such hairs give a characteristic reddish cast to the flower, e.g. *C. hispida* Sw. In the absence of hybridization corolla colour seems to be a reliable taxonomic character at species level.

To digress, coloured leaf-vesiture may also serve to attract pollinators to flowers of columneas, perhaps in conjunction with the red blotching of the leaves of some species in sects. *Ortholoma, Collandra* and *Columnea*. Seen from below the foliage, where the flowers hang and from which direction pollen vectors generally approach the flowers, the translucent red windows in leaves of species such as *C. sanguinolenta* (Kl.) Hanst. (sect. *Ortholoma*) are most conspicuous. Is there perhaps a parallel here with extrafloral nectary development?

Red sepal colours are observed to attract hummingbirds to columneas. *T. polytmus* attempted to feed from the red calyces of *C. hirsuta* which had lost flowers and set no fruit. Red sepals may therefore supplement the corolla in making the flowers of columneas conspicuous, especially in species with small or inconspicuously coloured corollas. Similarly the red, reddish or bronze floral vestiture found in species in sect. *Collandra* and elsewhere may also help attract pollinators. Elaboration of the sepal-margin by teeth increases the area of pigment containing tissue, as when the teeth are pectinate and also covered with pigment filled multicellular hairs, e.g. *C. major* Hanst. (sect. *Pterygoloma*). To support this notion, species with pectinate sepals usually have red sepals.

At fruiting time the colour of the sepals may serve to attract dispersal agents to the berries which may be white, pink, red, magenta or yellowish, and vary

---

PLATE I

A. *C. fawcettii* (Urb.) Mort. pollen.
  × 3850.
B. *C. fawcettii* pollen.
  × 1750.
C. *C. fendleri* Sprague pollen.
  × 7630.
D. *Alloplectus ambiguus* Urb. pollen.
  × 1540.

from about 5 to 20 mm in diameter depending on species. When surrounded by brightly coloured sepals the fruits are more conspicuous than if the calyx were green, chaffy or caducous. The fruits are presumed to be bird-dispersed in the absence of observation, but secondary agents are ants (Morley, 1968). Ridley (1930) noted that the combination of red and white colours is often associated with bird-dispersed fruits, and such a colour combination is expressed in many fruiting columneas. In *C. harrisii* (Urb.) Britton ex Mort. calyx-colours are correlated with fruiting time, the sepals being green at anthesis but becoming reddish as the white or pink berry ripens. Species such as *C. florida* (sect. *Collandra*) do not have red-white fruiting colours: the sepals are instead orange and fruits yellow to orange. The berries of Jamaican columneas are slightly sweet to taste but very seedy.

Sepal-characters have been used at sectional, specific and intraspecific levels in *Columnea* taxonomy. The usefulness of calyx-characters at and below the rank of species, despite variation in species such as *C. scandens*, is undisputed, but at sectional level there is as much variation within sections as between. Sepal characters seem to have responded to selection pressures relating to pollination and dispersal ecology.

The most helpful data for testing the reliability of corolla-morphology as a sectional character comes from the breeding work of Sherk and Lee (1967). Their findings are that interspecific hybrids with fertile pollen are obtained from crosses between sections *Pterygoloma* and *Ortholoma*, *Ortholoma* and *Collandra*, within *Columnea*, and within *Ortholoma*; crosses between sections *Columnea* and *Collandra* and *Columnea* and *Pterygoloma* were found to produce hybrids with poor pollen fertility. Since then, Saylor (1971) has succeeded in crossing and backcrossing species in sect. *Columnea* and *Pterygoloma*. The data are based on the relatively few species in cultivation and do not include results for species in sect. *Pentadenia*. Comparison between Figs 1 and 2 shows that columneas with similar or related corolla-morphology are interfertile and produce fertile

---

FIG. 2. Scatter diagram of corolla length against length of dissected limb. 1. *C. consanguinea* Hanst. (sect. *Collandra*); 2. *C. affinis* Mort. (sect. *Collandra*); 3. *C. crassa* Mort. (sect. *Collandra*); 4. *C. florida* Mort. type material (sect. *Collandra*); 5. *C. pectinata* Mort. type material (sect. Collandra); 6. *C. species* Koie 4778 (sect. *Collandra*); 7. *C. purpurata* Hanst. (sect. *Collandra*); 8. *C. pallida* Rusby type material (sect. *Collandra*); 9. *C. dissimilis* Mort. type material (sect. *Collandra*); 10. *C. inaequilatera* Poepp. & Endl. (sect. *Collandra*); 11. *C. rubida* (Mort.) Mort. type material (sect. *Collandra*); 12. *C. dimidiata* (Benth.) Leeuwenb. (sect. *Collandra*); 13. *C. perpulchra* Mort. type material (sect. *Pterygoloma*); 14. *C.*

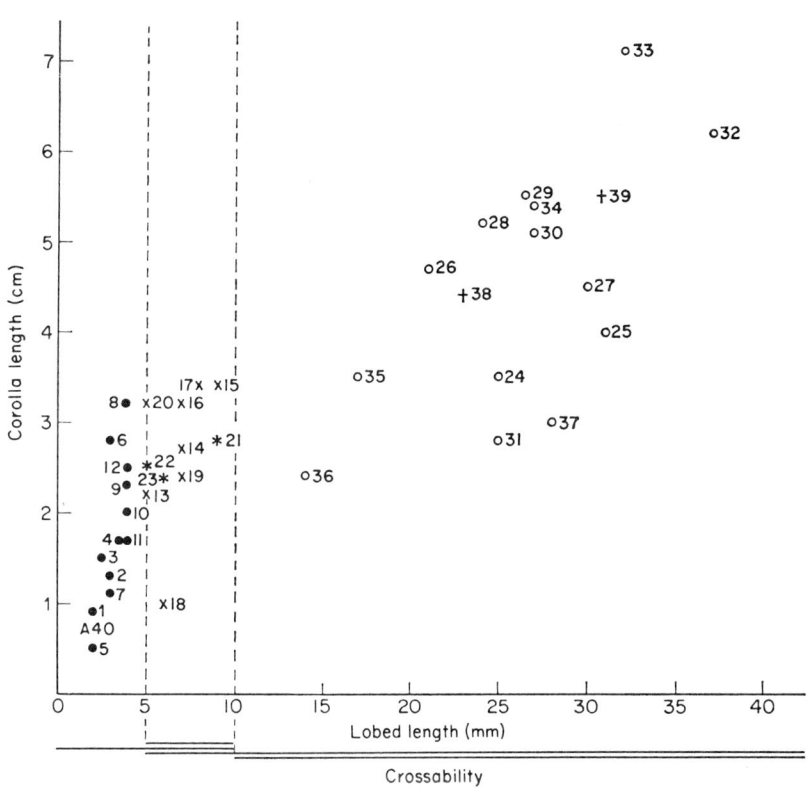

silvarum Mort. type material (sect. Pterygoloma); 15. *C. illepida* Moore (sect. Pterygoloma); 16. *C. species* Holm-Nielsen & Jeppesen 664 (sect. Pterygoloma); 17. *C. moorei* Mort. type material (sect. Pterygoloma); 18. *C. jamaicensis* Urb. (sect. Pterygoloma); 19. *C. rubrocincta* Mort. type material (sect. Pterygoloma); 20. *C. pulcherrima* Mort. type material (sect. Pterygoloma); 21. *C. warscewicziana* (Klotzsch ex Oerst.) Hanst. (sect. Ortholoma); 22. *C. grata* Mort. type material (sect. Ortholoma); 23. *C. sanguinolenta* (Klotzsch) Hanst. (sect. Ortholoma); 24. *C. urbanii* Stearn (sect. Columnea); 25. *C. picta* Karst. (sect. Columnea); 26. *C. dictyophylla* Donn. Smith (sect. Columnea); 27. *C. kalbreyeri* Hook. (sect. Columnea); 28. *C. wilsoni* Wiehler (sect. Columnea); 29. *C. hirsutissima* Mort. type material (sect. Columnea); 30. *C. rubricaulis* Standley (sect. Columnea); 31. *C. verecunda* Mort. (sect. Columnea); 32. *C. nicaraguensis* Oerst. type material (sect. Columnea); 33. *C. pilosissima* Standley (sect. Columnea); 34. *C. oerstediana* Klotzsch ex Oerst. type material (sect. Columnea); 35. *C. tomentulosa* Mort. var. *tulea* (Urb.) Morley (sect. Columnea); 36. *C. hispida* Sw. (sect. Columnea); 37. *C. tincta* Griseb. (sect. Columnea); 38. *C. incarnata* Mort. (sect. Pentadenia); 39. *C. magnifica* Klotzsch & Hanst. ex Oerst. type material (sect. Pentadenia); 40. *Alloplectus ambiguus* Urb. All measurements from single corollas.

progeny. Interfertility seems to be directly proportional to corolla-shape, and crosses between sections which are morphological extremes, such as *Collandra* and *Columnea*, give infertile progeny. Crosses between morphological intermediates, such as *Pterygoloma* and *Ortholoma*, or morphological extreme and intermediate, such as *Columnea* and *Pterygoloma*, give fertile progeny.

2. *Discussion*

The plot of corolla-length against length of the lobed portion (Fig. 2) shows that as corolla-length increases throughout the genus, so dissection of the corolla increases. It also suggests that species in sect. *Ortholoma* and *Pentadenia*, both with corollas more or less ventricose, may represent parallel developments arising from an ancestry in sect. *Pterygoloma* and *Columnea* respectively. This is supported by species with corollas which are only just ventricose such as *C. magnifica* Kl. & Hanst. ex Oerst. in sect. *Pentadenia* and *C. sanguinolenta* in sect. *Ortholoma*.

From what is known of the stamen retraction mechanism in sect. *Collandra* and *Columnea* it is logical to suggest that with increasing corolla-length and dissection through the genus, there is a trend away from autogamy. The ability to self for periods when outcrossing does not occur may explain some of the continuous variation patterns found in collandras. When outcrossing does not occur in species in sect. *Columnea* no fruits are produced: the genotype is unaltered by recombination.

On the basis of present knowledge relationships such as those discussed above lead me to think that *Columnea* sensu lato is a natural group with oligophilic or monophilic pollinators and not a polyphilic collection of groups. The number and diversity of *Columnea* species makes an intrageneric classification essential and at present there seems no better character than corolla-morphology. While the distinctions between sections *Pterygoloma* and *Ortholoma*, or *Pterygoloma* and *Collandra* are at times arbitrary, for the majority of species, corolla-morphology provides a useful classification in agreement with current systematic data.

ACKNOWLEDGEMENTS

I wish to thank the authorities at Kew, the British Museum, Missouri Botanical Garden, and Smithsonian Institution for enabling me to use herbarium material. Thanks are also due to Mr B. L. Burtt, Dr A. Skovsted, Dr S. Vogel, Miss C. Humphries, Mr I. Marks for the scanning electron micrographs, and not least, the late Mr C. V. Morton for his valuable observations. Part of the work was financed by a Science Research Council scholarship (London).

REFERENCES

BARKMAN, J. J. (1958). "Phytosociology and Ecology of Cryptogamic Epiphytes." Assen.

GRUBB, P. J. et al. (1963). A comparison of montane and lowland rainforest in Ecuador. *J. Ecol.* **51**, 567–601.

HARBORNE, J. B. (1966). Comparative biochemistry of flavonoids II. *Phytochemistry* **5**, 589–600.

HOSOKAWA, T. (1943). Studies on the life-forms of vascular epiphytes and the epiphyte flora of Ponape, Micronesia. *Trans. nat. Hist. Soc. Formosa.* **33**, 132 and 141.

HOSOKAWA, T. (1949). Studies on the life-form of vascular epiphytes and the spectrum of their life-forms. *Jap. J. Bot.* **24**, 41–45.

HOSOKAWA, T. (1953). On the nomenclature of aerosynusiae. Proceedings of the Seventh International Botanical Congress, Stockholm, 691–694.

KIRK, J. T. O. and TILNEY-BASSETT, R. A. E. (1967). "The Plastids" (frontispiece). London.

MORLEY, B. D. (1966). *Columnea* and aspects of its evolution. *Science Notes and News, Jamaica* **2**, 13–14.

MORLEY, B. D. (1968). "Biosystematics of *Columnea* in Jamaica." Unpublished thesis, University of the West Indies, Jamaica.

MORLEY, B. D. (1970). *Hohenbergia* and *Columnea* ecology. *Brom. Soc. Bull.* **20**, 130–132.

MORLEY, B. D. (1971). A hybrid swarm between two hummingbird pollinated species of *Columnea* (*Gesneriaceae*) in Jamaica. *J. Linn. Soc. (Bot.)*, **64**, 81–96.

MORLEY, B. D. (1972). The distribution and variation of some gesneriads on Caribbean islands, in *Taxonomy, Phytogeography and Evolution*. ed. Valentine, D. H., 239–257, Academic Press, London.

MORLEY, B. D. (in press). Notes on some critical characters in *Columnea* classification. *Ann. Mo. bot. Gdn.*

RICHARDS, P. W. (1952). "The Tropical Rainforest." Cambridge University Press.

RIDLEY, H. N. (1930). "The Dispersal of Plants Throughout the World." Reeve, Ashford.

ROSS, M. A. and HARPER, J. L. (1972). Occupation of biological space during seedling establishment. *J. Ecol.* **60**, 77–88.

SAYLOR, W. R. (1971). Gesneriad Cross Roads. *The Gloxinian* **21**, 30.

SCHIMPER, A. F. W. (1903). "Plant Geography." Pergamon, Oxford.

SHERK, L. C. and LEE, R. E. (1967). Interspecific hybridization in the genus *Columnea*. *Baileya* **15**, 89–96.

URBAN, I. (1901). *Columnea. Symb. Ant.* **2**, 358–364.

14 | Mode of Pollination as a Consequence of Ecological Factors

DAN EISIKOWITCH

Department of Botany, Tel-Aviv University, Israel*

Abstract: The influence of physical, biotic, and seasonal factors on reproductive dynamics of the seashore plants in Israel has been examined. The coastal plain of the Mediterranean sea in Israel is directly exposed to the sea breezes, continuous sea-water spray, strong solar radiation, and to summer droughts. Despite the arid conditions, most plants bloomed in the high summer.

Two factors have been investigated: (1) the physical effect of the wind on pollinators; (2) the effect of sea spray on pollen germination.

It has been found that the fluctuation of the wind (which is correlated with sea spray) is unpredictable except in midsummer. This led to the following adaptations: (1) blooming during the calmest hours of the day; (2) resistance to salt spray; (3) "ephemeral" blooming; (4) independence of pollinators.

It has been concluded that these extreme conditions decrease the spectrum of pollinators and that the plant–insect synchronization has to be much more intimate than in moderate conditions.

INTRODUCTION

Many living organisms have been investigated in their actual habitats and much of their morphology, anatomy and physiology has been found to be correlated with their environment. By knowing the life-form of a plant, we can normally guess the habitat in which it is usually found.

Life-forms of plants are usually described in terms of the morphology of their leaves or shoots, but although the flower is basically but a shortened shoot, it is less variable in structure than are leaves or shoots. The flower, like other components of a plant, is the result of continual exposure to its environment and much attention has been paid to its morphology and pollen-transfer mechanisms. Over the two hundred years (Kölreuter, 1761), during which the science of anthecology has developed, there has been a huge accumulation of records on

---

* Present address: Department of Zoology, South Parks Road, Oxford.

flower structure and pollinators. Since the time of Frisch (1919), the function and significance of floral structure, odour and colour have been investigated in depth, and we now have an enormous amount of information on flower–insect interrelationships.

One very important aspect of anthecology has been relatively neglected: the influence of physical, biotic and seasonal factors on reproductive dynamics (van der Pijl, 1961; Macior, 1971). Only in the last few years has there been a revival of this subject, especially due to the teams of Janzen, Linsley, Ehrlich, Raven and others. In order to understand the influence of the habitat on flower–insect relationships it is useful to choose an extreme habitat that will show its influence much more clearly than a moderate habitat.

THE HABITAT

The habitat we selected was the light soil belt of the coastal plain of Israel, which is in the immediate vicinity of the sea and is directly exposed to the sea breeze, to continuous sea-water spray, to radiation and to drought throughout the summer. One of the most striking phenomena on the seashore in Israel is the typical fluctuation of the winds, which blow from the sea toward the land by day and the reverse by night.

In summer the fluctuation is very predictable: at midnight the wind velocity is least ($0\text{--}1$ m s^{-1}); it rises toward dawn (2 m s^{-1}) and reaches a peak (up to 6 m s^{-1}) at the hours 12.00–16.00. Towards sunset the wind velocity decreases and again reaches a minimum at midnight. In the winter, the sequence is generally the same but it is not as predictable, the minimum wind velocity being almost always higher than 4 m s^{-1} (Rosenan, 1970).

OBSERVATION, EXPERIMENTS AND RESULTS

Since the habitat described in the previous section is arid or semiarid, it was very unexpected to find most plants blooming in the height of summer, which is the driest time of the year.

Two factors have been investigated in this habitat: (a) the physical effect of the wind on the pollinators; (b) the effect of sea spray on pollen germination.

As an example, I will discuss here three plants. (a) *Pancratium maritimum* L. (Amaryllidaceae)—grows along the seashore and regularly produces seeds. (b) *Glaucium flavum* Crantz (Papaveraceae)—grows along the seashore, mainly on "Kurkar" shore, and normally produces seeds. (c) *Urginea maritima* (L.) Baker (Liliaceae)—normally grows inland but introduced to the seashore where it grows fairly well, but does not produce seeds. All these three plants flower in midsummer, but each of them has a different mode of pollination. *Pancratium maritimum* is a typical hawkmoth flower (Faegri and van der Pijl, 1971): it blooms at night and depends almost absolutely on hawkmoths for pollination (Eisikowitch and Galil, 1971). Most of these hawkmoths are migratory insects which come to the Israeli coast only in the summer (Lane and Rothschild, 1961; Yathom and Rivnay, 1967).

The period of their stay generally coincides with the flowering period of *Pancratium* (and other seashore plants). Using infrared binoculars, and measuring windspeed with a scale anemograph at the height of the flowers, it was observed that pollination took place only on calm nights.

When the wind speed was above 3 m s^{-1} the moths could not reach the flowers and no seeds were produced. On the other hand, it was shown that the hawkmoths, when they could hover and pollinate, were inconstant in the species they visited. Sea spray on the open flower, whether put there artificially or naturally during storms, does not affect pollen germination on the stigma.

Glaucium flavum is a "dish-shaped blossom" (Faegri and van der Pijl, 1971) which opens in the early morning and closes the same afternoon. Usually it is bud-pollinated, but even so, seed formation takes place only on fairly calm days because sea spray can damage pollen grains on the stigma. Provided the windspeed does not reach the threshold of 4 m s^{-1} it can also be pollinated by honey bees and solitary bees; however, these conditions are not assured.

Urginea maritima is the only one of the three plants in which the flower is open for more than a day. It is dependent on day pollinators (honey and solitary bees) which can only reach the flower if the windspeed is less than 4 m s^{-1}. Because of the high sensitivity of its pollen grains to sea spray this plant does not produce seeds near the shore unless the flowers are pollinated and protected artificially.

The observations on the behaviour of the pollinating insects and the effect of sea spray on pollen grains shows that the main factor in pollination of the seashore plants is wind intensity. Two speed levels of wind were distinguished, each being a limit for the pollination activity of a certain group of pollinators: (a) windspeed of $0-3 \text{ m s}^{-1}$ which is the range for the activity of hawkmoths; (b) windspeed of $0-4 \text{ m s}^{-1}$ which is the range for bees.

The typical fluctuation of the wind that is so unpredictable all year, except in midsummer, causes selection in various directions: (a) blooming during the calmest hours of the day; (b) resistance to salt spray; (c) "ephemeral" blooming; (d) independence of pollinators.

The only plant that does not show any adaptation of its sexual reproduction to the seashore conditions is *Urginea maritima* which therefore does not produce seeds near the sea.

DISCUSSION

Pollination under unfavourable conditions is often solved by autogamy (Uphof, 1938; Hagerup, 1951). This has been discussed by Stebbins (1950) and by Grant and Grant (1965); it was therefore not unexpected to find that *Glaucium flavum* may be pollinated through an autogamic process. As has been stated above, the main physical force in the seashore habitat that limits pollination is the wind velocity.

The inhibiting influence of wind on the flight of winged insects is well known (Robertson, 1888; Hering, 1926; Larsen, 1943; Linsley, 1958; Lewis, 1970). Since the windspeed oscillates like ebb and flood tides, pollination can take place in the "ebb time" (i.e. low windspeed). There is synchrony between the physical conditions, time of flowering and the biological characters of the pollinators. Synchronization between certain conditions and insect behaviour has already been described by Linsley (1958, 1962), Baker (1961), Janzen (1967, 1968) and Linsley and Cazier (1970).

Pancratium maritimum, like *Ipomopsis candida* and *I. thurberi* (Grant and Grant, 1965), conforms in size, form, orientation, colour and time of blooming to the structural characteristics and behaviour of a long-tongued hawkmoth. We can agree therefore to Grant and Grant's hypothesis that these flowers have been moulded by natural selection to fit hawkmoths. Furthermore, they even flower at the time of the hawkmoths' migration to the seashore in Israel.

From Grant and Grant's hypothesis that the array of flower-visiting animals is a part of the plant environment, we emphasize that in our case this environment changes from month to month and even from hour to hour. Therefore the "lock" (Grant and Grant, 1965) in the relationships between floral mechanisms and the hawkmoths cannot be "open" all the time. Moreover, since there are no "non-adapted" keys (as has been found by Grant and Grant, 1965), the spectrum of coadapted pollinators is narrow.

In *Glaucium flavum*, in spite of its sensitivity to salt spray, pollination success is assured by its "combined pollination system" (Baker, 1961), i.e. bud and insect pollination.

If pollinators distribute their visits to flowers on the basis of the food reward they gain (Heinrich and Raven, 1972) synchronous blooming in the calmest warm nights may be advantageous through minimizing time and energy expenditure by pollinators.

In conclusion, shifting the flowering period into midsummer, when wind conditions are more predictable, decreases the spectrum of pollinators and favours intimate plant–insect synchronization. This leads to a certain risk in pollination, but perhaps to a saving in energy expenditure by the pollinator.

ACKNOWLEDGEMENTS

I wish to express my gratitude to Dr M. Smyth and Dr J. N. M. Smith for their helpful suggestions.

REFERENCES

BAKER, H. G. (1961). The adaptation of flowering plants to nocturnal and crepuscular pollinators. *Q. Rev. Biol.* **36**, 64–73.

EHRLICH, P. R. and RAVEN P. H. (1964). Butterflies and plants: a study in coevolution. *Evolution* **18**, 586–608.

EISIKOWITCH, D. and GALIL, J. (1971). Effect of wind on the pollination of *Pancratium maritimum* L. (Amaryllidaceae) by hawkmoths (Lepidoptera: sphingidae). *J. Anim. Ecol.* **40**, 673–678.

FAEGRI, K. and PIJL, L. VAN DER (1971). "The Principles of Pollination Ecology" (2nd edition). Pergamon Press, Oxford.

FRISCH, K. VON (1919). Über den Geruchsinn der Bienen und seine blütenbiologische Bedeutung. *Zool. Jahrb. Abt. allg. Zool. Physiol.* **37**, 1–238.

GRANT, V. and GRANT, K. A. (1965). "Flower Pollination in the Phlox Family." Columbia University Press, New York.

HAGERUP, O. (1951). Pollination in the Faroes in spite of rain and poverty in insects. *Dan. Bull. Medd.* **15**, 1–4.

HEINRICH, B. and RAVEN, P. H. (1972). Energetics and pollination ecology. *Science, N.Y.* **176**, 597–602.

HERING, M. (1926). "Biologie der Schmetterlinge." Verlag von Julius Springer, Berlin.

JANZEN, D. H. (1967). Synchronization of sexual reproduction of trees within the dry season in Central America. *Evolution* **21**, 620–637.

JANZEN, D. (1968). Reproductive behavior in the Passifloraceae and some of its pollinators in Central America. *Behaviour* **32**, 1–3.

KÖLREUTER, J. G. (1761). "Vorläufige Nachricht von einigen das Geschlecht der Pflanzen betreffenden Versuchen und Beobachtungen." Leipzig.

LANE, C. and ROTHSCHILD, M. (1961). Notes on migrant lepidoptera in Israel. *Entomologist* **94**, 295–304.

LARSEN, E. B. (1943). The importance of master factor for activity of noctuids. *Ent. Medd.* **23**, 252–374.

LEWIS, T. (1970). Patterns of distribution of insects near a windbreak of tall trees. *Ann. appl. Biol.* **65**, 213–220.

LINSLEY, F. G. (1958). The ecology of solitary bees. *Hilgardia* **27**, 543–599.

LINSLEY, F. G. (1962). Ethological adaptations of solitary bees for the pollination on desert plants. *Särtyck ur Meddelande nr 7 fran Sveriges Fröodlarefobund.*

LINSLEY, F. G. and CAZIER, M. A. (1970). Some competitive relationships among matinal and late afternoon foraging activities of *Caupolicanine* bees in Southeastern Arizona. (Hymenoptera. Colletidae). *J. Kansas ent. Soc.* **43**, 251–261.

MACIOR, L. W. (1971). Co-evolution of plants and animals—systematic insights from plant–insect interaction. *Taxon* **20**, 17–28.

PIJL, L. VAN DER (1961). Ecological aspects of flower evolution. II. Zoophilous flower classes. *Evolution* **15**, 44–59.

ROBERTSON, C. (1888). Effect of wind on bees and flowers. *Bot. Gaz.* **13**, 33–34.

ROSENAN, N. (1970). "Atlas of Israel" [1st edition, 1956–64] (2nd edition, in English, 1970). Section IV. Climate. Survey of Israel and Elsevier Pub. Co., Amsterdam.

STEBBINS, G. L. (1950). "Variation and Evolution in Plants." Columbia Univ. Press, New York.

UPHOF, I. C. TH. (1938). Cleistogamic flowers. *Bot. Rev.* **4,** 21–50.

YATHOM, SH. and RIVNAY, E. (1967). [Phenology of hawkmoth species in Israel.] *Ktavim* **17,** 57–72.

15 | Co-evolution of Plant Hosts and their Parasites as a Taxonomic Tool

A. D. J. MEEUSE

*Hugo de Vries-Laboratorium Voor Bijzondere Plantkunde,
University of Amsterdam, Netherlands*

Abstract: Some illustrative examples chosen from a much larger number of cases of host–parasite relations clearly indicate that there are certain connections between the taxonomy and the chemistry (and sometimes also the morphology) of the natural group of host plants on the one hand, and the taxonomy and the perception (and, in the case of animal parasites, the instincts) of their parasites on the other. An "old" association leading to a gradual co-evolution, the speciation of the hostal and the parasitic taxa proceeding along partly independent lines but held together by, presumably, mainly chemical links, evolved into a more or less characteristic, or even exclusive, bond between a host taxon above the species or generic level and its associated and phylogenetically diversified group of parasites. Such a specific association, when encountered at the present-day level of evolution, is often indicative of taxonomic affinities between taxa attacked by the same group of parasites, and between parasites found on the same group of host plants. There are exceptions to this rule partly caused by a switch-over to other (secondary) groups of hosts which contain the same, or at least related, phytoconstituents as the primary host taxon but acquired these constituents by a process of biochemical convergence. The problems appear to be complex and not always unequivocal, as is shown by the selected example of the pierid Butterflies and their food plants. Generalizations are permissible but corroborative evidence from other sources is required before far-reaching conclusions can be drawn: every indication of a taxonomic affinity between taxa emanating from a study of host–parasite relations must be followed up, and augmented by indications from independent sources of diagnostic taxonomic inquiry. Any "negative" taxonomic evidence obtained from host–parasite relations and seemingly arguing against established relationships of taxa may be explicable by the acquisition of a special biochemical constitution by one taxon which thus achieved indemnity from attack at an early stage of its phylogeny, whereas the related taxa did not develop a chemical repellant mechanism and retained their associated parasites (compare Solanaceae/Scrophulariaceae, Liliaceae/Amaryllidaceae). "Negative" evidence thus becomes more cogent whenever indications of a possible chemical repellency are not manifest.

INTRODUCTION

The host specificity of an appreciable number of parasites has been recognized for a considerable length of time. This has gradually led to a certain amount of understanding of the reasons why a given parasite is so often restricted in its occurrence. It is quite clear that certain interactions between the specific host and its parasite must be responsible for their association. In other words, characteristic, and apparently taxon-specific features of both host and parasite decide why the host is subject to attack by a particular parasite (and not by others even if they seem to be potentially dystrophic), and, conversely, why the parasite, at least in one of the stages of its life cycle, becomes dependent on its special host

FIG. 1. Some parasites are very selective in the choice of their source of food.

(and not on other, seemingly potential ones). Irrespective of which factors (chemical, biochemical, mechanical, physiological, anatomical or otherwise) actually determine the association of the host and its parasite, they all must be taxon-linked and presumably hereditary. The obvious corollary is that there must be some connection between taxonomically significant characters and the incidence of specific host–parasite associations. From here it is only a short step towards the application of specific host–parasite relations in taxonomic inquiry.

The definition of a parasite is not such an easy matter, but for the purpose of the present paper it will be taken for granted that parasites are: (1) detrimental to their host (even if the host receives *some* benefit in return); (2) completely dependent on their host, or hosts, at least during a part of their life cycle; and (3) more or less host-specific.

Parasites may be rather particular as far as their host is concerned (Fig. 1) and such types are called *monophagous* or, if the range of their hosts is restricted, *oligophagous*. *Polyphagous* parasites are capable of feeding on an appreciable number of different hosts, but that number is always limited: truly polyphagous parasites are rare if they exist at all. This is the logical consequence of the basic cause of the mutual relationship between host and parasite, namely, the actions and interactions of a set of taxon-specific characters. The diversity of taxon-associated characteristics simply sets a limit to the tolerance of the diverse potential host plants to a given parasitic taxon, and, conversely, the attacking power of a range of dystrophic forms in respect of a single plant species differs so much that the plant is immune to the majority of them. The necessary interaction—one might also call it the compatibility—between a host taxon and the associated parasitic species suggests a mutually regulating action during their speciation and evolution, i.e. integrated evolution or co-evolution of parasite and host.

It follows that the specificity of the host–parasite relationship is usually of taxonomic significance only if the number of hosts is strictly limited, although the host taxon may be specific even if it is as large as a genus or a family. A group of parasites may be polyphagous in the sense that the members are found on a widely divergent set of species, but these hosts may belong to the same major taxon (family or order), so that a certain mutual specificity is present, conceivably as the result of co-evolutionary speciation. This may be sufficiently indicative of specific taxonomic relationship to permit certain conclusions. One must bear in mind, however, that an analysis of such situations may easily lead to circular reasoning if one is not careful enough when drawing taxonomic conclusions from the ecological evidence. The taxonomic pointers gain in significance if the parasite is heteroecious (i.e. attacks at least two different hosts in different stages of its life cycle) and shows a high degree of specificity in its relationship with its alternating hosts. The association of a parasite with *two* taxonomically different host groups suggests a long-lasting co-evolution.

THE DEVELOPMENT OF SPECIFIC HOST–PARASITE RELATIONSHIPS

There are at least three possible pathways leading to special host–parasite relationships which, I think, yield as many valid, alternative explanations of the mutual specificity and co-evolution of taxa: (1) the "exploitation" theory, i.e. a change-over from a situation of mutual benefit to a dependence of the one life-form concerned (after having become a parasite) on the other; (2) the development of a certain degree of interdependence and mutual benefit from a primarily one-sided parasite–host relationship favouring co-evolution; and (3) the tolerance-adaptation principle of Dethier (1941, 1954), Thorsteinson

(1960), and Schoonhoven (1968), involves biochemical interactions as the origin of co-evolution.

1. *The Exploitation Theory*

The first of these explanations is usually of only little taxonomic value and need not be discussed in great detail. As an example, the exploitation of the instincts of visitors (potential pollinators) by orchids whose flowers are pollinated by pseudo-copulation may be mentioned. Conceivably, the flowers were originally dependent on pollinators which were "rewarded" for their efforts by the presentation of a source of food, as in the majority of entomophilous flowers including the bulk of the Orchidaceae. In the ultimate stage of evolution of this peculiar syndrome, the specific insect pollinator is a male specimen lured and deceived by the flower which simulates the female of that insect species. There is no close taxonomic connection between the groups of mimetic insects and their respective pollinators (which belong to the groups Scolicidae, Sphegidae, Ichneumonidae, and presumably also Apidae) of the Hymenoptera, for instance, representatives of the genus *Ophrys* alone being associated with, among others, sphegid and scolicid wasps.

Mimetism in the case of coprocantharophily and copromyophily (flowers or inflorescences exuding the odour of putrifying meat, rotting fish, carrion, dung, etc.) is also clearly a matter of convergence: it is known in widely divergent groups such as Araceae, Iridaceae (*Ferraria*), the stapelioid Asclepiadaceae, some Orchidaceae, *Desherainea*, and *Aristolochia*. In these, and in similar cases, there is clearly a convergence of adaptive types. The taxonomically more important examples mostly belong to the group of relationships explicable by means of the repellent-tolerance hypothesis.

2. *Interdependence*

The development of a certain degree of interdependence from a primarily one-sided parasite–host relationship is only a form or variant of the cases to be dealt with under (3), but for the sake of completeness it should be mentioned as a special and independent pathway of co-evolutionary advancement. The most striking and ancient example is the relation between the representatives of the genus *Ficus* and the figwasps, specialized braconid Hymenoptera, which must initially have been strictly parasitic. However, after the development of special gall flowers in the sycones, the relationship assumed the form of a true symbiosis (although, strictly speaking, the wasps retained the characters of phytophagous parasites). The fig trees are as much dependent on the wasps as the wasps on their specific fig species (the fig wasps are normally monophagous to oligophagous on one or a few closely related species of *Ficus*—barring a number of secondarily hyperparasitic forms which parasitize other fig wasps).

This is certainly not the only example. Tripetid flies are normally strictly parasitic (the larvae develop in the pistils of their host plants, destroying the ovules and young seeds), but there are some cases in which a situation of mutual benefit developed. Some tripetid flies, e.g. species living on South African Compositae, such as perennial species of *Vernonia*, not only complete their lifecycle on a group of host plants (so that not only the time of emergence of the imagines coincides with the time of full anthesis of the host), but also their activities usually remain confined to that group of plants (the imagines feeding on the pollen and/or nectar of the host plant). Pollination is assured by the activity of the females who lay their eggs in the flower heads and although many ovules may be destroyed, a sufficient quantity of viable achenes is produced as long as the number of individual flowers is large enough.

A comparable case is found in some species of *Drosophila* (and of other flies) that lay their eggs inside the corolla of a certain species of plant (e.g. a representative of the Acanthaceae or Solanaceae). The gravid females crawl into the corolla tube and effect pollination. These *Drosophila* species and other, comparable, dipterid taxa are not usually true parasites, however: the larvae normally develop after the corolla—which is expendable in any event—has been shed and starts decaying on the forest floor (Pipkin, 1964; Pipkin *et al.*, 1966; see the summary by Carson, 1971). The specialization and co-evolution must, at any rate, have proceeded along the lines of a chemical evolution of the host plant and the adaptive compliance of the semi-parasitic or truly parasitic species: the selection of the flowers of a specific taxon implies that the females recognize (identify!) and select the living substrata by chemoreception. This train of thought implies the capacity in the parasite of a *discriminating* sense of chemoperception. Anyone who has repeated the experiment at dusk with sweet-scented evening flowers (such as honeysuckle) in a perforated box, and watched the hawkmoths zig-zagging against the trail of scent carried along by the gentle wind of the summer evening, ultimately to fly through a hole into the box in search of the flowers they can smell but not see, becomes quite convinced of the fine power of olfactorial reception of insects. More cogent evidence concerning the recognition of particular scents is provided by the discovery of taxon-specific pheromones (sexual attractants) and by numerous laboratory experiments in which the specific chemoreception of insects was tested (cf. Schoonhoven, 1968).

The specific attraction of parasites by host plants is, of course, a very important aspect of applied entomology and phytopathology if these parasites attack crops. It is, therefore, not at all surprising that one of the centres of research in this field is at the Agricultural University of Wageningen (Professor J. de Wilde and Dr L. M. Schoonhoven).

There are two ways in which specific chemoreception is being studied: by means of selection-preference exhibited by insects in feeding and oviposition experiments; and by means of studies of the chemoreceptors and their reaction to specific chemical stimulii. Chemoreceptors are commonly situated in the antennae of adult insects, but in some groups, notably in Diptera and Lepidoptera, also in the legs. The chemoreception may be "tactile" in such cases, in that the direct contact of the chemoreceptory sense organs (when the animal sits or lands on the living substratum) causes a response. This is of course important if the substance that is specifically recognized is not or hardly volatile, and we may assume that in a number of cases the parasites are guided by such non-volatile compounds in their search for an adequate supply of food or a substratum to deposit their eggs. We may, therefore, assume that the host is identified by a phytophagous insect parasite by the latter's chemoreceptory organs which are capable of reacting to volatile compounds in air currents and to non-volatile substances by direct contact. The specificity of host–parasite relations is clearly explicable by the instinctive reactions of the parasite induced by the stimulation of its chemoreceptors by special, taxonbound chemical compounds (or groups of related compounds) which can be referred to as "attractants" or "chemical indicators". The interesting point is that such attractants were originally, and sometimes still are, substances with an insect-repellent action.

3. The Tolerance-Adaptation Principle

Most of the taxonomically more important cases of host–parasite relationships belong to the group explicable by means of the repellent-tolerant hypothesis, first developed by Dethier (1941), Fraenkel (1956, 1959), Thorsteinson (1960) and others and later summarized by Ehrlich and Raven (1965, 1967) and also by Schoonhoven (1968). Various aspects have also been discussed in several papers in Harborne (1972). In theory, a taxon may have a parasite which is so far adapted to its host that it is sufficiently resistant to toxic or incompatible (serologically active, e.g. antigenic) compounds contained in the host which would repel or even kill other potential parasites. A mutative modification of the genotype resulting in a change of the chemical build-up of the host may have an adverse effect upon the parasite. This is neatly illustrated by the sickle-cell mutation in humans (symbol: Hb^S, against the "normal" allele Hb^A). Homo- and heterozygous individuals with this factor (genotype Hb^S—) have only a single amino acid of the protein component of their haemoglobin molecules replaced by a different one, but such persons are much more resistant to the plasmodium of tropical malaria than individuals with the other (more particularly the normal Hb^A) alleles. Natural selection has resulted in allele frequencies of the Hb^S factor of over 20% (= 40% $Hb^A Hb^S$ heterozygotes) in areas where this form of malaria is endemic (e.g. parts of tropical Africa), in spite of the sublethality of the Hb^S allele in the heterozygous individuals. Conceivably, in similar cases a chemical mutation in individuals of a host population may cause an increased resistance of these individuals to some specific parasites of the species in question and lead to a selective pressure on the populations of those parasites. If, in the long run, practically all individuals of the host species possess the new resistance factor, only a few of the parasitic species may have survived, either by a "natural" resistance or by some fortuitous adaptive mechanism such as a chance mutation; since all the others have been eliminated, a greater specificity of the host–parasite relationship must result. That this mechanism functions is neatly demonstrated by the specificity of the so-called physiological races of our cultivars of economic plants: a given cultivar is usually much more resistant to the physiological races in the area of cultivation than to physio-races isolated in other parts of the world where different cultivars are grown, and in a number of cases the virulence and resistance factors have been shown to be genetically determined. In a number of cases the resistance particularly to fungal attack is associated with the presence of specific repellents and inhibitors called phytoalexins (see Deverall in Harborne, 1972, pp. 217-233). Although such substances are normally produced by the higher plants it is conceivable that they are sometimes synthesized in the so-called phyllosphere (see Last and Warren, 1972) by non-pathogenic micro-organisms.

In more strongly selective chemical mutants the resistance may have become so great that only a few parasitic species survive, or special resistant strains which may be quite "immune" to the phenotypic effect of the "chemical" mutation by an increased degree of tolerance. One of the most striking examples is the tobacco beetle which feeds on (uncut) tobacco, cigars and cigarettes, and thus absorbs quantities of nicotine which would be absolutely lethal to nearly all other insects. (Note that nicotine is a powerful insecticide, until recently used on a large scale in greenhouses until it was superseded by synthetic biocides!) Beetles related to the tobacco beetle are thus deprived of the opportunity of feeding on the tobacco plant, and owing to this isolation mechanism the relation of the tobacco beetle to its food must have become highly specific. One could visualize a situation in which a proto-*Nicotiana* still devoid of the biochemical mechanism to synthesize the insecticidal alkaloid nicotine was not protected against insect attack, but after a "chemical" mutation acquired an immunity against all but the nicotine-resistant insects. Parasites may become tolerant of highly toxic substances: the South African species *Dichapetalum cymosum* often contains monofluor-acetic acid, but is not likely to be free from parasites.

FIG. 2. Plants often contain obnoxious or toxic substances acting as repellents to potential parasites.

The incidence of poisonous or repellent substances in certain plant taxa (Fig. 2) and the complete absence in related forms may well explain the striking difference in the susceptibility of related taxa to the same group of dystrophic animals. If one starts from the assumption that the Amaryllidaceae are closely allied to the Liliaceae, a statement to which most phanerogamists would subscribe (some of them even uniting the two families), it appears strange that certain fungi and phytophagous insects discriminate quite clearly between liliaceous forms (including the Allieae as we shall see), which they attack, and the Amaryllidaceae, which are immune to these parasites. The same can be said about the conspicuous absence of weevils of the tribes Cionini and Mecinini, commonly found on such families as Scrophulariaceae, Labiatae, Verbenaceae,

Oleaceae, Labiatae, Orobanchaceae, Bignoniaceae, Pedaliaceae, Plantaginaceae, etc., as parasites on representatives of the Solanaceae. There are a number of reasons why in taxonomic classifications Solanaceae and Scrophulariaceae are placed in the same group (in "Tubiflorae", etc., see, e.g., Hartl, 1963), or even considered to be very closely related, so that the absence of the parasites must be for a special reason. Considering that the genus *Digitalis* is also remarkably free from phytophagous taxa characteristic of the *Scrophulariaceae* (such as some genera of the Mecinini, etc.), a phytochemical basis for the freedom from attack seems highly probable: Amaryllidaceae, Solanaceae, and *Digitalis* all contain taxon-specific, toxic alkaloids which are lacking in Liliaceae and in the other Tubiflorae mentioned above. It follows that the taxonomic relationships between Amaryllidaceae and Liliaceae, and Solanaceae and Scrophulariaceae, respectively, must not be judged by the cantharotaxonomic criterion alone: if there are strong indications of a specific chemical barrier mechanism one must carefully weigh all other positive and negative evidence of taxonomic proximity—taxonomy is a never-ending synthesis!

On the other hand, specific repellence may account for the more or less exclusive occurrence of phytophagous animals on plant groups containing obnoxious bioconstituents as we shall see.

Adaptive changes "protecting" a plant taxon against a potential dystrophic attacker need not necessarily be of a chemical nature. Certain morphological features may have a selective effect. The lesser susceptibility of certain cultivars of cotton to leaf virus transmitted by Jassids is associated with the degree of pubescence: the hairier the plant, the less the attack by the insect vector. In the case of fungal parasites the higher resistance of a form of a species may be due to a thicker leaf cuticle or of a thicker waxy layer over the cuticle, preventing the penetration of hyphae. The co-evolution of such special morphs and "adapted" phenotypes or mutants of the parasite may lead to a specialization combined with speciation in very much the same way as will be described in the case of the chemical mutants and adaptive parasites, namely by the gradual ecological isolation of the better adapted parasite populations, ultimately followed by a genetic barrier.

The chemical repellent-tolerance hypothesis must be augmented by the assumption that the tolerance exhibited by a parasite in respect of repellents or toxic compounds gradually becomes changed, if not into a preference at least into a (mostly olfactory) signal which guides the parasite to its adequate source of food. There is experimental evidence clearly showing the preference of insects to a specific host plant (or group of host plants), cf. the review by Schoonhoven (1968). The experimental set-up is fairly simple: a neutral substance or non-specific substratum (such as leaves of plants unrelated to the specific host or hosts) to which a quantity of the probable attractant is added is offered to the females (or larvae, as the case may be) with the appropriate controls. In the case of the Cabbage White family (Pieridae) the specific attractants, viz. the thioglucosides and the derived mustard oils so characteristic of the Capparidales *sensu* Takhtajan, are known exactly, as will be indicated in the following paragraphs.

In other plant taxa the host–parasite relationship is clear but the specificity of the attractant is not unequivocally established. Danaid butterflies are associated with the poisonous Apocynaceae and Asclepiadaceae, the caterpillars obviously being capable of accumulating the toxic principle without any adverse effects, but it does not follow that the toxic substances are the actual attractants (danaid caterpillars thrive on alkaloid-free mutants of *Asclepias*). In their compilation of food plant groups of rhopalocerid caterpillars, Ehrlich and Raven (1965, p. 594) state that numerous distinctive segments of other insect orders and groups feed on the two plant families concerned (and the same applies to such taxa as the Solanaceae), but it is striking that in a number of cases (danaids, also sphingids: Harris, 1972) the host spectrum includes (apart from Apocynaceae and Asclepiadaceae) a number of plant families containing latex: Moraceae, Euphorbiaceae, Caricaceae, Convolvulaceae, and even Sapotaceae. Apparently the danaids and sphingids concerned are tolerant of the toxic principles in Apocynaceae and Asclepiadaceae, but their primary host taxa are more likely to be the more ancient Moraceae (and Euphorbiaceae?), and the attractant is perhaps a factor associated with, or contained in, the milky juice.

According to a number of workers, who base their opinion on experimental evidence, specific host groups contain substances which paradoxically still act as repellents to the specific phytophagous insect (alkaloids in Solanaceae), for instance on the Colorado Beetle *Leptinotarsa*: Fraenkel, 1959; thioglucosides render food unpalatable for *Pieris* caterpillars in quite low concentrations: e.g. Lourens (1971). The same reasoning presumably applies to the papilionid butterflies often associated with the Polycarpicae or Ranales which contain benzylisoquinoline alkaloids: the attractants may conceivably be components of the ethereal oils also commonly occurring in these plants and in alkaloid-free related taxa. Apparently the repellent, or even toxic, principle sometimes acts clearly as an attractant by having become the signal for the chemoreceptive response of the insect (secondarily inducing an approach followed by oviposition or feeding as the case may be), but it need not always be the attractant—compare the case of the alkaloid-free Asclepiadaceae and danaid caterpillars.

Such evidence strengthens the case for the repellent-tolerance hypothesis, and it is highly probable that the acquisition of a sufficient degree of tolerance has enabled certain groups of parasites "to penetrate new adaptive zones in which they radiated" (Ehrlich and Raven). Certain irrefutable experimental data provide additional support for the tolerance hypothesis. In the case of the Pieridae, the attractants vary from species to species in that the isothiocyanate, sulphate and glucose units are coupled with a more or less taxon-specific radical (R), compare Kjaer (1960, 1963, 1966)

$$\begin{array}{c} R \\ \diagdown \\ C = N-OSO_3^- \\ \diagup \\ Glc-S \end{array}$$

$R = -CH_3, -C_2H_5, -CH_2C_6H_5,$
$-CH_2CH=CH_2,$ 3-indolyl-CH_2-, etc.

The females prefer a species with the characteristic compound on which to lay their eggs, as can be demonstrated in experiments with neutral substrata flavoured with specific and some non-specific thioglucosides: the specific agent is selected when the females are offered a number of substances simultaneously. Better still, the emended tolerance hypothesis (the coupling of the tolerance principle with conceivable evolutionary processes leading to co-evolutionary speciation) is neatly supported by experiments in which pierid caterpillars wer reared on non-specific substrata or on "neutral" plant material to which were added thioglucosides differing from the specific attractant of the pierid species in question, and also by a "natural experiment". The caterpillars usually take the offered food and thus become "conditioned" to the (unspecific) attractant and "the memory lingers", so that the females hatched from the pupae of the "conditioned" caterpillars prefer the nonspecific host plant in experiments in which they were offered several glucosides including the *original* taxon-specific one and the one to which they had, as caterpillars, become conditioned.

In this way certain strains of the common Cabbage White with a divergent chemotaxis must have originated in our European gardens, because populations of *Pieris brassicae* live as caterpillars on the alien nasturtium (*Tropaeolum majus*) which contains a thioglucoside (with $R = $ phenylmethylene or benzyl, $C_6H_5CH_2-$) different from the sinigrin (with $R =$ butenyl-1-2 or allyl, $CH_2=CH_2-CH_2-$) found in original host plants. A parallel case is that of the European weevil *Ceuthorrhynchus contractus* Marsh. and certain halticid beetles which originally fed on Cruciferae but have recently developed strains thriving on *Tropaeolum*.

Such phenomena may set up a chain of events which may culminate in isolation mechanisms followed by speciation. Conceivably, the aberrantly chemotactic female races or strains become subject to selective pressure by the different milieu of the "new" host plant, or by a different degree of predation of certain phenotypes within the population, or perhaps also by the different chemical composition of the new source of food itself; and if the original host plant is rare or restricted to special environmental niches, whereas the "new" chemotactic variant is not (or the secondary host prefers other sites), an ecological (niche) isolation promoting inbreeding and hampering outbreeding, or even some geographical separation, may result. The "change-over" of species of *Papilio* to the Umbelliferae-Apioideae in temperate regions (from ranalean forms as "primary" hosts) thus becomes more plausible: these butterflies must have "migrated" *directly* to the Apioideae, because there are no records of other groups of Umbelliferae or Araliaceae as food plants of Papilionidae.

Phenotypic changes induced by the cumulation of mutative genetic processes in the separate populations may enhance the evolutionary isolation, more particularly when the pattern of sexual behaviour (sexual display and mating) is affected (e.g. by a morphological divergence), so that males no longer outbreed. Migration of females to areas in which the host plant is rare or absent might easily result in a local population with a different chemotactic behaviour if the egg-laying urge supersedes other instincts and eggs are deposited on aspecific substrates, e.g. on "substitute" substrates sufficiently *resembling* the characteristic host plant to act as an "Auslöser" of the egg layer instinct. A generalization of this phenomenon explains at least some of the change-overs of parasite groups to "secondary" host taxa by a primarily phytochemical convergence of the primary and the secondary host groups. That is why one must be aware of the pitfall of circular reasoning: the fact that *Pieris brassicae* and *Ceuthorrhynchus contractus* have switched to *Tropaeolum* as a secondary host plant undeniably has something to do with the phytochemical resemblance (the thioglucosides occurring in both Capparidales and Tropaeolaceae), and the evidence *may* point to a common genetic basis of the shared chemical characteristic (i.e. to a close taxonomic relationship between the taxa concerned), but the host–parasite relationship and the chemodiagnostic data are not necessarily independent or mutually corroborative.

Nevertheless, some workers place so much reliance on host–parasite relationships that they reject taxonomic classifications based on other (usually morphophenetic) criteria. Harris (1972) came to the conclusion that on the basis of their host spectra the Sphingidae-Semanophorinae (a sub-family of the hawkmoth family) can be arranged in five groups, which he believes to be taxonomic entities (i.e. tribes or subtribes), and he suggested that the existing taxonomic framework contains so many anomalies that the family should be totally rearranged. Published comments of other workers, who studied anatomical criteria, seem to corroborate his conclusions. Whether Harris is right or not is irrelevant as far as the future progress of sphingid taxonomy is concerned; his important contribution provides a new impetus to systematic studies of the group by opening up interesting perspectives. It must also be noted that Harris made allowances for shifts to secondary host groups, and for chemical similarities (possible convergences, as in the case of latex families), which renders his account more convincing. After this general discussion, some selected examples will be given of the taxonomic application of host–parasite relationships. In these examples the hosts are always higher plants, but the parasites are animal, fungal, or spermatophytic.

FUNGAL PARASITES AS TAXONOMIC POINTERS

Host–parasite relations are most useful if they open up new perspectives or add corroborative evidence in more or less unequivocal cases. Parasitic fungi are excellent systematists and may discriminate more precisely than one might expect. Some interesting examples have been borrowed from a paper by Patzke (1966), but there are additional ones.

Not everyone follows Hutchinson in the severance of the Allieae from the family Liliaceae mainly on account of the umbellate inflorescence of the former. As pointed out a considerable time ago by Savile (1954), *Puccinia asparagi* DC., a fungal parasite of the indubitably liliaceous genus *Asparagus*, grows to some extent on *Allium* when transferred to this host. This supports chemodiagnostic evidence that the Allieae are not more closely related to the Amaryllidaceae than to the Liliaceae (in Hutchinson's opinion). Insect parasites also confirm this inclusion of Allieae in Liliaceae rather than in Amaryllidaceae as we shall see.

The parasitic fungi of the genus *Peronospora* are strictly monophagous as a rule, so that one may wonder whether the exceptions do not signify some error in the taxonomic evaluation of the technical diagnostic characters supposed to separate different hosts concerned.

Most phanerogamists dealing with Compositae recognize the genera *Matricaria*, *Anthemis*, *Achillea* and *Chrysanthemum*; some distinguish, in addition, *Chamaemelum*, *Tripleurospermum*, and perhaps more segregates. These genera are all susceptible to attack by *P. radii* De Bary and one may indeed query the generic delimitation based on technical characters. These characters do break down: *Matricaria sensu lato* for instance is supposed to differ from *Anthemis* in that it has no paleae, but *Matricaria recutita* f. *paleata* has a paleate common receptacle. To my mind, this splitting is excessive and some genera, at least, appear to be artificial rather than "natural". The cruciferous genera *Brassica* (*sensu lato*), *Raphanus* and *Sinapis* (*sensu lato*) are all attacked by the same parasitic fungus (*Peronospora brassicae* Gäum. = *P. parasitica* (Fr.) Tul.) and also in this case one must admit that the genetic relationship between *Brassica* and *Raphanus* is very close, because intergeneric hybrids can be obtained (× *Raphanobrassica*). For several reasons, even on account of practical considerations, taxonomists may prefer to maintain the chamomiles, mayweeds, marigolds, etc. as different genera, and also to separate *Brassica* from *Raphanus* and *Sinapis*, but it is important to remember that the genera concerned may not be very "good" and will, perhaps, have to be merged again with allied ones when additional evidence becomes available. One should at least be prepared to consider such proposals.

Some interesting cases involve heteroecious parasites (i.e. forms with a change of host during each complete life-cycle), particularly the bulk of the Uredinales. The taxonomic argument is more cogent if two host taxa are shared by the parasitic taxon, and it is of course also more significant as an indication of the mutual relationships of the members of one of the host groups. Although most agrostologists do not now combine *Bromus* with *Lolium* and *Festuca* (and some other genera) in the Festuceae, the mycotaxonomic evidence refuting the inclusion of *Bromus* in the *Festuceae* is still highly instructive. The genus *Bromus* is attacked by the uredo-telial stage of *Puccinia symphyti-bromorum* F. Müller (= *P. recondita* Rob. & Desm. f. sp. *bromina* Eriks.) which has the boraginaceous *Symphytum* (Anchuseae) as its first (aecidial) host. The closely related *Puccinia cerinthes-agropyrina* Franzschel (= *P. recondita* f. sp. *echii-agropyrina* Gäum. & Terr.) attacks representatives of other boraginaceous tribes (e.g. *Cerinthe* and *Echium*) and the grass genus *Agropyron* (*sensu lato*), but there is no related uredinalean species living in its aecidial phase on Boraginaceae and in its telial phase on *Festuca*, *Lolium*, or on other undisputed Festuceae. *Bromus* is clearly associated with such genera as *Triticum*, *Agropyron* (*Elytrigia*), *Hordeum*, *Elymus*, etc. which make up the Triticeae or Hordeae. The structure of the starch grains (i.e. compound in *Bromus*, simple in *Festuca*), the occurrence of glucoflavones in *Bromus* and in *Hordeum*, but not in *Festuca* or *Lolium*, and the parasitization of *Bromus* by *Ustilago bromivora* (Tul.) Fischer v. Waldh. and of *Agropyron*, *Elymus* and *Hordeum* by the very closely related, if not specifically identical, *U. bullata* Berk., are all clearly corroborative and confirm the alliance of *Bromus* with the group of genera centred around *Triticum*, *Agropyron*, *Elymus* and *Hordeum* beyond doubt.

A similar case is presented by the genus *Trisetum*. The superficial morphological resemblance of representatives of this genus with *Avena* (also reflected in vernacular or constructed vernacular names such as Yellow Oat, Goldhafer, Goudhaver) led to its original inclusion in the Aveneae, although the reference of the East Mediterranean species *Trisetum pumilum* (Desf.) Kunth to the genus *Koeleria* as *K. pumila* (Desf.) Domin (other synonyms are *Trisetaria pumila* (Desf.) Maire and *Lophochloa pumila* (Desf.) Bor) already indicates that *Trisetum* might belong to the Festuceae rather than to the Aveneae. This is clearly supported by host–parasite relationships. Related uredinalean fungi with species of *Sedum* (Crassulaceae) as the first (aecidial) host and *Trisetum* or *Koeleria* as the second (uredotelial) host are known: *Puccinia triseti* Eriksson ($=$ *P. recondita* f. sp. *triseti* Eriks.) is known from 8 or 9 species of *Trisetum* but after inoculation also grows on *Koeleria gerardii* Munro; and the related *P. longissima* Schroet. is known from several species of *Koeleria*. These rusts are not found on Aveneae. Other uredinological data (Holm, 1969) point to the inclusion of *Molinia* in the Arundineae (near *Phragmites*) rather than in Danthonieae or in a special tribe Molineae, and to the relationships between *Spartina* with the Chlorideae instead of the festucoid Phalaridae or Eragrostoideae.

The Salicaceae, at one time frequently included in the Monochlamydeae, Amentiferae (Amentiflorae), but nowadays usually considered to form a monotypic order Salicales of somewhat uncertain affinities, have been associated with the Flacourtiaceae (Violales) by some authors such as Hallier (1905, 1912) and Takhtajan (1959, 1969). Some workers regarded the evidence as extremely thin (e.g. Gilg, 1914) and the suggestion remained rather dubious or was at least not considered to be very convincing. However, the uredology points to a possible connection between the Salicaceae and the Flacourtiaceae-Idesieae: Holm (1969) showed that the occurrence of species of *Melampsora* with a Conifer (Pinaceae) as the aecidial host, and Salicales and one species on *Idesia* as their uredotelial hosts, is certainly a pointer to reckon with. This evidence agrees with the fact that the caterpillars of the Nymphalid genus *Atella* feed on *Salix* or on Flacourtiaceae, and recently the discovery of the identity of the benzoylated glucoside nigracin (previously only known from Salicaceae) with idesin (isolated from *Idesia*) and of related glucosides also found in both some Flacourtiaceae and Salicaceae makes an even stronger case for the taxonomic relationship between Salicales and at least some Violales.

Leppik (1966) has applied the host–parasite relationship in a phytogeographical study of gene centres of the genus *Cucumis*. He believes that the relative immunity to fungal parasites of the species with old, "established" areas of distribution is higher than that in secondarily invaded regions. This is a simplification, however: speciation must have proceeded during migration and the behaviour of the new taxa in newly invaded territory need not be a yardstick of the constitution of their older and more "sedentary" parent taxa. Still, it is an interesting idea which might be applied to other taxa.

INSECT PARASITES AND THEIR HOST PLANTS

If we consider the association of butterflies and plants, the case of the Papilionidae and the Polycarpicae is interesting if we assume the co-evolution of plant orders from the magnolialean–ranunculalean–lauralean nexus, and this butterfly group. Rutales–Sapindales, subsequently Umbelliferae–Araliaceae and ultimately Asterales are most probably derived from some ranalean–rutalean stock. Papilionid butterflies associated with Rutaceae, Umbelliferae, etc. are most probably derived from ancestors associated with ranalean progenitors of the Rutalean and Aralialean (*sensu stricto*) assemblies.

The association of two unrelated groups of beetles with Centrospermae has been discussed by Tempère (1967), cf. Table I. Both certain curculionid and certain chrysomelid beetles (which are specialized according to this French entomologist) show a host spectrum which includes the Caryophyllaceae. This family is on present evidence "aberrant" among the Centrospermae in that the red or blue to purple flower colours are caused by the presence of anthocyanidins, whereas nearly all Centrospermae *sensu lato* contain betacyanidins (and betaflavones, together termed betalains) instead of anthocyanidins; for this reason Mabry has excluded the Caryophyllaceae (and "Molluginaceae") from the centrospermous assembly. The affinities of the Caryophyllaceae to the betalain-containing families included in the Centrospermae *sensu lato* are quite clear, in my opinion, from morphological, anatomical, embryological and palynological data, whereas the chemical discrepancy may not be so great as it appears, but details will not be discussed here. The cantharotaxonomic data provided by Tempère confirm the close relations between the centrospermous families and indicate that the exclusion of the Caryophyllaceae from the assembly is distinctly erroneous.

The subfamily Criocerinae of the beetle family Chrysomelidae is almost always found on Liliaceae (and the associated *Smilax*, *Dracaena*, Dioscoreaceae), and, in addition, on Commelinaceae, Gramineae, and Orchidaceae *Lilioceris merdigera* (L.) (the lily beetle) attacks *Lilium* (*L. martagon* L. and related taxa particularly), but also *Convallaria*, *Polygonatum*, and wild and cultivated species of *Allium*, but has not been recorded from Amaryllidaceae. (N.B. The related asparagus beetle *Crioceris asparagi* L. attacks *Asparagus*). The inclusion of the tribe Allieae in the Amaryllidaceae is refuted by the evidence of this group of beetles. It is interesting that the most primitive Criocerid beetle with a known host plant is *Pseudocrioceris*, feeding on *Dracaena*. The taxonomic evidence provided by the tribes Cionini and Mecinini of the phytophagous Curculionidae (weevils) stresses, among other things, the relationships of *Buddleia* with Scrophulariaceae rather than with Loganiaceae, and the close affinity between the Plantaginaceae and several taxa of the Tubiflorae. The absence of these weevils on Solanaceae, a family often placed in juxtaposition to the Scrophulariaceae (and some related groups) is noteworthy. The explanation is conceivably that (as in the case concerning the most spectrum of the criocerinid beetles, which includes Liliaceae *sensu lato*, etc. but not the Amaryllidaceae) a chemical or biochemical repellent mechanism developed which prevented the changeover to the (related) families: both Solanaceae and Amaryllidaceae contain poisonous alkaloids.

Other examples, not to be discussed here, can be found in papers by Forbes (1958), Ehrlich and Raven (1965, 1967), Tempère (1967) and Crowson (1970). A special discussion will be devoted to the Pieridae and their hosts (see next section).

THE PIERIDAE AND THEIR HOST PLANTS

A survey of the representatives of the Cabbage White Family and their food plants (Ehrlich and Raven, 1965; Terofal, 1965; cf. the concise summary of their data in Table II) reveals that within the family there is a considerable divergence in host preference which roughly coincides with the principal subfamilies more or less generally accepted by lepidopterologists (cf. Klots, 1933; Ehrlich, 1958; Terofal, 1965). Although most of the present genera are associated with the thioglucoside families (and in this host group primarily, no doubt, with the Capparidales *sensu* Takhtajan), it is by no means certain that such thioglucoside plants constituted the primary host group of the Pieridae.

Unfortunately the biology of the forest-dwelling, monotypic West African subfamily Pseudopontiinae is unknown. Its habitat—tropical rainforest—suggests a host taxon other than Capparidaceae (which do not occur in dense lowland forest biotopes as a rule and prefer more open sites), but this evidence is inconclusive; moreover ancestral Capparidales might have preferred more wooded environments. If the Dismorphiinae are also more primitive than the remaining subfamilies, the only (incomplete) records based solely on the palaearctic genus *Leptidea* point to Leguminosae rather than Capparidales as their normal food plants, and might be taken as an indication of Leguminosae being the primary host group of the family Pieridae, but reliable life histories of the neotropical *Pseudopieris* and *Dismorphia* are apparently still lacking. This is surprising because the almost riotous incidence of mimicry and the, perhaps associated, development of large sex patches in the (appropriately named!)

TABLE I. Hosts of certain species of beetles: Centrospermae/Polygonales (data from Tempère, 1967)

Phytophagous Beetles	Chenopodiaceae (Betal.)	Amaranthaceae (Betal.)	Caryophyllaceae (Anthoc.)	Basellaceae (Betal.)	Portulacaceae (Betal.)	Polygonaceae (Anthoc.)
Curculionidae-Hypurini:						
Pseudophytobius acalloides Fairm.	N: Chenopodiaceae					
Hemiphytobius sphaerion (Bohem.)	E: *Chenopodium Atriplex*					
			N: *Silene nutans* L. (S. spec., *Melandrium album*. (Mill.) Garcke, *Arenaria montana* L.) E: other Caryophyllaceae			
Hypurus bertrandi (Perris)				E: *Boussaingaultia baselloides* H.B.K.	N: *Portulaca oleracea* L. E: other genera, e.g. *Claytonia*	

15. Co-evolution of Plant Hosts and Parasites

Chrysomeleidae:
Cassidinae

Cassida vittata Vill.	N: *Atriplex, Beta, Salsola*	N: *Spergula Spergularia*
C. nobilis L.		N: various taxa

Halticidinae:

Chaetocnema tibialis (Illig.)	N: *Atriplex, Beta, Salsola, Salicornia, Suaeda*	
Ch. concinna (Marsh.)	(N: *Beta*)	

Chrysomelinae:

Gastroidea polygoni L.	(N: *Beta*)	N: *Polygonum Rheum, Rumex*
		N: *Polygonum*

Names in parentheses: more exceptional occurrence.
Betal. = betacyanin family, Anthoc. = anthocyanidin family.
N = nominal host.
E = adequate host in experiments.

TABLE II. Principal food plant taxa of pierid butterflies (data from Terofal, 1965; Ehrlich and Raven, 1965; and other sources)

Subfamily	Number of genera	Food Plant Taxa (T indicates that the food plant group contains thioglucosides)	Remarks
1. Pseudopontiinae	1	unknown	
2. Dismorphiinae	3	Leguminosae (if not unknown)	Some leguminous taxa (e.g. Mimosaceae) contain other organic sulphur compounds
3. Coliadinae	11 or 12	7 genera (mainly) on Leguminosae, 3 (or fewer) unknown, representatives of these and remaining genera (such as *Gonepteryx*) on various other non-T hosts (except one recorded from Euphorbiaceae, and *Phoebis* also on Capparidales)	
4. Pierinae	39–42	Pierinae-Euchloini (7 gen.) almost always on Capparidaceae (T) or Cruciferae (T); Pierinae-Pierini on T-families (Capparidaceae, Cruciferae, Resedaceae, Tropaeolaceae, Salvadoraceae (about 20 gen.)), but also on Loranthaceae and Santalaceae (about 6 gen.) and about 5 genera (also) on various other T-hosts (Euphorbiaceae, Salvadoraceae) and non-T hosts (representatives of about 6 genera)	(4) Of very rare occurrence on Leguminosae, sometimes only occasionally and not consistently (e.g., *Ascia*, oligophagous on at least six families)
5. Teracolinae	9 (Terofal)	Nearly all on Capparidales	(5) Subfamily not recognized by Ehrlich and Raven

genus *Dismorphia* has certainly focussed the attention on this interesting and diversified taxon. The discovery of an association of even a single South American dismorphiinid taxon with a representative of the Capparidales (or the Tropaeolaceae) would change the picture completely. If both the neotropical *Dismorphiinae* and the palaearctic genus *Leptidia* descended from a tropical group of progenitors, now largely extinct and perhaps of palaeotropical origin, the migration of *Leptidia* to a more temperate climatic zone (where Capparidaceae are scarce) necessitated a change of host. Many palaearctic Pieridae live on Cruciferae (which replace the Capparidaceae in cooler regions) and this suggests a host substitution; if *Leptidia* changed its host taxon, it could have shifted to an altogether different group of food plants (outside the Capparidales), but it does not follow that the neotropical representatives (which did not migrate to cooler climatic regions) changed their host specificity.

The remaining tribes were at one time united into a single subfamily, but more recently at least two separate subfamilies are distinguished, viz., the Pierinae (subdivided into Euchloini, the Orange Tips tribe, and Pierini, the Cabbage Whites tribe) and the Coliadinae. As far as the entomological evidence goes, the basic Euchloinid genera are the most primitive of all the Pierinae. Unfortunately the food plant of the key genus *Eroessa* is unknown; the hosts of the other six genera belong almost always to the Capparidales (one genus also feeds on Loranthaceae). A generalization is always hazardous, so that it is wiser simply to conclude that if *Eroessa* is associated with a capparidalean host there is a strong case for the assumption that the primary host group of the *Euchloini* is the thioglucoside-containing group of the Capparidales, with the woody and more tropical Capparidaceae prior to the herbaceous and temperate Cruciferae. For the time being I believe that the Capparidales indeed represent the most likely primary host group, so that the association of the majority of the Coliadinae with Leguminosae and with other non-thioglucoside hosts must be of a secondary nature.

A few coliadinid genera feed (partly) on Capparidales and this might be taken as an indication of the original host group of the subfamily. The primitive Pierinae-Pierini, *Eronia*, *Pareronia*, *Colotis*, *Ixias*, and their associated genera, also typically associated with Capparidales, are supposed to have an Euchloine derivation which is certainly not contradicted by their host specificity. According to Klots (1933, p. 193) they may be closely allied to or derived from ancestral stock from which the Coliadinae are derived, which again suggests that the original host plants of the Coliadinae were capparidalean rather than fabalean. The genera constituting the *Delias* group of the Pierini are supposed to have been derived from a progenitor nowadays best approximated to by the genus *Aporia* (the latter with Berberidaceae and sometimes Rosaceae as food plants). One might think of a "shift" from such woody hosts to Loranthaceae parasitizing these primary hosts, but *Cepora*, also supposed to be derived from some early Aporiid stock and morphologically belonging to the *Delias* group, is clearly associated with Capparidaceae. These indications suggest that the transition to Loranthaceae (the principal host taxon of the *Delias* group) took place in the tropics, starting from Capparidales, and that the palaearctic *Aporia* independently shifted to other families in the temperate region (presumably also from Capparidales).

Although the evidence is not conclusive—life histories of Dismorphiinae and of the euchloinid *Eroessa* are badly needed—the common host group of at least

Fig. 3. Migration to secondary hosts.

the ancestral Euchloini, Pierini and Coleadinae may well have been of proto-capparidalean affinity, and, if so, most probably contained thioglucosides. This is at least a starting-point for deductions concerning host speciation and co-evolution, supported by a good many experimental data. For the taxa feeding as caterpillars on the Capparidales the guiding chemical compounds are undoubtedly the mustard oil glucosides, but Verschaffelt (1910), whose experiments were later repeated and confirmed (cf. Terofal, 1965), showed that oviposition by gravid females is, at least in some cases, evoked by degradation products of these glucosides (more particularly the aglycone, i.e. the mustard oil), which products are more volatile than their glucosides. In numerous experiments the relation between the host specificity and particular thioglucosides has been confirmed (Hovanitz *et al.*, 1963; Hovanitz and Chang, 1964; Terofal, 1965; Schoonhoven, 1968; Lourens, 1971).

This is a good starting-point to speculate on host transfer (Fig. 3), because in other experiments anomalous food plants offered to larvae of pierids associated with thioglucoside hosts were shown to be eaten in the majority of the cases as long as the substitute food plant contained mustard oil glucosides. The larvae discriminate between different plants manifestly on account of the different glucosides contained in these plants. In order to eliminate other and conceivably repellent properties of the plants, such as hairiness, consistency, fibre content, and other chemical factors, leaves of a "neutral" plant such as lettuce smeared with extracts of thioglucoside plants were used, and even altogether artificial substrates. By using standard media Lourens could work quantitatively and show a concentration effect. Higher concentrations of thioglucosides are suboptimal, most probably because a repellent or even toxic action prevails over the attraction as a "guiding" substance and potential host indicator. Some aspecific mustard oil glucosides are more readily taken than others, which may be associated with a lesser or a greater repellent action on pierid larvae.

Most interesting are the results of "conditioning" mentioned before; caterpillars reared on aspecific host plants from the egg stage give rise to females which normally prefer this aspecific host (and the corresponding glucoside) to the nominal food plant (and the nominal thioglucoside spectrum) of the pierid species in question. A shift to other thioglucoside families is thus explained, but this does not necessarily imply, as we have seen, that these plant families are closely related taxonomically.

The obviously secondary host group of the Santales is completely devoid of thioglucosides and the host specificity of the *Delias* group must have come about as the result of a different process of adaptive evolution. Conceivably, the proximity of a loranthaceous parasite growing on a primary host of Pierid caterpillars and the tissues of the common host containing mustard oil may have induced gravid females to lay eggs on the mistletoe instead of on the host plant proper (the foliage often intermingles), or, alternatively, the contamination of the parasitic plant with traces of thioglucosides may have been high enough to deceive the caterpillars. Perhaps a chemical stimulus of the *primary* host on caterpillars in a feeding mood was sufficient to cause incidental gnawing of loranthaceous leaves intermingled with the leafy twigs of the primary hosts. Reproducible experiments show that pierid caterpillars can be reared successfully on thioglucoside-free plants smeared with extracts of thioglucoside plants (and even with purified glucosides alone), and *Pieris brassicae* has repeatedly been recorded as a pest in crops of *Allium*. Apparently, the thioglucosides, though normally vital for normal oviposition and caterpillar development, are not altogether indispensable.

Nowadays the host specificity of the caterpillars of *Delias* is rather striking. I remember that as a boy I noticed broods of caterpillars of this genus which had completely eaten away all leaves and young twigs of their loranthaceous host plant and frantically crawled around on the remaining larger stems, but did not make attempts to feed on the host of the mistletoe (often *Citrus* as far as I remember), ultimately to pupate precociously; one of the smallest specimens of *Delias hyparete* (L.) I ever saw was reared from such a small pupa. It would be very interesting to study the behaviour of pierid caterpillars associated with Loranthaceae in such cases of starvation if the mistletoe in question is growing on a capparidaceous host: some simple laboratory experiments with detached almost bare branches of caterpillar-infested Loranthaceae and twigs of other (both thioglucoside and non-glucoside) plants might yield very interesting results.

The association of pierid taxa with Loranthaceae may well account for certain inconsistencies and even contradictions as regards the host specificity of certain genera. Caterpillars and pupae collected on a tree may actually have been associated with loranthaceous parasite of that tree (and the same holds for oviposition on a tree: not all observers, even if they are excellent entomologists, have had a good botanical training, and the loranthaceous parasites often resemble the foliage of their host at least in size and colour!), so that the host specificity may be greater than the records (as summarized by Ehrlich and Raven, 1965, and by Terofal, 1965) suggest, and discrepancies can be explained away. As an example, the genus *Pereute*, very close to *Delias*, is recorded from Loranthaceae and from Lauraceae (*Ocotea*), but it would not be surprising if it is exclusively associated with Loranthaceae.

As I pointed out in an earlier section of this paper, a "shift" to different hosts may lead to speciation, particularly if it is concomitant with an ecological and/or geographical isolation of the populations feeding on different taxa. An illustrative example is provided by the *Pieris napi* complex, which has a palaearctic and nearctic distribution and shows speciation at the intraspecific level. In the lowlands of western Europe the subspecies *Pieris n. napi* L. occurs which feeds on the somewhat succulent cruciferous genera *Nasturtium* and *Rorippa* growing in moist places, whereas the more or less clearly vicariant, montane *P. n. bryoniae* Huebn. feeds on smaller and less juicy genera such as *Arabis* and *Biscutella* which prefer drier sites. Similar conditions seem to obtain in the North American part of its range, where, e.g. *P. n. virginiensis* Edw. has a limited area of distribution, most probably attributable to its host plant speciation. The host speciation within the *P. napi* complex must already have resulted in an appreciable degree of genetic isolation, because there is a noticeable morphological distinction between the various subspecies. Attractant selection and host preference tests with different subspecies and with hybrids between subspecies with a clear-cut difference in host specialization will almost certainly yield important information (cf. Hovanitz and Chang, 1963a, b).

Summarizing, one may conclude that the Pieridae not only provide fine examples of various aspects of host–parasite relations applicable to taxonomic inquiry, but may also serve as objects for further experimental studies which will eventually elucidate some of the remaining problems.

PARASITIC HIGHER PLANTS AND THEIR HOST SPECIFICITY

A number of higher plants are dependent on specific hosts. Representatives of such families as Loranthaceae, Santalaceae, Rafflesiaceae, Cytinaceae, and Orobanchaceae, the genus *Cuscuta*, and many parasitic and semi-parasitic Scrophulariaceae are frequently associated with few hosts or even with a single one. There are interesting examples of monophagy and oligophagy. Certain Loranthaceae occur only on Conifers, one dwarf species is associated with a species of *Euphorbia*; species of *Rafflesia* are parasitic on Vitaceae *sensu lato*; and many of the representatives of the genera *Orobanche* and *Cuscuta* have only been recorded from single specific hosts. The explanation of the host specificity must be sought in a compatibility, especially in those cases in which there is protoplasmic continuity between the cells of the host and the haustorial cells of the parasite (plasmodesmata in cell walls between adjoining cells of host and parasite: *Cuscuta* is frequently used as a go-between to transfer a virus disease from one plant to another). Other experimental evidence indicates an influence of roots of potential hosts on the germination of the seeds of their parasites (e.g. of *Orobanche* and the scrophulariaceous genus *Striga*: cf. Edwards, in Harborne, 1972, pp. 235–248).

As regards the use of the relations between phanerogamic parasites and their hosts as taxonomic pointers one can conclude that there have been only a few practical results, perhaps because one has not looked deliberately for correlations. That taxonomically interesting relations are not altogether lacking is shown by observations of Wijnands (1972). The European species *Orobanche hederae* Duby normally occurs on *Hedera helix* L. and on *Hedera colchica* (C. Koch) C. Koch, but there are several records of its spontaneous occurrence on alien species of Araliaceae cultivated in botanic gardens in Europe such as *Fatsia japonica* (Thunb.). Dcne & Planch. and *Tetrapanax papyrifer* (Hook.) K. Koch. Presumably there are other potential araliaceous hosts, because the incidence of parasitization on secondary hosts is by no means low: in 1969 seven plants of *Orobanche hederae* were recorded on *Fatsia* in the Leyden botanic garden, and in 1972 twenty-three specimens on that anomalous host in the

Amsterdam botanic garden. The plants found growing on *Fatsia* appear to be vigorous and quite normal, so that the anomalous host is certainly adequate and not a second choice substitute. The most interesting record was the find of three normal plants of *Orobanche hederae* on the umbelliferous *Silaum silaus* (L.) Schinz & Thell. in the Amsterdam garden in 1972. Nearly all phanerogamists consider the two families Araliaceae and Umbelliferae (Apiaceae) to be very closely related, so that if a host of a parasite on araliaceous taxa outside the family might be expected, an umbelliferous taxon would be a most likely candidate.

ACKNOWLEDGEMENTS

The assistance of Dr J. A. von Arx (Centraal Bureau voor Schimmelcultures, Baarn) and of the staff of the Division of Entomology of the Institute of Taxonomic Zoology, Amsterdam, with points of nomenclature and with references is gratefully acknowledged.

REFERENCES

CARSON, H. L. (1971). The ecology of Drosophila breeding sites. *Univ. Hawaii H.L. Lyon Arbor. Lecture* **2**, p. 1-27.

CROWSON, R. A. (1970). "Classification and Biology." Heinemann, London, 350 pp. (Esp. Chapter 10: Hosts, parasites and classification, pp. 115-130).

DETHIER, V. G. (1941). Chemical factors determining the choice of food plants by Papilio larvae. *Am. Nat.* **75**, 61-73.

DETHIER, V. G. (1954). Evolution of feeding preferences in phytophagous insects. *Evolution* **8**, 33-54.

EHRLICH, P. R. (1958). The comparative morphology, phylogeny and higher classification of the Butterflies (Lepidoptera: Papilionoidea). *Univ. Kansas Sci. Bull.* **39**, 307-370.

EHRLICH, P. R. and RAVEN, P. H. (1965). Butterflies and plants: a study in co-evolution. *Evolution* **18**, 586-608.

EHRLICH, P. R. and RAVEN, P. H. (1967). Butterflies and plants. *Sci. Am.* **216**, 104-113.

FORBES, W. T. M. (1958). Caterpillars as botanists. *Proceed. 10th Intern. Congr. Entomol.* (*Montreal* 1956) **1**, 313-317.

FRAENKEL, G. (1956). Insects and plant biochemistry. The specificity of food plants for insects. *Proc. 14th Int. Congr. Zool.* 383-387.

FRAENKEL, G. (1959). The raison d'être of secondary plant substances. *Sciences* **129**, 1466-1470.

GILG, E. (1914). Zur Frage der Verwandtschaft der Salicaceae mit den Flacourtiaceae. *Engl. Bot. Jb.* **50**, *Suppl.* (*ENGLER-Fest-Band*), 424-434.

HALLIER, H. (1905). Provisional scheme of the natural (phylogenetic) system of flowering plants. *New Phytol.* **4**, 151-162.

HALLIER, H. (1912). L'origine et le système phylétique des angiospermes. *Arch. néerl. Sci. Exact. Nat.* III, B (*Sci. Nat.*) **1**, 146-234.

HARBORNE, J. B. (ed.) (1972). "Phytochemical Ecology." Academic Press, London and New York.

HARRIS, P. (1972). Food-plant groups of the *Semanophorinae* (Lepidoptera: Sphingidae): a possible taxonomic tool. *Can. Ent.* **104**, 71–80.

HARTL, D. (1963). Das Placentoid der Pollensäcke, ein Merkmal der Tubifloren. *Ber. Deut. Bot. Ges.* **76**, 70–72.

HOLM, L. (1969). An uredinological approach to some problems in angiosperm taxonomy. *Nytt Magas. f. Botanikk* **16**, 147–150.

HOVANITZ, W. and CHANG, V. C. S. (1963a). The effect of hybridization of host-plant strains on growth rate and mortality of Pieris rapae. *J. Res. Lepidopt.* **1**, 157–162.

HOVANITZ, W. and CHANG, V. C. S. (1963b). Change of food plant preference by larvae of *Pieris rapae* controlled by strain selection, and the inheritance of this tract. *J. Res. Lepidopt.* **1**, 163–168.

HOVANITZ, W. and CHANG, V. C. S. (1963c). Selection of allyl isothiocyanate by larvae of *Pieris rapae* and the inheritance of this tract. *J. Res. Lepidopt.* **1**, 169–182.

HOVANITZ, W. and CHANG, V. C. S. (1964). Adult oviposition responses in *Pieris rapae*. *J. Res. Lepidopt.* **3**, 159–172.

HOVANITZ, W., CHANG, V. C. S. and HONCH, G. (1963). The effectiveness of different isothiocyanates on attracting larvae of *Pieris rapae*. *J. Res. Lepidopt.* **1**, 249–259.

KJAER, A. (1960). Naturally derived isothiocyanates (mustard oils) and their parent glucosides. *Fortschr. Chem. Org. Naturst.* **18**, 122–176.

KJAER, A. (1963). The distribution of sulphur compounds. *In* "Chemical Plant Taxonomy" (T. Swain, ed.), pp. 453–473. Academic Press, London and New York.

KJAER, A. (1966). The distribution of sulphur compounds. *In* "Comparative Phytochemistry" (T. Swain, ed.), pp. 187–194. Academic Press, London and New York.

KLOTS, A. B. (1933). A generic revision of the Pieridae (Lepidoptera), together with a study of the male genitalia. *Ent. Am.* (N.S.) **12**, 140–254.

LAST, F. T. and WARREN, R. C. (1972). Non-parasitic microbes colonizing green leaves: their form and functions. *Endeavour* **31** (no. 114), 143–150.

LEPPIK, E. E. (1966). Searching gene centers of the genus *Cucumis* through host–parasite relationship. *Euphytica* **15**, 323–328.

LOURENS, J. H. (1971). Voedselspecialisatie bij Pieridae (Food plant specialization in Pieridae). Unpublished M.Sc. report, Dept. of Applied Entomology, Univ. of Amsterdam (in Dutch).

MABRY, T. J. (1966). The Betacyanins and Betaxanthins. *In* "Comparative Phytochemistry" (T. Swain, ed.), pp. 231–244. Academic Press, London and New York.

MACIOR, L. W. (1971). Co-evolution of plants and animals—systematic insights from plant-insect interactions. *Taxon* **20**, 17–28.

MERZ, E. (1959). Pflanzen und Raupen. *Biol. Zbl.* **78**, 152–188.

PARMELEE, J. A. and SAVILE, D. B. O. (1954). Life History and Relationships of the Rusts of *Sparganium* and *Acorus*. *Mycologia* **46**, 823–836.

PATZKE, E. (1966). Biologische Indizien zur Klärung von Verwandtschaftsverhältnissen. *Ber. Deut. Bot. Ges.* **79**, 489–497.

PIPKIN, S. B. (1964). New flower breeding species of *Drosophila* (Diptera: Drosophilidae). *Ent. Soc. Washington* **66**, 217–245.

PIPKIN, S. B., RODRÍGUEZ, R. L. and LÉON, J. (1966). Plant host specificity among flower-feeding Neotropical *Drosophila* (Diptera: Drosophilidae). *Am. Nat.* **100**, 135–156.

SAVILE, D. B. O. (1954). The Fungi as Aids in the Taxonomy of Flowering Plants. *Science, N.Y.* **120**, 583–585.

SCHOONHOVEN, L. M. (1968). Chemosensory bases of host plant selection. *A. Rev. Ent.* **13**, 115–136.

STEBBINS, G. L. (1970). Adaptive radiation of reproductive characteristics in Angiosperms, I: Pollination mechanisms. *In* R. F. Johnston *et al.*, *Annual Rev. Ecol. and Systematics* **1**, 307–326.

TEMPÈRE, G. (1967). Un critère méconnu des systematiciens phanérogamistes: l'instinct des insects phytophages. *Botaniste* **50**, 473–482.

TEROFAL, F. (1965). Zum Problem der Wirtspezifizität bei Pieriden (Lep.). Unter besonderer Berücksichtigung der einheimischen Arten *Pieris brassicae* L., *P. napi* L. und *P. rapae* L. *Mitteil. Münchener Entom. Ges.* **55**, 1–76.

THORSTEINSON, A. (1960). Host selection in phytophagous insects. *A. Rev. Ent.* **5**, 193–218.

VERSCHAFFELT, E. (1910). The cause determining the selection of food in some herbivorous insects. *Kon. Akad. Wetensch. Amsterdam, Proceed.* **13**, 536–542.

WIJNANDS, D. O. (1972). *Orobanche hederae* Duby op *Fatsia japonica* (Thunb.) Dcne. et Planch. en op *Silaum silaus* (L.) Schinz et Thell., een nieuwe waarneming voor Nederland. *Gorteria* **6**, 108.

WOHLPART, A. and MABRY, T. J. (1968). The distribution and phylogenetic significance of the betalains with respect to the Centrospermae. *Taxon* **17**, 148–152.

16 | Adaptive Significance of Major Taxonomic Characters and Morphological Trends in Angiosperms

F. EHRENDORFER

Institute of Botany, University of Vienna, Austria

Abstract: A survey is presented of several morphological (and anatomical) traits and trends in the vegetative and reproductive characteristics of Angiosperms that are important for their major taxonomic classification. Many of these traits and trends are clearly related to particular conditions of environment, pollination, fertilization, seed development, dispersal and seedling establishment. While they consequently appear to be of adaptive significance, such an interpretation is difficult in many cases and apparently impossible in others. Some of the reasons for this situation are discussed in general terms (p. 323 *et seq.*). In order to improve our still largely hypothetical knowledge about the macro-evolutionary differentiation of Angiosperms, several methods of approach are suggested (p. 326).

INTRODUCTION

It is still regarded as highly controversial whether the morphological and anatomical traits and trends used to characterize Angiosperms, their classes, orders and families have (or had) some adaptive significance: is this never, sometimes, often or always the case? As an example of the changing views on the subject one may briefly refer to that most significant character of the group, to their angiovulate condition, i.e. the enclosure of their ovules in carpels. As long as wind-pollination (anemophily) was regarded as a primitive feature in Angiosperms, such carpels, obstructing the direct passage for the pollen and pollen tubes with sperm cells to the ovules and egg cells had to be regarded as a useless, or even a disadvantageous character. Most students now believe that the earliest Angiosperms were beetle-pollinated, possibly somewhat similar to certain Annonaceae, a family of the generally primitive Magnoliales (Gottsberger, 1970). In species of *Guatteria*, for example, the flowers develop very slowly (over a period of several weeks to several months); as the greenish perianth leaves turn yellowish, an intense fruit smell is produced which particularly attracts beetles that normally inhabit and feed on fruits. The perianth finally folds together to form a pollination chamber where the beetles stay, feed from the softening perianth parts, deposit their eggs, pollinate the stigmas and carry away pollen to other flowers. As delicate ovules are likely to be devoured or damaged by such robust phytophagous flower visitors, protection of them evidently must have been advantageous (Grant, 1950). The angiovulate condition therefore can be interpreted as an adaptation to such forms of zoophily.

Other aspects of the zoophilous syndrome of Angiosperms are hermaphrodite flowers, which economize by the deposition and collection of pollen in a single step; and a perianth, useful first for protecting the buds, and later for attracting visitors optically as well as olfactorily. We can assume that it was this system of employing insects to carry pollen precisely and economically, and even over longer distances, from one plant and flower to other plants and flowers of the same population, which gave early Angiosperms, possibly growing quite scattered as an understory in tropical-montane mesozoic Gymnosperm forests, the decisive adaptive advantage over their less effectively wind-pollinated predecessors (Ehrendorfer, 1971, p. 664). It is remarkable in this connection that parallel adaptations and characters were formed in a now extinct sister group of early Angiosperms, the Bennettitales, where a perianth and hermaphrodite flowers are combined with a totally different female cone carrying single naked ovules protected by intercalated sterile bracts.

In this short paper it is impossible to give more than an outline of the salient features and selected examples to illustrate these complex problems. For some important vegetative and reproductive traits and trends in Angiosperms, I shall try to give possible functional interpretations, and suggest where adaptations seem to have come into play; for many others which may be non-adaptive, it may be asked if I shall have to confess our ignorance at present. Does this bring us into conflict with current ideas on synthetic evolutionary theory? In analysing examples for seemingly non-adaptive traits and trends, we may find an answer to this question. We may also ask what can be done in the future to advance our understanding of the functional aspects of Angiosperm macro-evolution and major systematics. This will be considered in the final section.

For detailed information and further literature on the nature of major taxonomic characters and trends in Angiosperms the reader is referred to more general publications, e.g. Stebbins (1951, 1967), Melchior and Werdermann (1954, 1964), Takhtajan (1959, 1969), Cronquist (1968), and Ehrendorfer (1971).

POSSIBLY ADAPTIVE TRAITS AND TRENDS

1. Vegetative Trends

The trends in growth- and life-form, from trees to lianas, shrubs and herbs, perennial and finally annual, are generally recognized today as important—though not always irreversible—guide lines in the evolution of Angiosperms, radiating from tropical-montane forests into the numerous habitats of our biosphere, and filling its many ecological niches. Growth is continuous or interrupted by drought or cold, with consequent development of protective bud scales, and the shift from evergreen to deciduous. There is evidence for repeated changes from little-branched monopodial and pachycaulous types with few, large shoot apices to strongly branched, sympodial leptocaulous growth-forms with numerous smaller apices (Corner, 1968). This evidently increases the speed of development; it is advantageous under periodically unfavourable conditions, and reduces the risk of damage to apical buds.

. Secondary growth of the eustelic bundle system is a prerequisite for the primitive tree form of the Angiosperms. Transformation of tracheids to vessels, and the shortening and broadening of vessels and sieve tubes, are generally recognized as progressions. These, as well as the different organization-levels in the arrangement of parenchyma and libriform elements in xylem and phloem (Braun, 1970), appear to be expressions of improved (and/or specialized) conductive capacities. The shift to a herbaceous life-form involves reduction of secondary growth and is often connected with the replacement of the main root by an adventitious root system. Maximum rate of development has been reached in annual herbs by increasing neoteny. The aquatic Lemnaceae represent a culmination of this trend.

2. Reproductive Trends

Differences in flower (and fruit) morphology represent the most important major taxonomic characters in Angiosperms. Several relevant trends have been widely accepted, e.g. numerical reduction (oligomerization) and progressive fusion of various flower parts, or the shift from poly- to bi- and monosymmetry. Yet functional interpretations of such differences or trends have hardly been considered or have even been denied. A different approach was initiated by Stebbins (1951), who indicated the number of Angiosperm families where such changes (and others) had become established as differential characters. Thereby he convincingly demonstrated that these features were not combined by chance but were clearly correlated. Reduced stamen number, fused petals and inferior ovary, for example, form a particularly common combination. These phenomena were interpreted as adaptive syndromes and as results of natural selection.

(a) *Oligomerization*. The shift from a spiral to a cyclic arrangement of numerous floral parts is clearly correlated in Angiosperms with their numerical reduction and definite attachment, and stabilization in definite numbers, usually in whorls of 5 or 3. Commonly one can observe that reduced flower size is compensated by an increase in flower number, leading to the formation of various inflorescence types. The most important selective pressures responsible for these correlated trends may be those favouring speed and economy of individual flower development, and reduced risk of damage or of lack of pollination and seed set (because of more numerous flowers developing over a longer period of time). The predominance of 5- or 3-merous flowers is possibly connected with the most common phyllotaxies of 2/5 and 1/3.

(b) *Pollination types*. It has been known for a long time that different modes of pollination are correlated with complex syndromes of floral features (inflorescences, etc.), but it was stressed particularly by Vogel (1954). Functional aspects of these syndromes have been recently reviewed by Stebbins (1970) and Faegri and van der Pijl (1971).

Many traits and trends in Angiosperm flower morphology are clearly linked to animal pollination, i.e. zoophily. The morphological differentiation of perianth into calyx and corolla, or the origin of petals from stamens, reflect functional differentiation relative to protection and pollinator attraction. Shifts from a superior to an inferior ovary or one raised on a gynophore appear to be a continuation of protective trends. Reduction in stamen number and pollen production in animal-pollinated outbreeders is usually linked with the origin of nectar-producing devices, and is often reflected in a change from pollen-feeding to predominantly nectar-sucking flower visitors. The selective advantage for the plant of this trend may result from the reduced energy expenditure involved. It appears likely that reversals of this trend have resulted in secondary polyandry (e.g. in primitive Rosidae, Dilleniidae and Caryophyllidae), and are connected with secondary pollen feeding.

Other prominent floral changes evidently linked to zoophily are the fusion of flower parts, particularly petals, and the origin of monosymmetric (usually zygomorphic) flowers (Leppik, 1972). These trends result in the better protection and fixed position of stamens and stigma, give the flower the appearance of depth, facilitate the landing of dorsiventral pollinators and standardize their method of entry into the flower. This makes pollen uptake and application more precise, and has often led to further staminal reduction. It is interesting in regard to such floral changes that experiments with flower models have shown that pollinators clearly prefer those with clearly lobed margins and depth (Leppik, 1956; Kugler, 1970).

It can be demonstrated that selection by pollinators is actually responsible for the shaping of floral structure by the numerous, often surprising, analogies between simple flowers (euanthia) where the same functions are accomplished by organs of a different morphological nature (e.g. optical attraction by corolla, staminodes, stamens, calyx, bracts, etc.), and between euanthia, meranthia (part flowers, e.g. *Iris*), and pseudanthia or even double pseudanthia (e.g. Proteaceae, Euphorbiaceae, Compositae, Araceae) (Troll, 1928).

Further evidence for the role of function and selection in shaping of flowers and inflorescences comes from the well known fact that different animal groups are connected with different flower types, characterized by complex character syndromes: it is sufficient to mention only cantharophily, myophily, melittophily, psychophily, sphingophily, ornithophily and chiropterophily. A number of major taxa of Angiosperms are characterized by the predominance of one of these types, e.g. Aristolochiaceae, Araceae and Rafflesiaceae by myophily; Lamiaceae by melittophily; Fouquieriaceae and Liliaceae-Aloineae by ornithophily; while many others are more or less strongly differentiated in their floral biological specialization and radiation.

While shifts to autogamy are limited to species and species groups, changes from primary zoophily to secondary anemophily are characteristic of many major taxa of Angiosperms. In both instances "superfluous" devices for animal pollination are reduced, e.g. petals and nectaries. In addition, flower size reduction combined with unisexuality and numerical increase of male flowers (compensation for decreased probability of pollination), smooth and small pollen grains, enlargement of stigmas, and numerical reduction in number of ovules per ovary (pollination usually by single grains) are trends typically linked to the syndrome of anemophily. It is found in many families of Hamamelididae (= Amentiferae), but also in other subclasses, e.g. in Sanguisorbeae (Rosaceae) *Fraxinus* (Oleaceae), *Antemisia* (Asteraceae), etc.

(c) *Fertilization and seed development.* Trends towards the elaboration and numerical increase of apertures in Angiosperm pollen grains, often characteristic of major taxa, may be connected with improved germination. The cohesion of grains in tetrads, polyads or pollinia is more or less limited to groups with many ovules per ovary. Widespread and taxonomically very important tendencies are fusion of carpels, styles and stigmas (choricarpy → coenocarpy), and shifts from laminal to submarginal, and from axillary to parietal and central placentation. These changes may be of adaptive importance because with limited amounts of pollination they will tend to ensure the fertilization of as many ovules per flower as possible; furthermore, the increased and better integrated vascular supply between the carpels and placentas may improve the nutrition and development of ovules, embryos, endosperms and seeds.

The ovules and gametophytes of the Angiosperms are clearly more reduced than those of the Gymnosperms, and generally develop much faster. Various trends contribute further to this accelerated development, e.g. the change from crassi- to tenuinucellate ovules, the switch from monosporic to di- and tetrasporic embryo sacs in several orders and families, or the origin of pseudomonad pollen (from tetrads) in Cyperales.

Nutritional functions serving the embryo are gradually transferred to later stages of development, probably also for reasons of economy and reduction of risk. The primary (haploid) endosperm of Gymnosperms develops before fertilization, the secondary (usually triploid) endosperm of Angiosperms only after fertilization. In many derived Angiosperm families nutrient storage has been shifted from the secondary endosperm to the cotyledons of the embryo, and in Orchidaceae even the development of ovules is delayed until after successful pollination.

(d) *Seed dispersal and seedling establishment.* Differences in fruit and seed structure rank high as differential characters in Angiosperm macrosystematics. Their adaptive nature often is quite evident: hydrochorous and anemochorous seeds in Nymphaeaceae and Salicaceae, endozoochorous berries in Grossulariaceae, autochorous capsules in Balsaminaceae, anemochorous nuts in (more primitive) Compositae, etc. Functional aspects have recently been reviewed by van der Pijl (1969) and Stebbins (1971a). Relationships (and compromises) between seed size, dispersal, life form and habitat point to the complex nature of such adaptations. Examples include Pyrolaceae and Orchidaceae, where a clear connection exists between mycotrophy and enormous numbers but extremely small size of rather undifferentiated seeds. In the Fagaceae, relatively large, heavy, one-seeded fruits with a good nutrient supply are evidently a prerequisite for successful establishment of seedlings under the unfavourable conditions of shady deciduous and long-lived climax forests. In contrast, the related Betulaceae (*sensu stricto*) develop light, winged fruits with little nutrient supply; they have much better dispersal facilities but allow seedling establishment only in open, short-lived pioneer woodland. The Corylaceae occupy a somewhat intermediate position in many respects.

An obvious trend in Angiosperm reproduction is the transference of dispersal function from seeds to fruits, and sometimes to fruit assemblages. This trend is often connected with the reduction of seed numbers per ovary, with few- or one-seeded fruits which often remain closed. The change from seed to fruit also seems to confer the advantage of increased morphological versatility and adaptability; a comparison between the seeds of Hydrophyllaceae, Helleboreae (Ranunculaceae) or Spiraeoideae (Rosaceae) with the fruits of the related Boraginaceae, Anemoneae or Rosoideae will serve as an illustration.

Up to this point the survey seems to indicate that various differential characters, both vegetative and reproductive, in Angiosperm macrosystematics, do have adaptive significance. As most of these characters also appear at the microsystematic level (Stebbins, 1971b), their establishment may very well have been due to selective advantages in the early stages of evolution of the various groups. Coadaptation and co-evolution with animal symbionts and parasites must have been of great importance in many instances.

Most of the trends discussed can be associated with an advantageous acceleration of the developmental processes, usually resulting in neoteny and reduction. The consequence has usually been standardization of organs, compensation by numerical increase, further plasticity in differentiation and addition by integration of organs of lower into those of higher levels of organization. Risk reduction, improved economy, and exploitation of new resources by radiation are general aspects of these processes.

SEEMINGLY NON-ADAPTIVE TRAITS AND TRENDS

1. Examples

For quite a number of major traits and trends in Angiosperms not even satisfactory working hypotheses concerning their possible adaptive importance can be advanced as yet. Examples are: dicotyledony versus monocotyledony and many other seedling characters; alternate versus opposite (or whorled) leaves; simple versus compound leaves; elaboration of leaf margins; differences in nodal anatomy; presence or absence of internal phloem; scattered bundles in monocotyledons; open (polytelic) versus closed (monotelic) inflorescences; various aspects of structure, number, position and aestivation of floral parts; position of ovules; details of pollen structure; etc. My assumption is that even detailed and experimental studies will often fail to demonstrate *direct* adaptive significance of such characters. One may then ask if this negates the applicability of micro-evolutionary concepts to macro-evolutionary events? For various reasons I believe that this is not so.

2. Origin

Nearly all the features we utilize in Angiosperm macrosystematics have also been found as spontaneous or induced mutations within species (see e.g.

Schwanitz, 1959; Gottschalk, 1970). The recessive alleles involved in these differences may be carried along in the populations, particularly if they have some positive heterozygote or complementary effect. But even if they are neutral or momentarily slightly detrimental, their incorporation into the population may occasionally be possible, especially in early phases of rapid evolution. Under such conditions, small population size and exclusionary interaction of incipient species (Bock, 1972) will favour genetic drift, and selection pressure will be directed more towards divergence than towards close adaptation.

3. Opportunistic Evolution

The way in which an established structure may suddenly acquire new importance has recently been demonstrated in the case of extrafloral nectaries on leaves and calyces of Malvaceae. While they may only function as "sap valves" in many species, they have led to a symbiotic relationship with aggressive ants (*Camponotus* sp.) in the Brazilian *Hibiscus henningsianus*; this species is efficiently protected from phytophagous insects as long as its symbionts are present (Gottsberger, 1972). One may recall in this connection that extrafloral nectaries (probably functionally connected with pollinators) are a regular feature in Malpighiaceae. A parallel case is the origin and perfection of staminal nectaries in Ranunculaceae: starting with minimal and diffuse nectar secretion from still fertile stamens (e.g. *Pulsatilla*), they have developed through very simple nectar staminodes (e.g. *Trollius*) to enormously elaborate nectar containers (e.g. *Nigella*). In other members of the family this function has been changed again, to petaloid nectaries (e.g. *Ranunculus*), and to true petals (e.g. *Adonis*). All these examples tend to demonstrate the opportunistic and sometimes erratic course of evolution, where reconstruction of the original adaptive importance of certain traits will often be very difficult.

There are evidently numerous groups whose differential characters have become established as adaptations favoured by selection, under conditions different from those prevailing at present, so that today they even appear as detrimental. Such "obsolete adaptations" include: dioecy in entomophilous *Salix*; the perianth in aquatic Ceratophyllaceae; sympetaly in anemophilous Plantaginaceae; etc.

4. Unrecognized Correlations

Another reason for the seemingly non-adaptive nature of certain traits and trends may be found in their unrecognized correlation with adaptive features. A recently analysed example is the trend from 1- to 3- and multilacunar cotyledon nodes in Juglandaceae (Conde and Stone, 1970); we now know that this is connected with an adaptive progression from small and anemochorous fruits with epigeic germination to large, nutrient-rich, animal-dispersed fruits with hypogeic germination.

Similar correlations between adaptive and seemingly non-adaptive traits and trends (e.g. unisexuality, seed size) mark the evolution of (hemi)parasitic Angiosperms (Kuijt, 1969). This aberrant mode of life has been established in only a few lineages but was usually connected with very drastic and complex divergence, necessitating classification as separate subfamilies, families or even orders. Common adaptive trends are reduction of foliage and photosynthetic apparatus, replacement of roots by haustoria, neoteny, and sometimes unisexual flowers. Otherwise strategies diverge widely. Starting from the twining life form, the fully parasitic Cassythaceae (close to Lauraceae) and the Cuscutaceae (close to Convolvulaceae) have developed haustorial contacts with stems and leaves of their host plants; generally, both families have evolved in a remarkably parallel fashion. The other groups seem to have started as terrestial hemi-parasites, forming haustoria on the roots of their hosts. The Santalales, culminating in epiphytic members of Loranthaceae and Viscaceae, have retained autonomous germination of their relatively large seeds. Extreme neotenic reduction and fusion of flower parts as well as unisexuality may be a matter of economy here. Only one seed is produced per ovary, and neoteny may have served as a precondition for extreme specialization, leading from zoochory to autochory (as in *Arceuthobium*). The remaining orders and families of parasitic Angiosperms have switched to very small neotenic seeds whose germination has to be induced by the host. Two lineages develop large numbers of such seeds per ovary in large, mostly hermaphrodite flowers: Scrophulariaceae-Rhinantheae and Orobanchaceae, Rafflesiaceae and Hydnoraceae; Lennoaceae may belong to this type as a third lineage. Another type is represented by Balanophoraceae and Cynomoriaceae, where the production of only one tiny seed per ovary is linked to very small and (consequently?) unisexual flowers.

It seems that morphological novelties, even if neutral or slightly detrimental, may be maintained, particularly in early evolutionary phases. Nevertheless, even if major taxonomic traits and trends in Angiosperms have no direct adaptive importance, they often appear to be correlated in one way or another with the adaptive radiation of the group involved. Former adaptations, obsolete under present environmental conditions, and the opportunistic or even erratic evolution of morphological features and their function add to the difficult interpretation of seemingly non-adaptive traits and trends.

OUTLOOK

It is clear from the foregoing paragraphs that most of our functional interpretations of major morphological differences and trends in Angiosperms today are still no more than working hypotheses. As a conclusion, a few approaches may be mentioned that might improve this situation:

(a) Modifications and mutants, approximating major taxonomic differences, might be induced experimentally in larger numbers. Furthermore, closely related taxa that differ in such traits and trends should be selected.

(b) This material should be studied comparatively, using modern methods, particularly in regard to developmental aspects and energetic economy.

(c) Evolutionarily important aspects of competition at germination, seedling and adult stages, and of plant–animal relationships in parasitism (phytophagy, etc.), and symbiosis (pollination, seed dispersal, etc.) should be subjected to comparative studies under field and experimental conditions. Correlations between various phenomena involved would deserve particular attention.

(d) Multidisciplinary surveys and statistical evaluations of morphological traits and trends in different habitats, climates and other environmental conditions might give a clue to adaptive syndromes as yet unknown or unexpected.

It is evident that such approaches have as yet been applied very inadequately or not at all to the basic problem. Many new fields of cooperation should open up here between systematics and ecology. The data we can expect from such studies should make a fundamental contribution to our still very imperfect understanding of macro-evolution in Angiosperms.

REFERENCES

Bock, W. J. (1972). Species interaction and macroevolution. *Evol. Biol.* **5**, 1–24.

Braun, H. J. (1970). "Funktionelle Histologie der sekundären Sproßachse, I, Das Holz." Handbuch der Pflanzenanatomie, 2. Aufl., **IX/1**, Berlin.

Conde, L. F. and Stone, D. E. (1970). Seedling morphology in the Juglandaceae, the cotyledonar node. *J. Arnold Arbor* **51**, 463–477.

Corner, E. J. H. (1968). "The Life of Plants." Mentor Books, New York.

Cronquist, A. (1968). "The Evolution and Classification of Flowering Plants." Nelson, London and Edinburgh.

Ehrendorfer, F. (1971). Spermatophyta, Samenpflanzen. In "Lehrbuch der Botanik für Hochschulen" (D. Denffer et al., eds), pp. 584–741. G. Fischer, Stuttgart.

Faegri, H. and van der Pijl, L. (1971). "The Principles of Pollination Ecology" (2nd edition). Pergamon Press, Oxford.

Gottsberger, G. (1970). Beiträge zur Biologie der Annonaceen-Blüten. *Öst. bot. Z.* **118**, 237–279.

GOTTSBERGER, G. (1972). Blütenbiologische Beobachtungen an brasilianischen Malvaceen. II. *Öst. bot. Z.* **120**, 439–504.
GOTTSCHALK, W. (1970). Möglichkeiten der Blattevolution durch Mutation und Rekombination. *Z. Pfl physiol.* **63**, 44–54.
GRANT, V. (1950). The protection of the ovules in flowering plants. *Evolution* **4**, 179–201.
KUGLER, H. (1970). "Blütenökologie." 2. Aufl. G. Fischer, Stuttgart.
KUIJT, J. (1969). "The Biology of Parasitic Flowering Plants." University of California Press, Berkeley and Los Angeles.
LEPPIK, E. E. (1956). The form and function of numeral patterns in flowers. *Am. J. Bot.* **43**, 445–455.
LEPPIK, E. E. (1972). Origin and evolution of bilateral symmetry in flowers. *Evol. Biol.* **5**, 49–85.
MELCHIOR, H. and WERDERMANN, E. (eds) (1954, 1964). "A. Engler's Syllabus der Pflanzenfamilien," Band I u. II. Gebrüder Bornträger, Berlin.
PIJL, L. VAN DER (1969). "Principles of Dispersal in Higher Plants." Springer, Berlin–Heidelberg–New York.
SCHWANITZ, F. (1959). Genetik und Evolution bei Pflanzen. *In* "Evolution der Organismen", 2. Aufl. **1**, 425–551. G. Fischer, Stuttgart.
STEBBINS, G. L. (1951). Natural selection and the differentiation of Angiosperm families. *Evolution* **5**, 299–324.
STEBBINS, G. L. (1967). Adaptive radiation and trends of evolution in higher plants. *Evol. Biol.* **1**, 101–142.
STEBBINS, G. L. (1970, 1971a). Adaptive radiation of reproductive characters in Angiosperms, I: Pollination mechanisms, II: Seeds and seedlings. *A. Rev. Ecol. Syst.* **1**, 307–326; **2**, 237–260.
STEBBINS, G. L. (1971b). Relationships between adaptive radiation, speciation and major evolutionary trends. *Taxon* **20**, 3–16.
TAKHTAJAN, A. (1959). "Die Evolution der Angiospermen." G. Fischer, Jena.
TAKHTAJAN, A. (1969). (transl. C. Jeffrey). "Flowering Plants, Origin and Dispersal." Oliver and Boyd, Edinburgh.
TROLL, W. (1928). "Organisation und Gestalt im Bereich der Blüte." Springer, Berlin.
VOGEL, S. (1954). Blütenbiologische Typen als Elemente der Sippengliederung. *Bot. Stud.* (Jena) **1**, 338 pp.

17 | Ecological Data in Practical Taxonomy

V. H. HEYWOOD

Department of Botany, The University of Reading, Reading, England

Abstract: Ecologists are among the major users of taxonomy but frequently express dissatisfaction at the ways in which taxonomic units are constructed and defined. Taxonomists, on the other hand, aim to produce classifications which will satisfy the requirements of many different classes of user, including ecologists, and the results consequently represent a compromise. The nature of the taxonomic process and the limitations imposed by inadequate information and resources are discussed. Traditional methods of publication such as Floras are considered and ways in which they might be improved to meet the needs of ecologists are proposed. The relationships between populations and taxonomic units, and the problems of sampling are surveyed and consideration is given to the problem of employing special or auxiliary categories in classification.

INTRODUCTION

An unusual feature of taxonomy is that its results are brought together and published in a number of highly specialized and stylized forms such as Floras and monographs. There is nothing quite comparable in other branches of biology. This is not altogether surprising if one considers that one of the major roles of taxonomy is to provide a data-processing and retrieval service for biology. Ecologists are amongst the major users of this service and it is entirely reasonable that they should question the efficiency of the system for their particular needs. Both Floras and monographs vary widely in their style and presentation but their general layout and content were established in a period when the main data available for synthesis and presentation derived from morphology and anatomy, and ecology had not even begun its development as a modern science. Yet Floras and other identification manuals provide the basic tools available to the ecologist for floristic information and identification.

The aim of this paper is to consider the various instruments of formal taxonomic publication such as Floras, the nature of the taxonomic process and its limitations, and how far these are compatible with the needs of ecologists. At the same time it must be stressed that there are pressures from other users (including taxonomists) for a rethinking of these questions in the face of the enormous increase in the amount and nature of data which taxonomists are required to handle. Apart from various attempts to use electronic data processing and storage, computer-generated keys, and graphic or symbolic methods of data presentation, little effort has been expended on devising ways and means in which these new data can be economically assembled and effectively made available in non-traditional formats. There are obvious limits to what Floras and monographs can include and as a consequence a large proportion of taxonomically valuable information remains uncorrelated and effectively lost in countless papers in journals and reviews.

PRACTICAL TAXONOMY

Many of the criticisms laid at the door of taxonomy by the ecologist are due to a misunderstanding of the taxonomic process, its aims and its limitations. Part of the difficulty stems from the avowed attempt of taxonomists to produce (*inter alia*) "general purpose" classifications which presumably are designed to serve the ecologist as well as other classes of user. It would be more accurate, however, to regard them as serving a wide variety of purposes rather than a "general purpose". Even this is more of a pious hope than a deliberate conscious attempt to consider the needs of the different classes of user. For one thing, these needs are seldom clearly enunciated (cf. Heywood, 1966); for another, those requirements that *are* expressed are often either unrealistic or are mutually incompatible. There is, moreover, a widely held belief by taxonomists that their classifications represent "how it must be", stemming from a belief in the underlying evolutionary (and therefore factual) basis of classifications. The term "natural" classification (whether it is used in the phenetic or phylogenetic sense —cf. Heywood and McNeill, 1964) has a similar connotation.

By their nature, the general classifications of the taxonomist must be a compromise as far as the users are concerned, for it is only the taxonomist who is aware of the limitations of information and resources and who is forced to make decisions in situations that are far from ideal. It is, for example, simply not practicable to produce classifications in a vacuum as some critics have proposed. Why not, the argument runs, restrict classifications to those organisms which are of actual or potential value? Surely it is not necessary to devote resources to the study of exotic groups of plants (or animals) which have no conceivable economic or practical value (meaning in fact those in which the critic is not interested at that time). Such a view has received support from very eminent biologists, such as Sokal (1970) who writes:

> "Do we need to describe and classify every species of organism on earth? Should we even attempt to do so? Is the day of the descriptive naturalist ecologist, or systematist

over? What are we to do with the large body of systematists and ecologists who find in mid-career that their subject matter has changed so dramatically as to make much of their work irrelevant?"

Certainly *ad hoc* classifications can be prepared for restricted groups of organisms in special circumstances but to apply this principle generally would be to destroy the whole value of biological classification which is to provide a content within which relationships can be expressed. Whatever else the taxonomist may or even should be concerned with, he is, to quote from Fosberg's (1972) recent statement:

> "engaged in making a directory for finding our way around in the amazing complexities of the plant and animal kingdoms. Finding means of identification and placing in proper context the myriad of kinds of living things with which we share and that are part of the environment that we inhabit. . . . We are engaged in the construction of a framework on which to hang or arrange the total available biological information, the data about life, whether on the molecular or the organismal or the population levels . . ."

It may seem extraordinary that these ideas, which have been expressed many times in the past by many systematists, need repeating today; yet, as reference to Fosberg's paper will show, many of our scientific colleagues are unconvinced of the need for such an activity in systematics, at least on a wide scale. It is all the more extraordinary that this should be so when so much attention is being paid to the environmental crisis and the need for as accurate and detailed information as possible on the floristic resources of the various regions of the world without which conservation studies would founder. As I have indicated elsewhere (Heywood, 1971a) it seems surprising that the problems of conservation of resources have been regarded as primarily the concern of the ecologist while the taxonomist's role has remained somewhat vague. Taxonomists in fact occupy a key position in relation to our knowledge of the environment and this is only somewhat belatedly being recognized.

Many voices have been raised in support of the view that taxonomists today should not concern themselves primarily with such mundane matters as floristics and practical taxonomy or that such activities should be left to a small body of herbarium taxonomists who can be left to get on with their useful but unexciting "pigeon-holing". However, while not disagreeing with Smith (1969) that every taxonomist is obliged to contemplate four fundamental questions about his taxa: "what is it? when did it arise? where did it originate? and how did it acquire its present distribution?", I feel that there has been a tendency to neglect the methodology of basic descriptive taxonomy in our attempts to persuade ourselves and others that we are all evolutionary biologists. This is not to suggest any restriction of our overall role but rather to propose that we look again into the balance of our efforts in the whole field of systematics. This general point has been lucidly discussed by Jacobs (1969) in relation to the problems arising from the neglect of monographs of large families. His section headed "Fragmentation and Frustration" is a fair summary of the situation and one cannot but agree with his depressing conclusion that "Taxonomists are nibbling at the crumbs of their work, without being able to digest the substantial chunks" despite the fact that "a plethora of sophisticated methods for processing and simplifying information about taxa has been developed (Jardine and Sibson, 1970)".

It is important to clarify the relationships between decision-making and the consequent presentation of these decisions in the form of Floras and monographs, usually as keys and descriptions. The taxonomic description is in fact the major method still employed by the taxonomist for presenting the "distillate of his research to other biologists" (Kruckerberg, 1969). Descriptions in Floras, monographs and revisions, are how the generalizations of the taxonomist are *systematically* packaged for the consumer. Other forms of publication are of course employed in research papers but the Flora in particular is as Shetler (1971) puts it "a time-honoured information retrieval system; it is as old as taxonomy itself and equally indispensable".

FLORAS—THEIR NATURE AND ROLES

A question that has to be asked today is just how effective a Flora is as an information storage system since what can be retrieved from it depends on the amount of information that is actually stored (which is quite different from that used in making the decisions or descriptions contained therein) and how it is organized, since, as Shetler also says, many Floras and other reference works "contain much apparent information but little real or effective information, because most of the data are for all practical purposes inaccessible".

What, however, are the functions of Floras and monographs? Let us consider Floras since it is here that perhaps the greatest confusion lies. Two rather different answers have been proposed recently: I have myself suggested that the function of a Flora up to the present is to establish how many groups there are in the area concerned that merit the rank of genera, species, subspecies, etc.; indicate how they may be recognized by means of keys and descriptions, where they may be found and under what ecological conditions. The descriptions are given to facilitate or confirm identification, not to provide comparative data for taxonomists or other biologists (Heywood, 1973). This definition is similar to that of Bentham who stated that the principal object of a Flora was to afford the means of determining any plant growing in the area concerned.

Clearly, however, many Floras, if not most, can also serve as a source of taxonomic/descriptive information of various sorts, whether or not this was the intention of the writer. This depends largely on the kind of Flora concerned, as is discussed below, but has led taxonomists whose main orientation is towards computerized data-processing to regard the Flora as

> "a physical repository of descriptive data about plants which are organized and formatted, usually in book form, so as to answer a time-tested set of prescribed questions. Other questions often can be answered but only if the user is willing to expend much time and effort to extract the embedded information. The data elements of a typical Flora are well known to us; in general, they cover the basic certainties of plant diagnosis, description, and documentation" (Shetler, 1971).

It is when one regards the Flora as a repository of descriptive data that one must to a large extent agree with Watson (1971) that "perusal of the average taxonomic-descriptive work (such as Floras) usually reveals that *as a source of comparative data it is hopeless*".

This comes as no surprise to the practising taxonomists, for the conventions employed in Floras to ensure brevity and ease of identification virtually guarantee a loss of comparative descriptive information. It follows that to attempt to use Floras as a source of character data for numerical taxonomy is largely doomed to failure; indeed if one attempts to make, say, a generic key on the information given in generic descriptions in practically any Flora, the result is hopeless. The reasons for this are discussed below.

Before considering what remedies, if any, might be sought for such a situation, it is worth looking at the different types of Flora which are produced since in aims and achievements they vary a great deal.

Apart from the classic discussions of last-century authors such as Bentham, little has been written on the methods and techniques of writing Floras. Van Steenis's (1954) review, given at the Paris International Botanical Congress, on the general principles in the design of Floras is a notable exception, and the chapter "Presentation of Data" in Davis and Heywood's (1963) text on Angiosperm Taxonomy brings together much relevant information.

The aims and scope of a Flora are usually related to the state of our knowledge of the area or region concerned. For some areas it may be the only source of comparative data for taxa apart from original descriptions (although new taxa may be included in the Flora). This is especially true of research or "creative" Floras, as Van Steenis (1954) calls, them which bring together a large amount of information for an area or region for the first time. Examples are Boissier's

"Flora Orientalis", Rechinger's "Flora Iranica", Davis's "Flora of Turkey", Willkomm and Lange's "Prodromus Florae Hispanicae", Maire's "Flore de l'Afrique du Nord" and tropical regional Floras such as "Flora Malesiana", "Flore du Congo Belge", "Flora of Tropical East Africa", etc. Although, as Davis and Heywood (1963, p. 296) point out, they occupy a position between the shorter, concise Floras of well-known regions and monographs, and are self-contained in the sense that original material and collections are cited, they do not all contain detailed descriptions and keys. Thus, while it is true that they provide a basis for further research, they do not necessarily embody enough descriptive information about the plants concerned to act as a useful source of descriptive data.

From the point of view of identification a Flora with keys but no descriptions (e.g. Rechinger's "Flora Aegaea") is easier to use than one with descriptions but no keys (e.g. Post and Dinsmore's "Flora of Syria, Palestine and Sinai"). The latter is more useful, however, if one is seeking information about the characters of the plants.

Fisher (1968) and Watson (1971) make a case for a distinction to be made between the observational and classificatory aspects and the identificatory aspects of Floras (and to some extent monographs, as is discussed later). It seems to me that we should recognize three activities:

1. Decision-making, which is reflected in the Floras as the classification into those families, genera, species and infraspecific taxa actually adopted.
2. Identification, which is provided by keys, synopses and comparative short descriptions, in Floras and monographs principally.
3. Comparative observation and data storage, which are poorly represented in most Floras except imperfectly through the descriptions and comments.

Floras contain the results of large numbers of decisions made about the identity, circumscription, variability and presence of taxa, but seldom do they give much information about the ways in which they have been based.

1. Decision-making in taxonomy

Decision-making in taxonomy is a complex process and the fact that the results are presented in a morphological framework often misleads the non-taxonomist who sees only statements about classificatory position and status of organisms expressed or justified in terms of morphological characters. Most taxonomists today would agree that all kinds of available evidence—anatomical, cytological, chemical, cytogenetical, etc.—should be taken into account in deciding on the recognition and rank to be afforded to a group of organisms, especially species (cf. Heywood, 1958), but for practical purposes it is accepted that the resultant groups have to be recognizable in terms of morphology so far as higher plants (and animals) are concerned. This is primarily because man is a visually gifted animal and depends on eyesight for recognition, his other senses being used in a subsidiary capacity.

Not only is there inevitably, therefore, a morphological bias in presenting the decisions, but also to some extent in actually making them. Morphological features are always available and can be directly assessed whereas other kinds of data are seldom available for more than a few members of a group and it is often not possible to obtain them easily as a result of further research due to special requirements (living material, viable seeds, etc.). Moreover, morphological features can be continually tested and re-tested by generations of taxonomists although precisely how this experience can be utilized and recorded is difficult to explain and set out.

In most Floras no provision is made for indicating what evidence, other than morphological and occasionally cytological, has been used in decision-making. The very condensed and specialized format of most Floras precludes this. It would, however, be misleading to believe that even the morphological data employed are fully documented in Floras. In fact a very clear distinction has to be made between the decision-making and the way in which the decisions are documented. This is very clearly highlighted by Taylor (1971) who comments that once we have made a decision regarding how many taxa to recognize in a group "we tend to organize all our information to support this decision and end by making the assumption we have made our most important contribution by making the taxonomic decision itself. Clearly, much of the information about the organism is lost in such a process because in the final communication, we have transmitted only an abstraction premised by a taxonomic decision and have lost the goal of constructing a bank of information about the organism." There is much truth in Taylor's claim that we organize our information so as to justify the decisions we have arrived at. This is inevitable so long as we continue to employ the traditional taxonomic method of unanalysed entities (Heywood, 1971b)—groups are recognized as a consequence of the taxonomist's assessment of whole organisms, organ systems and individual organs in terms of overall similarity and the complex body of information handled by the taxonomist normally is not consciously broken down into characters until it is necessary to communicate the decision—in the form of a key or description.

The decision-making process involves not only this highly personal mental assessment of sensory impressions about organisms or parts of them, but a complex series of model-testing whereby various hypotheses are proposed (often involving conscious selection or weighting of particular lines of evidence or characters) and tested to see whether we are satisfied with the groups or classifications that result. There is a considerable amount of feedback in this process and only the final result is communicated as our decision. It is the vague nature of the process, or rather our inability to explain precisely what the mental processes involved are, that leads to the often voiced criticism as to the unscientific basis of taxonomy. There can be no doubt that this side of taxonomy is highly subjective and the results can only be tested by trying them out. There is all the more reason, it would seem, therefore, for setting down on paper the evidence used even though we cannot state exactly how we have "processed" it.

Another factor that has to be taken into account is the conceptual framework employed by the individual taxonomist as regards the definition of species and other taxa: different traditions and the unorthodox usage of categories in different European Floras which affect the decisions made can lead to considerable difficulties of interpretation (Heywood, 1967).

It is not often appreciated, even by taxonomists, that a consequence of this "unanalysed entities" approach is that there is no guarantee whatsoever that individual features which might *a posteriori* be shown to play an important role in classification are considered at all. The approach is in other words highly selective. This may be a desirable or even necessary feature in that it speeds up the whole process, but it highlights even further the distinction that has to be made between the data involved in the decision-making process and the communication via description afterwards. There may be little direct connection between the two. It is as though one says, "here are the groups we have decided to recognize, now let us see what their characters are". These points are discussed further in the next sections. The actual explanation and justification of the decisions is a further aspect that in Floras is usually omitted altogether although it is often covered in monographs.

2. Identification

The main sources of identification are Floras, monographs, illustrations and herbaria. In the first two, the principal instruments of identification are the key and comparative description although, as noted above, some Floras either contain no keys or no descriptions, or the two are combined in descriptive keys.

I shall not concern myself here with the mechanics of key construction or with the different kinds available except to mention two recent developments: (1) the use of multiple-entry keys, especially for "difficult" groups (cf. key to the Umbelliferae, in Davis, "Flora of Turkey", Vol. 4 (1972)); (2) the introduction of computer-generated keys and automatic identification by computer (cf. Boughey et al., 1968; Hall, 1970; Pankhurst, 1970; Morse, 1971). These developments may go a long way to permitting the use of a wider range of characters, especially those of low diagnostic value, i.e. polythetic instead of monothetic keys. They may also include assessments of the relative convenience or ease of observation of the characters employed and of the expected abundance or frequency of occurrence of the taxa concerned (cf. Morse et al., 1971). Both these types of indicator would clearly be of considerable use in ecological work if they could be more generally applied.

There may, surprising though this may seem, not be an exact relationship in practice between the information given in a key and that given in the corresponding description in the text of a Flora. There are three reasons for this: (1) it may be deliberate policy to eschew in descriptions those features used in keys and vice versa so as to avoid repetition and save space; (2) there may be serious discrepancies between the information given in keys and descriptions due either to incompetence or error, the two being written separately; (3) the keys and descriptions may be written by different authors.

Clearly, as a source of comparative information, a key is not only an abomination, as Watson (1971) has noted, but often an inaccurate one at that!

To avoid confusing the user (and partly to save space) the number of characters used at any entry in a conventional key is usually limited to two, three or four. As a consequence, features, say vegetative ones, which may be of special value to the ecologist, are often omitted.

Descriptions are a necessary part of the identificatory process—if only to confirm or refute the conclusion reached by using the keys. For this purpose it is essential that descriptions be comparable, at least between related taxa, and desirable that they should be brief. Over a wide range of taxa, for example species in a large genus, complete comparability is not practicable since there is a shift in characters as one goes, for example, through a genus containing groups of closely related species. To describe these features for each species would lead to very long descriptions containing large amounts of information which were redundant for identification, although conversely valuable as a source of comparative data for storage and processing.

3. Recorded observations and data storage

As already noted descriptions in Floras are designed primarily to facilitate identification, not to provide a complete repertoire of comparative data for the taxonomist or other biologists. Descriptions are abstracts for practical convenience, or as Taylor (1971) puts it, "end-point abstractions". All practising taxonomists are aware of the enormous wastage in man hours in the preparation of Floras in relation to the published result in terms of *recorded* descriptive data. Although practical reasons largely dictate the length and nature of descriptions it has to be recognized that many of the characters not described *are* observed by the taxonomist during the decision-making processes mentioned above and in the preparation of descriptions, but are discarded. In other words, information is thrown away: the amount of this discarded information is of staggering proportions and it is no consolation to suggest that some of it is retained mentally by the taxonomist as part of his experience.

The more flexible format of the revision or monograph often permits certain kinds of information to be presented in tabular form or in other ways in addition to formal descriptions but basically the same considerations apply as in Floras.

Taxonomy as at present organized is an outstandingly successful information storage and retrieval system through the use of a hierarchical, nested-box type of classification and a nomenclatural system based on the binomial which serves as a key to the literature (i.e. data). Despite this it is obvious that the actual amount of information that is stored and therefore retrievable is only a small proportion of that actually handled or considered by the taxonomist in setting up the system!

Although in the past complaints about this lack of information in the system have come mainly from ecologists, largely because insufficient ecological data were recorded, the situation has become much more serious in recent years and has reached near-crisis proportions (Heywood, 1973) for two main reasons. The first is a consequence, in a sense, of the successful growth and development of systematics leading to the production of a vast corpus of information on comparative anatomy, embryology, palynology, micro-morphology, and, more recently, biochemistry (mainly micro-molecular). It would be idle to pretend that this information is processed and digested by taxonomists on a substantial scale and it tends, therefore, to circulate within the more or less closed system of its own particular discipline. We neither have the mechanism for consistently retrieving this information from pages of the journals nor the mechanism for presenting it in an adequate format were the difficulties in retrieving and synthesizing it to be overcome. This is not to say that valiant attempts are not being made to meet some of these problems by individual taxonomists with their private indexing systems and voracious reading habits, and by many abstracting journals or systems of varying degrees of effectiveness. It could, moreover, be argued that the same dilemma faces virtually every branch of learning but the point that has to be stressed is that systematics is specifically intended to be and is used as an information processing and retrieval system.

The second reason for the seriousness of the present situation arises from the development during the past decade or so of methods for the use of computers in systematics. Most of the attempts to date have been concerned with ways of setting up classifications (cf. Hall, 1972) and have necessitated precise comparative data in the form of taxonomic characters for all the units under consideration. Not only have numerical taxonomists had to face the problems of seeking these data out themselves from the plants or animals concerned in view of the unsuitability and lack of comparability of published descriptions as already discussed, but when they have assessed their data matrices they have soon realized that there is no ready means of publishing or otherwise making them available. Thus the lists of characters used, but not the individual character expression for each unit, and the resultant classification, in whatever form, alone are published, while the raw data remain on file.

Another, more recent, development is the use of computers and information systems in systematics for the summarizing and communicating of taxonomic information. Data about the specimens or taxonomic units are stored in the computer and constitute a data bank and with appropriate programmes the desired information can be retrieved as and when needed. Much work has been devoted recently to devising a suitable data format, especially in connection with the Flora North America project (cf. Morse et al., 1971). To meet the problem of general availability of such a data bank it is important to set up a network system, as is feasible using current information-processing technology, so that any user would contribute to and have in turn access to a centralized data bank. For this to be effective standardized data formats are necessary which "would make one investigator's data immediately available and comprehensible to all others using the same format. It would also make any information in the proper format available for one's own use without local modifications" (Morse et al., 1971).

There is still a long way to go before any agreement can be envisaged for the setting up of a standardized general taxonomic information retrieval system. This would be highly complex and sophisticated and by present standards of available finance for systematics, inordinately expensive. For the present much more modest packages are being devised and as Hall (1972) points out the main proposed applications are concerned with specimen-label data and associated taxonomic and bibliographic information (Crovello and MacDonald, 1970; Greene, 1972). These would certainly be extremely valuable as regards storage and communication of some kinds of ecological data.

A problem that has not been adequately faced up to is the format for publication of computer-stored data files or selected combinations of them. Unless and until, one can envisage widespread indeed commonplace, availability of on-line access to centralized data banks, and clearly this must be so far off as not to constitute a serious possibility for the present generation of biologists, there will be a major and continuing need for some form of publication that will present comparative data for both general and specific use by various classes of biologists. The need for a series of encyclopaedias of systematic groups was voiced as far back as 1954 at the Paris International Botanical Congress. The nearest approaches have been of two sorts: the second edition of Hegi's "Illustrierte Flora von Mitteleuropas", which contains, in addition to long and detailed morphological descriptions, extensive ecological and geographical data and what are virtually essays and bibliographies on topics such as pollen morphology and phytochemistry (cf. Heywood, 1971c). Although highly commendable, the result is an unhappy compromise and in any case the non-morphological data are summarized but not systematically detailed in a comparative form. Other approaches are volumes such as Harborne, Boulter and Turner's volume on the "Chemotaxonomy of the Leguminosae" (1971) which brings together a vast amount of information on various classes of chemical compounds, and Heywood's "Biology and Chemistry of the Umbelliferae" (1971) which does the same for a wide range of disciplines. Neither is a systematic account of the family concerned but they are rather source books of information from which data for synthesis can be obtained.

The family Umbelliferae, and in particular the tribe Caucalideae, is being intensively studied by a group of British, French and Spanish laboratories with a view to producing a detailed systematic survey of the morphology, anatomy, embryology, cytology, floral biology, palynology and phytochemistry of every genus and species possible and it is planned as a major goal of this programme to devise a suitable format for the publication of the processed information as a model for discussion and possible adoption more generally.* It is to be hoped that more workers will devote some attention to these problems in addition to their basic research.

POPULATION, SPECIES AND ECOLOGY

The conventional taxonomic units, especially the species, subspecies and variety, have been regarded by many ecologists as unsatisfactory for their work. The reasons advanced are various, such as lack of precision, inadequate biological or populational basis, unspecified nature of the samples on which they are based, etc.; in fact, most of the criticisms advanced by proponents of the "biological"

---

* (Conseil National de la Recherche Scientifique: Recherche coopérative sur Programme No. 286, Ombellifères, and Science Research Council grant B/SR/1923.)

species concept against the taxonomic (typological, nominalistic, phenetic, etc.) species concept. Despite this, most ecologists do not appear to be seriously handicapped in practice by handling the species prepared by taxonomists: were this not so, a workable alternative system would surely by now have been introduced.

My own view agrees closely with that advocated by Guinochet (this volume, p. 121) who considers that for general phytosociological work, conventional taxonomic species should be employed (indeed there is no alternative) while recognizing at the same time that within them ecological races, ecotypes and other kinds of more or less differentiated local populations occur in particular ecological conditions. On the other hand, experimental ecologists may find that taxonomic species or subspecies are too broadly defined and not sufficiently sensitive units with which to work, preferring to employ local populations. It is when one attempts to redefine taxonomic units on such a detailed population basis rather than to employ an *ad hoc* system that serious difficulties arise.

Some of these problems and possible solutions will be discussed in the next sections.

1. Populations and taxonomic units

It is a tenet of present-day taxonomy that species and subspecies are intended to be population units. If we consider the sequences:

$$\text{individual} \begin{array}{l} \rightarrow \text{population} \longrightarrow \text{(species)} \longrightarrow \text{community, etc.} \\ \searrow \text{population} \longrightarrow \text{species} \longrightarrow \text{genus, etc.} \end{array}$$

and ask the simple question, what is meant by a population, the answer almost inevitably is a population of a taxon, i.e. representatives of a group already defined by taxonomists and into whose definition the idea of a population is bound to come. It is not practicable to progress from individuals to communities via anonymous populations—they must be populations belonging to or believed to occur within or represent named taxa.

Apart from a few hundred cases where extensive experimental work has been undertaken, taxonomic species are models which have been set up, on the basis of field samples (unsystematically assembled, see below), which are considered to represent populations.

The fact that any named taxonomic species (binomial) is not *known* to represent a population is not so much a criticism of taxonomy as a reflection of our state of knowledge at that time. The taxonomist describing a species is setting up a hypothesis: that there will be found to exist in nature hundreds or thousands of individuals with similar characteristics to those in his model description derived from a more or less limited sample. The testing of this hypothesis is done by countless checking and rechecking in the field, herbarium, laboratory, etc. by later taxonomists, ecologists and other biologists who attempt to use the taxonomist's model (cf. Davis and Heywood, 1963, p. 351 et seq.)

There are many ways in which the term population can be defined, depending on the interests and purposes of the definer. For the general taxonomist a population is any group of individuals considered together at any one time because of features they have in common. This tends to mean in practice a group of plants growing together which look similar, or a series of such groups as they occur over a particular area or region. Such populations are not necessarily gene pools or breeding populations in the sense that the individual members can exchange genes. Gene-flow in the majority of cases is a likely inference not an established fact, based on small samples and extrapolation from similar morphology.

The ecologist is often concerned with the local population and it is likely that such local populations of taxonomic species will in fact be breeding populations and will often possess certain characteristics that can be correlated with local environmental conditions. They may, however, not be sufficiently distinctive morphologically to permit taxonomic recognition, the variation usually being of a quantitative rather than qualitative kind. Local ecologically related populations with some conspicuous morphological differences (habit, hairiness, etc.) have sometimes been recognized as taxonomic varieties (var. ecol.), an idea later extended by Turesson with his ecotypes (see below), but this practice is not widespread. The dangers of using the formal taxonomic categories for describing the population structure of a species have been pointed out by Mayr (1959) and Heywood (1963) even if the necessary analysis, collateral cultivation and experimental work has been undertaken. Variation at the intraspecific level is often multidimensional and not susceptible to discrete recognition. Polytopism, as is mentioned below, is a further complication.

As a generalization it can be said that while well-studied species have a population basis, the analysis should not be pursued too closely at the infraspecific level within formal taxonomic classification.

2. Sampling and taxonomy

Taxonomy is by its very nature *ad hoc* in the sense that no matter what the basis of sampling or range or quantity of evidence available in a particular case, a decision has to be made by the taxonomist about identity and status, subject to continual revision and reassessment as further information becomes available.

This means that taxonomists very often (perhaps normally) have to make decisions in circumstances and on evidence which other scientists would consider inadequate. Were this not so, much of the present-day fabric of classification would have to be discarded and taxonomy as a practical service would come to a halt.

I suspect that failure to realize this peculiar feature of taxonomy is responsible for much of the misunderstanding of the subject by non-taxonomists who see classifications expressed in terms of binomials or even trinomials thus giving an impression of uniformity and consistency of knowledge. There is no easy means of finding out how reliable is the information hiding behind these scientific names. Such is the magic of names that even fictitious species have masqueraded in the literature and been taken at face value by all too trusting users of taxonomy!

The majority of species recognized and employed today have been based on unsystematically sampled material—specimens collected by different people at different times and found in different herbaria, and in no way representing adequately what overall variation pattern may eventually be found. This is discussed in detail by Davis and Heywood (1963, p. 351 *et seq.*) However, it is not possible, even given unlimited resources, to base taxonomic units on objective, statistical sampling without some degree of circularity since the sampling has to be undertaken within populations which are only recognizable in terms of formally or informally recognized species.

Sampling in taxonomy is in practice a continuous process, mostly taking place after initial decisions have been made on the recognition of taxa and leading in many cases to a revised circumscription of them. Again this sampling is unsystematic and is seldom intensive except on a local scale. It is unrealistic to envisage intensive population sampling throughout the range of a species: even in experimental or genecological studies this is rarely feasible in terms of the effort required and the likely return from a scientific viewpoint. There is, of course, nothing to prevent an ecologist undertaking intensive sampling of local populations as needed, but in general ecologists, like other users of taxonomic classifications, will have to accept that the system is highly imperfect.

3. Special and auxiliary categories

The genecological system of categories introduced by Turesson and modified by later workers (for discussion see Heywood, 1959) might appear to hold out most attraction to ecologists seeking an alternative to the normal units of taxonomy. However, apart from some attempts to equate ecotypes with taxonomic varieties or subspecies, the higher categories of the genecological system—

ecospecies and coenospecies—have almost invariably been employed as descriptive of the micro-evolutionary status of taxonomic units in discussions, rather than as alternatives to such units.

There appears to be a general consensus of opinion that the use of special categories to accommodate particular kinds of variation, as a substitute for the formal categories of taxonomy, is of only limited practical value. The more specialized the category, the less general value it possesses, and since one is often concerned with different aspects of variation occurring within populations at the same time, a series of different, overlapping categories would be needed. In the taxonomy of cultivated plants, the introduction of a parallel set of categories such as specioid, subspecioid, etc. has met with little favour and only the cultivar is officially accepted and used. In a recent discussion of this question Hanelt (1972) rejects the use of very detailed morphological classification of infraspecific variation and of special categories, and proposes instead a triple system (he in fact calls it a dual system) in which, for example, *Vicia faba* is divided into two subspecies, subsp. *faba* and subsp. *minor*, each of which is divided artificially into varieties and subvarieties on the basis of legumes, seed and growth habit; these two formal classifications are then supplemented by an informal system based on evolutionary considerations, including a number of geographical races, which are not given formal Latin names, which may be distinguished by morphological, phenological and other criteria. For ecological purposes, such informal systems have many advantages: they are flexible, they are *ad hoc*, and since they do not require formal names the problems of nomenclature, synonymy, priority, etc. are avoided. On the other hand, such informal systems have the disadvantage that they are not well suited for ease of communication in that vernacular names are used and they are not "registered" or typified in the sense that formal categories are.

Similar considerations apply to the *-deme* terminology (Gilmour and Gregor, 1939; Gilmour and Heslop-Harrison, 1954; Gilmour, 1960) and the following comments I made ten years ago (Heywood, 1963) are still valid today.

> "When applied to ecotypic situations the *deme* terminology may be employed as suggested by Gregor (1963). Thus individual investigators would record such genoecodemes, for example, as they wished in a particular species and these could be recorded in Floras by listing the categories to which the examples had been assigned under their respective taxa, adding references to the literature where the relevant details are recorded, e.g. under *Plantago lanceolata*; genoecodeme (Gregor and Watson, 1961). The *deme* terminology is deliberately not designed for formal communication, but it is likely to have to serve a much greater role in communication when more examples of similar *demes* are recorded. Up to the present the system has been little used outside Britain, and furthermore the number of ecotypes (geno-

ecodemes, etc.) recorded has been comparatively few. With the more widespread adoption of the system and the continued description of further examples it is inevitable that some more formal method of recording and distinguishing between them will be needed, and in the light of similarities or parallels between others of them, there will be a need for some kind of informal hierarchization. For communication and meaningful interpretation of results systematization of knowledge is necessary.

"It may well be found in the future that a whole series of similar genoecodemes with similar phenotypic characteristics have been recorded within a species by different authors in different countries. Countless situations of this nature will arise. The need for international registration of ecodemes and for a means of distinguishing between them and of ascertaining which, if any, are 'the same' might well have to be considered. We are only just beginning to see what kinds of problems of classification and communication may lie ahead. One thing is clear, the orthodox classification system will come to play an increasingly important role as more and more kinds of information are attached to its framework."

CONCLUSIONS

Practical taxonomy has as one of its major goals the preparation of classifications to serve a range of different purposes. Such multi-purpose classifications have proved to be generally useful to ecologists who constitute a major group of consumers, although they may have to be supplemented by additional information in the form of auxiliary or special categories or by symbolic or graphic devices. Automatic data processing may eventually provide alternative, more flexible means of classification through the establishment of data banks, but the application of such methods is not yet feasible except on a very limited scale. In the meantime a dialogue should be established between ecologists and taxonomists so that each other's terms of reference, working concepts and requirements can be better understood. This is particularly important since both use largely the same data and raw material and even employ the same terms such as population, species, sampling, etc. although with different meanings.

REFERENCES

BOUGHEY, A. S., BRIDGES, K. W. and IKEDA, A. G. (1968). An automated biological identification key. *Univ. California (Irvine) Mus. Syst. Biol.*, Res. ser. no. 2.

CROVELLO, T. J. and MACDONALD, R. D. (1970). Index of EDP-1R projects in systematics. *Taxon* **19**, 63–76.

DAVIS, P. H. and HEYWOOD, V. H. (1963). "Principles of Angiosperm Taxonomy." Oliver and Boyd, Edinburgh and London.

FISHER, F. J. F. (1968). The role of geographical and ecological studies in taxonomy. *In* "Modern Methods in Plant Taxonomy" (V. H. Heywood, ed.), pp. 241–259. Academic Press, London and New York.

FOSBERG, F. R. (1972). The value of systematics in the environmental crisis. *Taxon* **21**, 631–634.
GILMOUR, J. S. L. (1960). The deme terminology. *Rep. Scott. Pl. Breed. Stn.* **1960**, 99–105.
GILMOUR, J. S. L. and GREGOR, J. W. (1939). Demes: a suggested new terminology. *Nature* **144**, 333–334.
GILMOUR, J. S. L. and HESLOP-HARRISON, J. (1954). The deme terminology and the units of micro-evolutionary change. *Genetica* **27**, 147–161.
GREENE, D. M. (1972). A taxonomic data bank and retrieval system for a small herbarium. *Taxon* **21**, 621–629.
GREGOR, J. W. and WATSON, P. J. (1961). Ecotypic differentiation: observations and reflections. *Evolution* **15**, 166–173.
HALL, A. V. (1970). A computer-based system for forming identification keys. *Taxon* **19**, 12–18.
HALL, A. V. (1972). Computer-based data banking for taxonomic collections. *Taxon* **21**, 13–26.
HANELT, P. (1972). Die infraspezifische Variabilität von *Vicia faba* L. und ihre Gliederung. *Kulturpflanze* **20**, 75–128.
HEYWOOD, V. H. (1958). "The Presentation of Taxonomic Information: a short guide for contributors to Flora Europaea." Leicester University Press.
HEYWOOD, V. H. (1959). The taxonomic treatment of ecotypic variation. *In* "Function and Taxonomic Importance". (A. J. Cain, ed.) Systematics Association Publication **3**, London.
HEYWOOD, V. H. (1963). Biosystematics and classification. *Rep. Scott. Pl. Breed. Stn* **1963**, 1–7.
HEYWOOD, V. H. (1966). How many taxonomies? *Rev. Roum. Biol. sér. Bot.* **11**, 101–106.
HEYWOOD, V. H. (1967). Variation in species concepts. *Bull. Jard. bot. nat. belg.* **37**, 31–36.
HEYWOOD, V. H. (1971a). Preservation of the European flora. The taxonomist's role. *Bull. Jard. bot. nat. belg.* **41**, 153–166.
HEYWOOD, V. H. (1971b). The characteristics of the scanning electron microscope and their importance in biological studies. *In* "Scanning Electron Microscopy" (V. H. Heywood, ed.), pp. 1–16. Academic Press, London and New York.
HEYWOOD, V. H. (1971c). The new Hegi Compositae. *Taxon* **19**, 937–938.
HEYWOOD, V. H. (1973). Taxonomy in crisis? Or Taxonomy is the digestive system of biology. *Acta Bot. hung.* (in press).
HEYWOOD, V. H. and MCNEILL, J. (eds) (1964). "Phenetic and Phylogenetic Classification." Systematics Association Publication **6**, London.
JACOBS, M. (1969). Large families—not alone. *Taxon* **18**, 253–262.
JARDINE, N. and SIBSON, R. (1970). Quantitative attributes in taxonomic descriptions. *Taxon* **19**, 862–870.
KRUCKERBERG, A. R. (1969). The implications of ecology for plant systematists. *Taxon* **18**, 92–120.
MAYR, E. (1959). Trends in avian systematics. *Ibis* **101**, 293–302.
MORSE, L. E. (1971). Specimen identification and key construction with time-sharing computers. *Taxon* **20**, 269–282.

MORSE, L. E., PETERS, J. A. and HAMEL, P. B. (1971). A general data format for summarizing taxonomic information. *BioScience* **21**, 174–181.

PANKHURST, R. J. (1970). A computer programme for generating diagnostic keys. *Computer J.* **13**, 145–151.

SHETLER, S. G. (1971). Flora North America as an information system. *BioScience* **21**, 524–532.

SMITH, A. C. (1969). Systematics and appreciation of reality. *Taxon* **18**, 5–13.

SOKAL, R. R. (1970). Another new biology. *BioScience* **20**, 152–159.

TAYLOR, R. L. (1971). The Flora North America project. *BioScience* **21**, 521–523.

VAN STEENIS, C. G. G. J. (1954). General principles in the design of Floras. *8me Congr. Int. Bot. Paris Rapp. & Comm.* sect. 2, 4, 5, & 6, 59–66.

WATSON, L. (1971). Basic taxonomic data: the need for organization over presentation and accumulation. *Taxon* **20**, 131–136.

Author Index

The numbers in *italic* indicate the pages on which names are mentioned in the reference lists

A

Abbayes, H. des, 33, 48, *65*
Abrol, Y. P., 214, 231, *237*
Adams, J., 8, *24*
Adams, M. S., 55, *65*
Adelberg, E. A., 152, *172*
Ahmadjian, V., 32, 33, 48, 61, *65*, 161, 164, 167, *168*
Ahti, T., 50, 53, *65*
Alexander, M., 168, *168*
Allan, P. B. M., 228, *238*
Allard, R. W., 8, 9, *24*, *27*, *29*
Almborn, O., 33, *65*
Amadon, D., 15, *24*
Amberson, J. M., 174, *187*
Anderson, D. J., 14, *24*
Anderson, E., 17, 19, *24*
Anderson, G., 165, *168*
Anderson, L., 216, *239*
Andrews, J. T., 59, *65*
Antonovics, J., 7, 8, *24*, *28*, 233, *238*
Apinis, A. E., 149, *150*
Asahina, Y., 57, *65*
Auclair, J. L., 245, *262*

B

Bachelard, G., 135, *138*
Baddeley, M. S., 59, *65*
Bailey, R. W., 227, *238*
Baker, 190, 191, 196
Baker, H. A., 89, *91*
Baker, H. G., 244, 258, 262, *262*, 286, *287*
Baldwin, C. L., Jr., 245, *264*
Bänziger, H., 246, *262*
Barkman, J. J., 32, 61, *65*, 145, 146, 147, 148, *150*, 161, *168*, 266, *281*

Barnes, H. F., 236, *238*
Bartlett, E. M., 165, 166, *169*
Barton, L. V., 114, *117*
Batiste, W. C., 228, 229, *238*
Beardmore, J. A., 7, 8, *24*, *25*
Behr, L., 32, *66*
Bendall, D. A., 225, *238*
Bennett, I., 202, *211*
Ben-Shaul, Y., 51, *65*, *66*
Bequaert, J., 174, *188*
Bernie, A. C., 48, *69*
Berrie, A. D., 176, 177, 183, 184, 185, *187*
Bidault, M., 127, *138*
Bishop, J. A., 237, *238*
Björkman, O., 4, *25*
Blaedel, W. J., 216, *239*
Blaise, S., 127, *138*
Bliss, L. C., 71, *91*
Bock, W. J., 324, *326*
Boer, R. de, 192, *199*
Bongers, J., 204, *210*
Bonner, W. D., 225, *238*
Bormann, F. H., 167, *171*
Borriss, H., 114, *117*
Borsos, Olga, 232, *238*
Boughey, A. S., 336, *345*
Bouman, F., 192, *199*
Bowen, G. D., 164, 165, 167, *169*
Bradshaw, A. D., 4, 7, 12, 17, 19, *24*, *25*, *27*, 62, *65*, 152, *169*, 233, *238*
Braun, H. J., 319, *326*
Bridges, K. W., 336, *345*
Briggs, B. G., 19, *25*
Briggs, D., 134, *140*
Briquet, J., 136, *138*
Brodo, I. M., 32, 33, 59, 61, 64, *65*

349

Brooks, M. A., 152, *169*
Brougham, R. W., 7, *25*
Brower, L. P., 12, *25*
Brown, D. S., 175, 180, 181, 182, *187*
Buchner, P., 152, *169*
Buck, J., 225, *241*
Burch, J. B., 175, 180, 182, *187*
Burges, N. A., 98, *118*
Butler, G. W., 216, *242*

C

Cain, A. J., 214, *238*
Cain, R. F., 61, *65*
Carpenter, L., 259, *262*
Carrodus, B. B., 163, *169*
Carson, H. L., 6, *25*, 293, *314*
Cartier, D., 127, *138*
Cavers, P. B., 114, 115, *117*
Cazier, M. A., 286, *287*
Chadefaud, M., 135, *139*
Chang, V. C. S., 310, 313, *315*
Chapman, D. S., 54, *67*
Charles, A. H., 7, *25*
Cherkasskaya, V. S., 59, *65*
Chiang, H. C., 53, *69*
Churchill, E. D., 20, *26*
Ciereszko, L. S., 167, *169*
Clapham, A. R., 20, *25*
Clarke, B., 8, *25*
Clarke, B. C., 231, *238*
Clatworthy, J. N., 16, *26*
Clausen, J., 12, 16, *25*
Clauss, E., 216, *238*
Cochrane, V. W., 163, *169*
Coe, M. J., 72, 87, *91*
Combes, R., 214, *238*
Conard, H. S., 127, *138*
Conde, L. F., 325, *326*
Conn, E. E., 214, 216, 231, *237*, *242*
Connell, J. H., 202, *210*
Constance, L., 1, 10, 20, *25*
Cook, C. D. K., 4, 5, *25*
Coppins, B. J., 45, 46, 61, *65*, *67*
Corkill, L., 223, *238*
Corner, E. J. H., 319, *326*

Cosgrove, D. J., 165, *169*
Craig, A., 227, *238*
Crawford-Sidebotham, T. J., 215, 219, 227, 236, 237, *238*
Crepet, W. L., 192, *199*
Cronquist, A., 318, *326*
Crovello, T. J., 11, 13, *25*, 339, *345*
Crowson, R. A., 305, *314*
Cuatrecasas, J., 87, *91*
Culberson, C. F., 52, 53, *65*, *66*
Culberson, W. L., 33, 52, 53, *65*, *66*

D

Daday, H., 217, 218, 236, 237, *238*
Daft, M. J., 165, *169*
Dale, M. B., 12, 13, *27*, *29*
Darwin, C. R., 215, 233, *238*
Datta, S. C., 95, *117*
Dautzenberg, P., 143, *150*
Davey, K. G., 245, *262*
Davidson, J. F., 20, *25*
Davies, M. S., 7, 17, *25*, *28*
Davis, P. H., 2, 6, 10, 11, 13, 20, *25*, 62, *66*, 253, *262*, 333, 334, 342, 343, *345*
Dayhoff, M. O., 231, *238*
Degelius, G., 46, *66*
Delevoryas, T., 192, *199*
Dempster, E. R., 6, *25*
De Sloover, J., 46, *66*
Dethier, Y. G., 291, 294, *314*
De Varigny, H., 214, *238*
De Vries, W., 232, *239*
De Waal, D., 231, 235, *239*
Dibben, M. J., 32, *66*
Dobzhansky, Th., 4, 6, 15, 19, *26*
Dobzhansky, T. L., 7, *25*
Dodson, C. H., 261, *263*
Donald, C. M., 4, *26*
Doudoroff, M., 152, *172*
Drew, E. A., 158, *169*
Dronamraju, K. R., 257, *262*
Droop, M. R., 158, 159, 160, 164, *169*
Duncan, U. K., 162, *169*
Durand, B., 132, *138*
Du Rietz, G. E., 45, *66*, 147, 148, *150*

E

Easty, D. B., 216, *239*
Edwards, C. A., 228, *239*
Eggerman, W., 204, *210*
Ehrendorfer, F., 133, *139*, 190, 191, 198, *199*, 318, *326*
Ehrlich, P. R., 15, 16, 19, 20, 24, *26*, 214, *239*, 284, *287*, 294, 298, 305, 308, 312, *314*
Esikowitch, D., 285, *287*
Eisner, H. E., 226, *239*
Eisner, T., 226, *239*
Emberger, L., 135, *139*
Endress, P., 197, *199*
Engelmann, F., 245, *262*
Euw, J., von, *242*
Evenari, M., 95, *117*
Ewers, W. H., 178, *187*

F

Fabiszewski, J., 61, *66*
Faegri, H., 320, *326*
Faegri, K., 191, *199*, 244, 246, 259, *262*, 285, *287*
Fairbrother, D. E., 9, *26*
Farrar, J., 55, *66*
Fedoseeva, L. L., 228, *239*
Feeny, P. P., 213, 214, 236, *242*
Feinburg, E. H., 8, *28*
Ferry, B. W., 59, *65*
Finegan, E. J., 59, *65*
Fischer, H., 143, *150*
Fisher, F. J. F., 334, *345*
Flor, H. H., 213, *239*
Flück, H., 231, *239*
Fogden, M. P. L., 207, *210*
Forbes, W. T. M., 305, *314*
Ford, E. B., 8, *26*, 246, *263*
Fosberg, F. R., 331, *346*
Foulds, W., 221, 222, 230, 237, *239*
Fraenkel, G., 216, 219, 220, *241*, 294, 298, *314*
Fraenkel, G. S., 213, 214, *239*
Frisch, K. von, 284, *287*

G

Gadgil, P. D., 167, *169*
Gadgil, R. L., 167, *169*
Galil, J., 285, *287*
Galinou, M. A., 161, *169*
Galun, M., 32, 48, 50, 51, *65*, *66*
Gardou, C., 127, 133, *139*
Geddes, P., 160, *169*
Gerdemann, J. W., 164, 165, *169*, *171*
Gibbs, R. D., 232, *239*
Gibson, J. B., 19, *28*
Gibson, L. P., 203, *210*
Gilbert, L. E., 246, 254, 256, 257, *263*
Gilbert, O. L., 58, *66*
Gilg, E., 303, *314*
Gillett, J. B., 16, *26*
Gilmour, D., 225, *239*
Gilmour, J. S. L., 2, *26*, 134, *139*, 344, *346*
Goddard, D. R., 225, *241*
Good, R., 31, *66*
Goodall, D. W., 11, 12, 13, 14, *26*
Gooday, G. W., 159, *169*
Goreau, N. I., 156, 157, *170*
Goreau, T. F., 156, 157, *170*
Gorenflot, R., 127, *139*
Gottsberger, G., 317, 324, *326*, *327*
Gottschalk, W., 324, *327*
Graham, J. H., 227, *239*
Graham, W. L., 54, *66*
Grandtner, M., 128, *139*
Grant, K. A., 259, *263*, 286, *287*
Grant, V., 134, *139*, 259, *263*, 286, *287*, 318, *327*
Grant, V. E., 190, *199*
Grant, W. F., 214, 232, *239*
Gray, E., 236, *241*
Greene, D. M., 339, *346*
Greenwood, R. M., 227, *238*
Gregor, J. W., 96, *117*, 344, *346*
Greig-Smith, P., 12, 13, 14, 20, *26*
Griffiths, D. J., 18, *26*
Grime, J. P., 221, 222, 230, 237, *239*
Grove, W. B., 64, *66*
Grubb, P. J., 266, *281*
Grummann, V. J., 33, 51, *66*
Guérin, P., 216, 227, *239*

Guinochet, M., 125, 127, 129, 133, *139*
Gutterman, Y., 95, *117*
Gyrisco, G. G., 226, 228, 229, *242*

H

Hagerup, O., 286, *287*
Hainsworth, F. R., 244, *264*
Hale, M. E., 32, 51, 54, *66*, 161, 164, 167, *170*
Hall, A. V., 336, 339, *346*
Hall, F. R., 226, *240*
Hallier, H., 303, *314*
Hamel, P. B. A., 337, 339, *347*
Hand, M. M., 114, *119*
Hanelt, P., 232, *240*, 344, *346*
Hansen, H. L., 229, *240*
Hanson, H., 20, *26*
Harberd, D. J., 7, 12, *26*
Harborne, J. B., 277, *281*, 294, 295, 313, *314*
Harley, J. L., 152, 153, 156, 156, 164, 168, *170, 171*
Harper, J. L., 4, 16, *26*, 111, 112, 114, 115, *117*, 270, *281*
Harrington, J. F., 114, *117*
Harris, G. P., 55, 58, *67, 68*
Harris, P., 298, 300, *315*
Harris, W., 7, *25*
Hartl, D., 297, *315*
Hastings, S., 39, *68*
Hawksworth, D. L., 32, 34, 45, 46, 47, 49, 51, 52, 53, 54, 63, *67*
Haydak, M. H., 245, 255, *263*
Hayes, J. T., 8, *28*
Hayman, D. S., 156, 166, *170, 171*
Haynes, F. H., 64, *67*
Head, W. B., 6, *25*
Heath, G. W., 228, *239*
Hedberg, O., 72, 74, 77, 79, 81, 82, 83, 84, 87, 88, 89, *91*
Hegi, G., 137, *139*
Hegnauer, R., 214, 231, 232, 237, *240*
Heikkilä, H., 32, *65*
Heinrich, B., 244, 258, *263*, 286, *287*
Heise, H., 152, *170*

Henry, S. M., 152, *170*
Hering, M., 286, *287*
Hertel, H., 61, *67*
Heslop-Harrison, J., 4, 10, 19, *26*, 96, *117*, 134, *139*, 192, *200*, 344, *346*
Hewitt, E. J., 225, *240*
Heywood, V. H., 2, 6, 10, 11, 13, 20, *25*, 27, 62, *66*, 98, *117*, 134, *139*, 253, *262*, 330, 331, 332, 333, 334, 335, 336, 338, 340, 342, 343, *345, 346*
Hiesey, W. H., 16, *25*
Hiesey, W. M., 10, 17, *27*, 96, *117*
Higgins, V. J., 227, *241*
Highton, R. B., 177, *188*
Hill, D. J., 49, 50, *67*, 158, *170*
Hocking, B., 244, 246, 260, *263*
Hollingworth, R. M., 226, *240*
Holm, A., 87, *91*
Holm, L., 303, *315*
Holm, R. W., 15, 20, 24, *26*,
Holmgren, P., 4, *25*
Honch, G., 310, *315*
Hosokawa, T., 266, *281*
House, H. L., 250, 258, *263*
Hovanitz, W., 310, 313, *315*
Hubbard, C. E., 5, *27*
Hughes, Monica A., 216, 232, *240, 241*
Hultén, E., *91*
Huneck, S., 52, *67*
Hunter, P. J., 186, *188*
Hurd, 190, 191, 196, *199*
Hurd, P. D., Jr., 244, *262*
Hurst, J. J., 226, *239*

I

Ikeda, A. G., 336, *345*

J

Jackobs, J. A., 165, *171*
Jacobs, M., 332, *346*
Jain, S. K., 4, 7, 16, 17, 19, *27*
James, P. W., 45, 51, 59, 61, *67*
Jamieson, C. A., 245, *262*
Janzen, D. H., 202, 203, 204, 205, 206, 208, 209, *210, 211*, 244, *263*, 286, *287*

Jardine, N., 332, *346*
Jelinkova, E., 46, *68*
Johnson, C. D., 207, *211*
Johnson, L. A. S., 15, *27*
Johnson, M. P., 16, *27*
Jones, D. A., 213, 214, 215, 216, 218, 219, 220, 223, 224, 225, 226, 227, 228, 229, 230, 234, 235, 236, 237, *240*.
Jones, F. G. W., 227, 228, *241*
Jones, M. G., 227, 228, *241*
Juhren, G., 94, *119*
Juhren, M. C., 94, *119*

K

Kafatos, F. C., 226, *239*
Kannenberg, L. W., 9, *27*
Kappen, K., 62, *67*
Kärenlampi, I., 41, 42, *67*
Keck, D. D., 16, *25*
Keilin, D., 225, *241*
Keister, Margaret, 225, *241*
Kershaw, K. A., 32, 55, 58, *68*, 160, *170*
Kettlewell, H. B. D., 233, *241*
Khan, A. G., 164, *170*
Kilby, B. A., 245, *263*
Kimura, Motoo, 231, *241*
Kirk, J. T. O., 275, *281*
Kjaer, A., 299, *315*
Klots, A. B., 246, *263*, 305, 309, *315*
Kölreuter, J. G., 283, *287*
Koller, D., 95, *117*
Korn, M. E., 237, *238*
Kristinsson, H., 41, *68*
Kruckeberg, A. R., 1, 16, *27*, 332, *346*
Krywolap, G. N., 167, *170*
Kugler, H., 191, *200*, 321, *327*
Kuijt, J., 325, *327*
Kurland, C. G., 226, *241*
Kurokawa, S., 38, *68*

L

Lacoste, A., 131, *140*
Lambert, J. M., 12, 13, *27*
Landers, R. Q., 111, *118*
Lane, C., 225, 228, 237, *241*, 285, *287*
Lang, A., 94, *118*
Lang, K., 226, 228, *241*
Larsen, E. B., 286, *287*
Larsen, K., 232, *241*
Last, F. T., 295, *315*
Laundon, J. R., 39, 40, 46, 47, 50, *68*
Leblanc, F., 46, *66*
Lebrun, J., 128, *140*
Lee, R. E., 278, *281*
Leeuwen, C. G., van, 223, *241*
Lemée, G., 127, *139*
Léon, J., 293, *315*
Leppik, E. D., 190, 194, 195, *200*
Leppik, E. E., 190, 191, 194, 195, 196, 198, *200*, 303, *315*, 320, 321, *327*
Leroux, P. L., 177, *187*
Levene, H., 6, *27*
Levin, D. A., 216, *241*, 244, 257, *263*
Levine, L., 8, *25*
Levins, R., 6, *27*
Lewis, D., 154, 157, 158, *172*
Lewis, D. H., 152, 153, 155, 156, 158, 159, 160, 161, 164, 165, 166, 167, 168, *169*, *170*, *171*
Lewis, H., 16, 19, *27*
Lewis, M. C., 11, *27*, *29*
Lewis, T., 286, *287*
Lewontin, R. C., 6, 8, *27*
Linskens, H. F., 245, 256, *263*
Linsley, F. G., 286, *287*
Lippmaa, T., 147, *150*
Lo, C. T., 180, *187*
Looman, J., 50, *68*
Lourens, J. H., 298, 310, *315*
Lovatt, Evans, C., 226, *241*
Löve, A., 71, *91*
Loyd, R. C., 236, *241*
Lucretius Carus, Titus, 215, *241*
Lüdi, W., 128, *140*
Lunan, D. A., 59, *68*
Lundeberg, G., 163, *171*
Lundegardh, H., 225, *241*
Luria, S. E., 215, *241*
Lüttge, U., 247, 249, *263*

M

Maarel, E., van der, 146, *150*, 223, *241*
MacArthur, R., 6, 27
McClelland, W. F. J., 185, *187*
McCullough, F. S., 177, *187*
MacDonald, R. D., 339, *345*
Macior, L. W., 244, *263*, 284, *287*, *315*
McLaughlin, J. J. A., 160, 164, *171*
McMillan, C., 1, 12, 20, 27
McNaughton, I. H., 16, *26*, 114, *117*
McNaughton, S. J., 111, *118*
McNeill, J., 330, *346*
McVean, D. N., 12, 27
McWilliams, E. L., 111, *118*
Mabry, T. J., *315*, *316*
Maher, E. P., 216, 232, *241*
Mahlstede, J. P., 111, *118*
Mandahl-Barth, G., 174, 180, 183, 186, *187*
Mangenot, G., 129, *140*
Marks, P. C., 167, *171*
Marsden-Jones, E. M., 4, 27
Marsh, P. B., 225, *241*
Marshall, D. R., 4, 27
Marton, K., 32, *66*
Maurizio, A., 247, *263*
May, V., 202, *211*
Mayer, A. M., 94, *118*
Maynard Smith, J., 6, 27
Mayr, E., 15, 16, 20, *28*, 342, *346*
Medwecka-Kornas, A., 128, 129, *140*
Meeuse, A. D. J., 190, 191, 195, 196, 198, 200
Meinwald, J., 226, *239*
Melchior, H., 318, *327*
Melin, E., 163, *171*
Merz, E., *315*
Michener, C. D., 12, *28*
Millar, R. L., 227, *241*
Millbank, J. W., 32, *68*, 160, *170*
Milner, H. W., 10, 17, 27, 96, *117*
Mitchell, B. D., 48, *69*
Montgomery, R. D., 225, *241*
Moore, D. M., 22, *28*
Morgan-Huws, D I., 64, *67*
Morley, B. D., 269, 270, 271, 272, 273, 278, *281*
Morley, F. H. W., 95, *118*
Morris, M. J., 228, *241*
Morse, L. E., 336, 337, 339, *346*, *347*
Mosse, B., 156, 165, 166, *170*, *171*
Müller, H., 246, 247, *263*
Murdoch, C. L., 165, *171*
Murray, S. A., 55, *68*
Muscatine, L., 154, 156, 157, 158, *171*, *172*

N

Natarajan, R., 175, *187*
Nayer, J. K., 216, 219, 220, *241*
Negbi, M., 95, *117*
Nelson, A. P., 4, *28*
Nelson, G. S., 177, *188*
Neunzig, H. H., 226, 228, 229, *242*
New, J. K., 114, *118*
Nicholas, D. J. D., 225, *240*
Nicolson, T. H., 164, 165, *169*, *171*
Nilsson, H., 163, *171*
Nordhagen, R., 128, *140*

O

Oberdorfer, E., 137, *140*
Oberholzer, G., 180, *187*
Odum, E. P., 2, *28*, 160, 164, *171*
Odum, H. T., 160, 164, *171*
Ohta, Tomoko, 231, *241*
Oliver, E. G. H., 89, *91*
Olson, J., 111, *118*
Ooststroom, S. J., van, 143, *150*
Orloci, L., 123, *140*

P

Pankhurst, R. J., 336, *347*
Paran, N., 51, *65*, *66*
Paris, R., 232, *242*
Parkin, J., 192, *200*
Parmelee, J. A., *315*
Parsons, J., 225, 226, 228, 229 237, *242*
Pate, J. S., 164, *171*
Paterniani, E., 19, *28*
Patzke, E., 301, *315*
Paulson, R., 39, *68*

Pavlovsky, O., 19, *26*
Pearse, V. B., 156, *171*
Pelkonen, M., 41, 42, *67*
Percival, M., 247, 249, *263*
Peters, J. A., 337, 339, *347*
Petterson, B., 88, *91*
Peveling, E., 51, *68*
Phillips, F. M., 165, *171*
Phillipson, W. R., 23, *28*
Pilsbry, H. A., 174, *188*
Pimental, D., 8, *28*
Pipkin, S. B., 293, *315*
Pisut, I., 46, *68*
Pitchford, R. J., 177, 180, *188*
Poelt, J., 44, 61, *68*
Poljakoff-Mayber, A., 94, *118*
Price, J. R., 231, *242*
Price, R. D., 260, *264*
Pryce-Jones, L., 247, *263*

R

Ratcliffe, D. A., 12, *27*
Rauh, W., 86, *91*
Raunkiaer, C., 81, *91*
Raven, P. E., 244, *263*
Raven, P. H., 16, 19, *26*, *27*, 214, *239*, 284, 286, *287*, 294, 298, 305, 308, 312, *314*
Read, C. P., 152, *171*
Reid, C. P. P., 153, 164, *171*
Richards, P. W., 266, *281*
Richardson, D. H. S., 32, 38, 51, *68*
Ridgway, R. L., 228, *242*
Ridley, H. N., 278, *281*
Rivnay, E., 285, *288*
Robertson, C., 243, 257, *263*, *264*, 286, *287*
Robinson, Muriel E., 215, *242*
Rocci, V., 225, *242*
Rodriguez, R. L., 293, *315*
Rogers, D. J., 124, *140*
Rose, C. R., 178, *187*
Rose, F., 45, 46, *67*
Rosenan, N., 284, *287*
Rosenthal, O., 226, *242*
Ross, M. A., 270, *281*

Rothmaler, W., 137, *140*
Rothschild, M., 225, 226, 228, 229, 237, *240*, 285, *287*
Rovira, A. D., 164, 165, 167, *169*
Ruijgrok, H. W. L., 232, *240*, *242*
Rundel, P. W., 49, 54, *68*
Runemark, H., 43, *69*
Ruszkowski, A., 227, 228, *242*

S

Sagar, G. R., 16, *26*
Salisbury, E. J., 20, *28*, 94, 115, *118*
Salomon, H., 161, *172*
Salt, G., 72, 87, *92*
Sanders, F. E., 116, *172*
Sanderson, A. R., 228, *242*
Sandholm, H. A., 260, *264*
Santesson, R., 61, 62, *69*
Sarkar, P., 71, *91*
Savile, D. B. O., 71, *92*, 301, *313*, *316*
Saylor, W. R., 278, *281*
Schade, A., 48, *69*
Schimper, A. F. W., 266, *281*
Schneider, A., 216, *242*
Schneiderman, H. A., 226, *241*
Schoonhoven, L. M., 292, 293, 294, 297, 310, *316*
Schrauwen, J., 245, 256, *263*
Schuette, H. A., 245, *264*
Schultz, W. M., 8, *28*
Schulze, E.-D., 62, *67*
Schutte, C. H. J., 175, 180, *187*
Schwanitz, F., 324, *327*
Schwarz, E., 174, *187*
Schwickerath, M., 146, *150*
Scott, G. D., 153, 154, *172*
Seaton, A. P. C., 8, *28*
Selander, R. K., 1, *28*
Sernander, R., 33, 35, *69*
Shankland, D. L., 226, *240*
Sharma, B. D., 192, *200*
Sherk, L. C., 278, *281*
Shetler, S. G., 332, 333, *347*
Shibata, S., 53, *69*
Shipe, E., 236, *241*

Sibson, R., 332, *346*
Sidhu, B. S., 214, 232, *239*
Simpson, G. G., 135, *140*
Siple, P. A., 47, *69*
Smith, A., 7, *28*
Smith, A. C., 331, *347*
Smith, D., 154, 157, 158, *172*
Smith, D. C., 156, 158, 159, 160, 161, 162, 167, *169*, *170*, *171*, *172*
Smith, F. A., 163, *172*
Smith, I., 248, *264*
Snaydon, R. W., 7, 17, *25*, *28*
Sneath, P. H. A., 13, *28*
Sokal, R. R., 13, *28*, 330, *347*
Solbrig, O. T., 22, *28*
Soó, R., 137, *140*
Soukatchev, V. N., 133, *140*
Southwood, T. R. E., 206, *211*
Spielman, A., 260, *264*
Stahl, E., 215, *242*
Stalker, H. D., 6, *28*
Stanier, R. Y., 152, *172*
Stark, N. M., 168, *172*
Starr, M. P., 152, *170*
Stearns, F., 111, *118*
Stebbins, G. L., 62, *69*, 133, *140*, 286, *288*, *316*, 318, 319, 320, 322, 323, *327*
Stebbins, J. L., 15, 16, *28*
Steiner, M., 61, *68*
Stent, G. S., 213, *242*
Stiles, F. G., 244, *264*
Stoker, J. R., 231, *237*
Stokes, P., 94, *118*
Stone, D. E., 325, *326*
Suneson, C. A., 8, *28*
Swinscow, T. D. V., 44, 52, *69*
Syers, J. K., 48, *69*
Syrett, P. J., 157, *172*

T

Takhtajan, A., 88, *92*, 318, *327*
Takhtajan, A., L., 196, 198, *200*, 303
Tapper, B. A., 216, *242*
Taylor, D. L., 159, *172*
Taylor, R. L., 335, 337, *347*

Teesdale, C., 177, *188*
Tempère, G., 304, 305, 306, *316*
Terofal, F., 305, 308, 310, 312, *316*
Theodorou, C., 165, *169*, *172*
Thien, L. B., 260, *264*
Thoday, J. M., 4, 19, *28*
Thompson, L. S., 227, *242*
Thompson, P. A., 98, 112, 113, 114, *118*
Thompson, R. C., 114, *117*
Thomson, J. W., 32, 45, 47, 54, 57, *69*
Thomson, T. E., 202, *211*
Thorpe, W. H., 10, *28*
Thorsteinson, A., 292, 294, *316*
Thurston, J. M., 111, 114, *118*
Tilney-Bassett, R. A. E., 275, *281*
Tinker, P. B., 166, *172*
Trass, H., 38, *69*
Trench, R. K., 157, 159, 160, *172*
Treub, M., 215, *242*
Troll, C., 82, *92*
Troll, W., 321, *327*
Tschiersch, B., 232, *240*
Turesson, G., 4, *28*
Turner, B. L., 9, 22, *29*
Turner, R. G., 233, *238*
Turrill, W. B., 4, 10, *27*, *29*, 96, *118*
Tutin, T. G., 20, *25*, 98, *118*

U

Uphof, I. C. Th., 286, *288*
Urban, I., 270, *281*
Usanis, S. A., 8, *28*

V

Valentine, D. H., 98, *118*
Van der Pijl, L., 190, 191, *199*, *200*, 244, 246, 259, 261, *262*, *263*, 284, 285, *287*, *288*, 320, 322, *326*, *327*
Van Eeden, J. A., 180, *187*
Van Steenis, C. G. G. J., 333, *347*
Van Valen, L., 6, *29*
Varasova, N. N., 114, *118*
Vegis, A., 94, 96, 111, *119*
Verschaffelt, E., 310, *316*

Vogel, S., 245, *264*, 320, *327*
Von Abrams, G. T., 114, *119*

W

Waddington, C. H., 4, *29*
Walters, M., 134, *140*
Walters, S. M., 20, *29*, 98, *118*
Warburg, E. F., 20, *25*
Wareing, P. F., 94, *119*
Warren, R. C., 295, *315*
Waterhouse, F. L., 228, *242*
Watson, L., 333, 334, 337, *347*
Watson, P. J., 344, *346*
Webb, D. A., 98, *118*
Webber, P. J., 59, *65*
Weber, W. A., 33, 39, 43, 44, 48, *69*
Weil, J. W., 236, *238*
Wells, P. V., 4, *29*
Went, F. W., 23, *29*, 94, 95, *119*
Werdemann, R., 318, *327*
Wesson, G., 94, *119*
Westergaard, M., 94, *119*
Wetherbee, R., 53, *69*
Wetmore, C. M., 32, 50, *69*
White, J. W., 247, *264*
Whitehead, F., 4, *29*
Whittaker, R. H., 137, *140*, 143, *150*, 213, 214, 236, *242*
Wigglesworth, V. B., 250, *264*
Wijnands, D. O., 313, *316*
Wilkins, D. A., 4, 11, 12, *29*, 235, *240*
Williams, S. N., 186, *188*
Williams, W. T., 13, *29*
Willis, C. B., 227, *242*
Wilmanns, O., 148, *150*
Wohlpart, A., *316*
Wolf, L. L., 244, *264*
Wood, P. W., 8, *28*
Woods, F. W., 153, 164, *171*
Woolhouse, H. W., 49, 50, *67*, 165, *172*
Workman, P. L., 8, *29*
Wright, C. A., 175, 176, 177, 180, 181, 182, 183, *187*, *188*

Y

Yathom, Sh., 285, *288*
Yonge, C. M., 156, 157, *170*, *172*

Z

Zahl, P. A., 160, 164, *171*
Zahlbruckner, A., 38, *69*
Zak, B., 167, *172*
Zertová, Anna, 232, *241*
Ziegler, H., 247, *264*
Zilg, H., 216, *242*
Zimmerman, M., 249, *264*
Zohary, D., 133, *140*

Subject Index

Numbers in **bold** type indicate Plates

A

Abies, 35
Acanthaceae, 292
Acarospora, 44, 61
  sinopica, 48
Acer, 61
Achillea, 133, 301
Actinomycetes, 155
Adaptation, 22, 71–90, 94, 317–326
Adaptive radiation, 23, 88
  evolution, 71–90
Adonis, 324
Aegilops ovata, 95
Agrestia, 43
Agropyron, 302
Agrostemma githago, 113, 114
Agrostis sclerophylla, 79
Akebia, 198
Alchemilla, 86
  argyrophylla, 86
  elgonensis, 86
  johnstonii, 86
Alectoria, 36, 51
  capillaris, 36, 47
  chalybeiformis, 36
  fuscescens, 36, 47, 62, 63
  intricans, 36
  minuscula, 43
  nigricans, 36
  ochroleuca, 36
  sarmentosa, 34–36, 45, 49, 51
  setacea, 36
  subcana, 47
  virens, 36, 49
Algae, 21, 31–64, 155
Alliances, 124
Allieae, 296

Allium, 301, 304, 311
Alnus, 61
  glutinosa, 60
Alloplectus, 275
  ambiguus, **276**, 279
Alpine environment, 71–90
Amaranthaceae, 306
Amaranthus retroflexus, 111
Amaryllidaceae, 253, 296, 297, 301, 304, 305
Amborella, 198
Amentiferae, 303, 321
Amentiflorae, 198
Amino acids, 243–262
Ammophila, 127, 128
  arenaria, 128
  arundinacea, 128
  baltica, 128
  breviligula, 128
Amphidinium chattonii, 159
Anaptychia ciliaris, 38
A. mamillata, 38
A. melanosticta, 38
Anatomy, 50, 94, 192, 304, 317–326, 334, 338, 340
Anemone, 81
Anemoneae, 323
Anemopsis, 198
Angraecum sesquipedale, 193
Angiosperms, 189–199, 317–326
Anisophylly, 271
Annonaceae, 317
Anthecology, 189–199, 243–262, 283, 284
Anthemis, 301
Anthocorms, 197
Anthospermum, 81
Anthoxanthum odoratum, 6–9, 17–19

Subject Index

Antibiosis, 167
Antigonon leptopus, 247
Apanteles, 226
   zygaenarum, 226, 229
Apatura iris, 246
Apidae, 292
Apion loti, 228
Apocynaceae, 298
Apomixis, 12
Aporia, 310
Arabidopsis thaliana, 221
Arabis alpina, 89
Araceae, 196, 292, 321
Araliaceae, 300, 304, 313, 314
Arceuthobium, 325
Arenaria montana, 306
Ariolimax reticulatus, 227
Arion ater, 227
A. hortensis, 227
A. subfuscus, 227
Aristolochia, 292
Aristolochiaceae, 259, 321
Armeria, 40, 60
Arthonia, 61
   glaucomaria, 60
Arthopyrenia, 61, 162
   halodytes, 44
Arundineae, 303
Ascia, 308
Asclepiadaceae, 252, 259, 292, 298
Asclepias, 275, 298
   curassavica, 257, 261
Asparagus, 300, 304
Aspicilia calcarea, 43
A. cinerea, 43
Association, 121, 145, 148, 153, 154
Atella, 303
Aterales, 304
Atriplex, 132, 306, 307
Avena, 111, 302
Aveneae, 303

B

Bacellaceae, 306
Bacidia, 61

Bacteria, 155, 215, 227
Baeamyces uncialis, 41
Balanophoraceae, 325
Balsaminaceae, 322
Bartsia, 81
Bennettitales, 194, 318
Berberidaceae, 310
Bergenia crassifolia, 250
Beta, 307
Betula, 60
   pendula, 63
Betulaceae, 322
Bignoniaceae, 297
Biocoenology, 141–150
Biomass, 147, 149
Biomphalaria, 173, 178
Biosystematics, 24, 138
Biscutella, 312
Biston betularia, 233
Blaeria, 81, 89
Bombyx mori, 245
Boraginaceae, 302, 323
Borrera ciliaris, 38
Boussaingaultia baselloides, 306
Brachypodium phoenicoides, 138
B. pinnatum, 138
Brachyrhinus ligusti, 228
Brassica, 302
Breeding system, 9, 195, 267, 300, 321
Bremiola, 228
Bromeliaceae, 269
Bromus, 302
Bruchophagus kolobovae, 228, 229
Bryophytes, 21
Buddleia davidii, 247
Buellia, 61
   canescens, 50
Bulinus, 173–186
   africanus, 174–186
   alexandrina, 186
   forskali, 174, 186
   globosus, 182–184
   nasutus, 182–185
   natalensis, 180
   pfeifferi, 186
   senegalensis, 174

Bulinus—contd.
  sericinus, 180
  tropicus, 175
  truncatus, 175–186
  ugandae, 186
Burmanniaceae, 152

C

Candelariella vitellina, 50
Calicium, 61
  corynellum, 60
  curtisii, 60
Calluna vulgaris, 249, 250
Caloplaca aurantiaca, 51
C. heppiana, 46
C. saxicolum, 48
C. thallincola, 60
Caltha, 246
Calycanthus occidentalis, 245
Campanulaceae, 253
Camponotus, 324
Capparidaceae, 310
Capparidales, 298, 300, 305, 308–310
Caprifoliaceae, 253
Caralluma chrysostephana, 260
Cardamine pratensis, 127
Carduus, 86
  chamaecephalus, 79
Carex glacialis, 71
Caricaceae, 298
Caryophyllaceae, 97, 304, 306
Caryophyllidae, 320
Cassida nobilis, 307
C. vittata, 307
Cassidinae, 307
Cassythaceae, 325
Castanea, 61
Catillaria, 61
  lightfootii, 46
Cattle, 228
Caucalideae, 340
Centaurea jacea, 127, 133
Centrospermae, 304
Cepaea hortensis, 227
  nemoralis, 227
Cepora, 310

Ceratophyllaceae, 324
Cercidiphyllum, 197
Cestrum parqui, 250
Cetraria ciliaris, 54
C. ericetorum, 41
C. islandica, 41
Ceuthorrhynchus contractus, 299, 300
Chaetocnema concinna, 307
C. tibialis, 307
Chamaemelum, 301
Chemistry, 52, 63, 145, 204, 213–237, 243–262, 277, 289–314, 334–338
Chenopodiaceae, 21, 306
Chenopodium, 132, 306
Chlamydospermae, 193, 197
Chlorideae, 303
Chrysanthemum, 301
Chrysomeleidae, 307
Chrysomelidae, 304
Chrysomelinae, 307
Chuquiraga spinosa, 259
Cionini, 296, 305
Citrus, 312
Cladina subtenuis, 49
Cladonia, 41
  cariosa, 53
  cervicornis, 41
  chlorophaea, 53
  coniocraea, 41
  furcata, 39, 48
  impexa, 41, 55
  ochrochlora, 41
  pacifica, 55
  pocillum, 39
  polycarpia, 53, 54
  polycarpoides, 53, 54
  polydactyla, 41
  pyxidata, 39
  rangiferina, 55
  squamosa, 41
  subtenuis, 49
  tenuis, 49
  uncialis, 41, 42, 43, 45, 49
Classification, 1, 5, 6, 10–15, 20, 134, 142, 145, 154, 165, 273, 317, 329, 338, 393, 342

Subject Index

Claytonia, 306
Clerodendron paniculatum, 247
Climate, 17, 71–90, 93–117, 127, 182, 218, 223, 267, 309
Coccomyxa, 159
Coelenterates, 155–157, 164
Coenogonium, 33
Coenospecies, 344
Coevolution, 8, 193, 213–237, 289–314, 323
Coleadinae, 310
Coleoptera, 193, 198, 228
Coliadinae, 308
Colotis, 309
Columnea, 265–280
   affinis, 278
   argentea, 268–270
   arguta, 266, 272
   aurantiaca, 274
   aureonitens, 272, 274
   billbergiana, 272
   brevipila, 268
   consanguinea, 278
   crassa, 278
   dictyophylla, 279
   dimidiata, 278
   dissimilis, 278
   fawcettii, 268, **276**
   fendleri, **276**
   florida, 272, 278
   grata, 279
   harrisii, 268, 278
   hirsutissima, 279
   hirsuta, 266–268, 270, 277
   hispida, 268, 277, 279
   inaequilatera, 278
   incarnata, 274, 275, 279
   illepida, 279
   jamaicensis, 268, 274, 279
   kalbreyeri, 279
   kucyniakii, 274
   magnifica, 279, 280
   major, 277
   microphylla, 277
   moorei, 279
   nicaraguensis, 271, 279
   oerstediana, 272, 279
   pallida, 278
   pectinata, 278
   perpulchra, 278
   picta, 279
   pilosissima, 279
   proctorii, 268
   pulcherrima, 279
   purpurata, 272, 278
   rubida, 278
   rubrocincta, 279
   rubricaulis, 279
   rutilans, 266, 268, 269, 271
   sanguinea, 272
   sanguinolenta, 279
   scandens, 271, 278
   silvarum, 279
   subcordata, 268
   superba, 274
   teuscheri, 277
   tincta, 279
   tulae, 274, 279
   urbanii, 266, 267, 269–271, 274, 279
   verecunda, 279
   warscewicziana, 279
   wilsoni, 273, 279
Commelinaceae, 304
Competition, 201–210, 270
Compositae, 195, 247, 253, 293, 301, 321, 322
Computers, 24, 336, 339
Condylactis gigantea, 160
Conifers, 204, 313
Contarinia loti, 228
Convallaria, 304
Convoluta ruscoffensis, 158
Coral, 161
Cornicularia muricata, 45
Corylaceae, 322
Corylus, 61
Crepis, 84
Cretaceous, 89
Criocerinae, 304
Crioceris asparagi, 304
Cruciferae, 237, 308, 309
Cryptogams, 141–150

Cryptothallus, 152
  mirabilis, 147
Cucumis, 303
Curculio, 203
Curculionidae, 305, 306
Cuscuta, 313
Cuscutaceae, 225
Cyanogenesis, 213–237
Cycadales, 191
Cycadeoidales, 191
Cycadeoidea, 192, 194
Cynomoriaceae, 325
Cyperales, 322
Cytinaceae, 313
Cytocoleus, 33
Cytogenetics, 334
Cytology, 175, 334, 340
Cytotaxonomy, 22, 214

D

Dactylis glomerata, 133
Danaids, 298
Danthonieae, 303
Data processing, 329, 334, 337, 339
Delias, 309–311
  hyparete, 312
Demes, 151
Deme terminology, 134, 344
Dendriscocaulon bolacinum, 33
D. umhausense, 33
Dendrosenecio, 74, 83, 87
Dermatina quercus, 61
Dermatocarpon dessertorum, 39
D. hepaticum, 39
D. lachneum, 39, 41
D. rufescens, 41
Deserts, 95
Desherainea, 292
Dianthoseris schimperi, **80**, 84
Dianthus barbatus, 257
Dichapetalum cymosum, 296
Dichromena ciliata, 193
Digitalis, 294
Dilleniidae, 320
Dioecy, 195, 324

Dioscoriaceae, 304
Diptera, 193, 198 228, 260, 294
Discontinuity, 185
Disjunction, 127, 128
Dismorphia, 305, 309
Dismorphiinae, 305, 308–310
Dispersal, 176, 178, 322
Distribution, 56, 57, 83, 88, 90, 93–117, 127, 128, 145, 178–186, 218, 223, 224
Ditylenchus dipsaci, 227
Dracaena, 304
Drimys, 198
  winteri, 252
Drosophila, 293
Dysdercus, 205

E

Echium, 302
Ecotype, 20, 62, 144, 341, 342
Electrophoresis, 176
Elleanthus, 275
Elymus, 302
Elytrigia, 302
Embryo, 322
Embryology, 192, 304, 338, 340
Empetrum, 35
Endemism, 116
Endogone, 156, 162
Endogyne, 153
Ephedra, 196
Epichloe typhina, 152
Epidendrum radicans, 261
Epilachna varivestis, 219
Epiphytism, 265–280
Eragrostoideae, 303
Erica, 89
  arborea, 89
  mediterranea, 250
Ericaceae, 21
Eroessa, 309, 310
Eronia, 309
Espletia, 87
Establishment, 322
Euchloini, 309, 310
Euphorbia, 313

Euphorbiaceae, 193, 298, 308, 321
Euphrasia, 4
Euptelea, 197–199
Euryops, 81
   dacrydioides, **80**
Euryurus leachii, 226
Evernia prunastri, 49, 59
Evolution, 16, 71–90, 189–199, 243–262 265, 317–326
   convergent, 261
   parallel, 84
   see also coevolution; speciation
Extinction, 209, 318

F

Fagaceae, 203
Fagus, 61, 81
Fatsia, 314
   japonica, 313
Fertilization, 321
Festuca, 127, 302
   abyssinica, 84
   ovina, 127
   pilgeri, **73**, **78**, 84, 86
   rubra, 7, 8
Festuceae, 302, 303
Ficus, 193, 196–198, 292
Flacourtiaceae, 303
Floral biology, 321, 340
Floras, 329, 332, 336, 337
Floristics, 121, 142, 145
Flower, 190–199, 243–262, 319
Fossils, 195
Fouquieriaceae, 321
Fraxinus excelsior, 114
Freycinetia, 193
Fulgensia fulgens, 60
Fungi, 146, 152, 155, 227, 301

G

Galerina, 145, 149
Gastroidea polygoni, 307
Gautteria, 317
Gene exchange, 42
   flow, 270, 342
   pools, 89
Généalogie, 135
Genecology, 10, 24, 343
Genetic constitution, 111, 121–138
   drift, 16, 23, 87, 88, 324
   system, 88
Genetics, 1–24, 202, 216, 231, 232, 235, 237, 300, 312, 324
   microbial, 215
Genotype, 3, 6–10, 32, 34, 58, 114, 280, 294
Gentianaceae, 152
Geranium, 11
Germination, 93–117, 321, 325
Glaucium flavum, 284–286
Gnetum, 196
Gonepteryx, 308
Graminae, 304
Graphina anguina, 44
Graphis scripta, 44
Grazing, 213–327
Grossulariaceae, 322
Gymnodinium microadriaticum, 159
Gymnosperms, 190–199, 318

H

Haematomma
   coccineum, 50
   lapponicum, 54, 56, 57
   ochroleucum, 50
   ventosum, 49, 50, 54, 56, 57, 60
Hamamelidales, 198
Hamamelididae, 321
Haplosciadium, 86
   abyssinicum, 79
Heavy metals, 233
Hedera colchica, 313
   helix, 313
Hedychium, 275
Helichrysum, 81
Heliconius, 246, 253, 256, 257
Helix aspera, 227
Helleboreae, 323
Hemerocallis fulva, 251
Hemiphytobius sphaerion, 306
Hemiptera, 228

Herbivores, 201–210
Hibiscus henningsianus, 324
Hierarchy, 2, 6, 21, 134, 143
Hohenbergia, 269
Homoptera, 228
Hordeae, 302
Hordeum, 302
Hosackia, 232
Host–parasite relationships, *see* Coevolution; Parasites
Host specificity, 201–210
Houttuynia, 198
Hyalophora cecropia, 226
Hybridization, 17, 132, 268, 269, 273, 277, 280, 313
Hydnoraceae, 325
Hydrophyllaceae, 323
Hymenoptera, 193, 196, 198, 207, 228, 229, 292
Hypera plantaginis, 225, 228
 variabilis, 228
Hypogymnia physodes, 55
Hypurini, 306
Hypurus bertrandi, 306

I

Ichneumonidae, 292
Icmadophila ericetorum, 60
Idesia, 303
Idiotaxonomy, 141–150
Ilex, 61
Immunology, 175
Indicator species, 20
Insecta, 228, 229, 304, 324
Ipomoea alba, 193
Ipomopsis candida, 286
I. thurberi, 286
Iridaceae, 292
Iris, 321
solation, 16, 18
 ecological, 300
 genetic, 312
 geographical, 84, 87, 271
 reproductive, 15, 19, 22, 136, 271, 300
Ixias, 309
Ixora, 193

J

Jacaranda acutifolia, 245
Jasione, 127
Juglandaceae, 203, 325

K

Kentranthus ruber, 250, 257
Keys, 5, 330, 332, 336, 337
Koeleria, 303
 gerardii, 303
 pumila, 303
Koenigia, 89, 90
 islandica, 71, 89, 90

L

Labiatae, 296
 (*see also* Lamiaceae)
Lactuca sativa, 114
Lamiaceae, 321
Lantana, 269
 camara, 261
Lardizabalaceae, 197
Lauraceae, 198, 312, 325
Laurales, 197
Lecania erysibe, 46
Lecanora
 calcarea, 42, 60
 chlarotera, 60
 cinerea, 42
 conizaeoides, 46
 dispersa, 39, 40, 60
 esculenta, 39
 expallens, 45
 gangaleoides, 51
 intricata, 45
 populicola, 60
 radiosa, 51
 reptans, 50
 rupicola, 60
 soralifera, 45
 varia, 46
Lecidea
 atrata, 48
 dicksonii, 48
 granulosa, 46

insularis, 60
lapicida, 48
lithophila, 48
mactocarpa, 48
oligotropha, 39
orosthea, 45
silacea, 49
tumida, 45
turgidula, 60
uliginosa, 39
Leguminosae, 253, 305, 308
Lempholemma vesiculiferum, 44
Lennoaceae, 325
Lepidoptera, 193, 225, 228, 245, 250, 294
Lepraria incana, 60
Leptidea, 305, 309
Leptinotarsa, 298
Leptorhaphis epidermidis, 60
Lichens, 31–64, 148, 155, 156, 160, 164
Life-forms, 81–83, 147, 201–210, 283, 319
Light, 49
Liliaceae, 253, 296, 297, 300, 304, 305, 321
Lilioceris merdigera, 304
Lilium martagon, 304
Limonium, 40, 60
Linum usitatissimum, 213, 231
Liriodendron tulipifera, 252
Littorina obtusata, 143
Lobaria amplissima, 33
L. pulmonaria, 34
Lobelia
  deckenii, 84
  keniensis, **76**, 85
  telekii, **73**, 74, **76**, 88
  wollastonii, 88
Loganiaceae, 305
Lolium, 302
  perenne, 18
Lophochloa pumila, 303
Loranthaceae, 308–310, 312, 313, 325
Lotus corniculatus, 213–237
L. uliginosus, 230
Lychnis flos-cuculi, 106–108, 113, 115
Lycopodiaceae, 152
Lysimachia, 4
Lythrum salicaria, 246

M

Magnolia, 198
Magnoliales, 317
Malacosoma neustria, 226
Malvaceae, 324
Malvales, 205
Malpighiaceae, 245, 324
Mammalia, 228
Man, effect of, 113, 130, 131, 186
Maritime factors, 35
Matricaria, 301
  recutita, 301
Mecinini, 296, 305
Melampsora, 303
  lini, 213
Melandrium album, 306
Menispermaceae, 197
Mercurialis annua, 132, 133
M. hueti, 132
Mesochorus temporalis, 226, 229
Microbial taxonomy, 22
Micropterygidae, 246
Microthelia, 61
Microtus agrestis, 219, 220, 228
Migration, 89
Molinia coerulea, 127
Molineae, 303
Molluginaceae, 304
Mollusca, 155, 227
Monimiaceae, 197, 198
Monochlamydeae, 303
Monographs, 329, 332, 336
Monotropa, 147, 152
Moraceae, 298
Morphology, 17, 39, 63, 72, 94, 196, 265, 283, 304, 317–326, 335, 338
Mountains, 17–90, 127, 206, 319
Mutation, 9, 294, 300, 323
Mycena iodiolens, 145
M. vitrea, 145
Mycoblastus sanguinarius, 60
Mycorrhiza, 152, 155, 156, 162, 164
Myosotis keniensis, 79
Myristicaceae, 197
Myrmecia, 33
Myrtaceae, 253

N

Nasturtium, 311
Nectar, 243–262
Nectarinia johnstonii, 87
Nematoda, 227
Neoteny, 319, 325
Nicotiana, 295
Nigella, 324
Nitrogen, 157
Noctuidae, 257
Nomenclature, 144, 338
Nostoc, 33, 34
Numerical taxonomy, 14, 333
Nymphaceae, 196
Nymphaeaceae, 322
Nymphaeidae, 261

O

Ochrolechia
  androgyna, 45
  geminipara, 45
  tartarea, 45
Ocotea, 312
Oleaceae, 297
Oligomerization, 320
Oncomelania, 173
Opegrapha atra, 44
Ophioglossaceae, 152
Ophioglypha, 159
Ophrys, 292
Orchidaceae, 245, 292, 304, 322
Orchids, 152, 215
Oreophyton, 86
  falcatum, 79, **80**
Oreotrochilus estella, 259
Orobanchaceae, 21, 297, 313, 325
Orobanche, 313
  hederae, 313, 314
Orthoptera, 227
Osteospermum jucundum, 231
Outbreeding, 81, 280

P

Palmae, 252
Pancratium, 285
  maritimum, 284, 286
Pangium edule, 215
Papaver, 114
Papilio, 300
Papilionidae, 261, 300, 304
Parasites, 4, 16, 146, 173–186, 213, 289–314, 323
Paratylenchus penetrans, 227
Pareronia, 309
Parmelia, 48
  borreri, 54
  britannica, 35, 37, 38
  caperata, 55, 58
  discordans, 55
  glabratula, 47
  mamillata, 38
  omphalodes, 55
  perlata, 45, 55
  revoluta, 35, 37–39, 45
  saxatilis, 58
  subrudecta, 54
  sulcata, 60
Parmeliella atlantica, 46
P. plumbea, 46
Pavetta, 193
Pedaliaceae, 297
Peltigera, 47
  aphthosa, 34
  polydactyla, 161, 162
  spuria, 46
Peltula obscurans, 50
Pereute, 312
Peronospora, 301
  brassicae, 302
  parasitica, 302
Pertusaria corallina, 60
Petrorhagia prolifera, 114
P. velutina, 105
Petunia hybrida, 256
Phalaridae, 303
Phaseolus
  aureus, 225
  limensis, 219, 220
  lunatus, 219
  mungo, 225
  nanus, 219

Phenotype, 3, 216, 221, 295, 300
Phenotypic control, 114
  factors, 34
  plasticity, 4, 41, 58, 62, 176
Philippia, 81, 89
Phlox paniculatus, 257
Phoebis, 308
Phosphorus, 164
Photoperiodism, 94
Phragmites, 303
Phylogeny, 2, 12, 22, 23, 190–199, 243–262, 330
Physcia
  ascendens, 45
  caesia, 45
  tenella, 45
  wainioi, 45
Physconia pulverulenta, 48
Physiology, 5, 17, 55, 93–117, 121–138, 142, 153, 208, 213–237
Physopsis, 174
Phyteuma, 127
Phytocénose, 122
Phytochemistry, 192, 340
  (*see also* Chemistry)
Phytogeography, 95, 303
Phytosociology, 13, 62, 141–150, 152
Phytotron, 32
Picea abies, 35
P. excelsa, 35
Pieridae, 261, 298, 305–313
Pieris, 298
  brassicae, 299, 300, 311
  napi, 312
Pinaceae, 303
Pinus taeda, 153
P. radiata, 165
Piperales, 197, 198
Placopsis gelida, 34
P. lepidophora, 34
P. venosa, 34
Planorbidae, 176
Plantaginaceae, 297, 305, 324
Plantago
  alpina, 127
  coronopus, 127
  lanceolata, 344
  serpentina, 127
Platyhelminthes, 155
Platymonas convolutae, 158
Pleistocene, 182
Pleuraloma flavipes, 226
Plumbaginaceae, 21
Poa kilimanjarica, 84
P. leptoclada, 84
Pollination, 18, 190–199, 227, 243–262, 273, 278, 283–286, 292, 293, 317–326
Porina, 44
  chlorotica, 52
Porites divaricata, 159, 160
Portulaca oleracea, 306
Portulacaceae, 306
Polycarpicae, 298, 304
Polychidium, 34
  umhausense, 33
Polygonaceae, 306
Polygonatum, 304
Polygonum, 132, 307
Polyommatus icarus, 225, 228
Polyploidy, 133, 145, 176, 180–182, 216
Polypodium vulgare, 143
Populus, 60
Potamogetonae, 21
Prey–predator relationship, 16
Proteaceae, 321
Protozoa, 155
Prunus javanica, 215
P. laurocerasus, 226, 233
Pseudevernia furfuracea, 47, 54
P. olivetorina, 54
Pseudocrioceris, 304
Pseudophytobius acalloides, 306
Pseudopieris, 305
Pseudopontiinae, 305, 308
Pseudotsuga, 35
Psychodidae, 193
Puccinia
  asparagi, 301
  cerinthes-agropyrena, 302
  longissima, 303
  radii, 301
  recondita, 302

Subject Index

Puccinia—contd.
  symphyti-bromorum, 302
  triseti, 303
Pulsatilla, 324
Puya raimondii, **85**, 87
Pyralidae, 257
Pyrenula nitidella, 61
Pyrgophysa, 174
Pyrolacaceae, 322

Q

Quercus, 61

R

Racodium, 33
Rafflesia, 313
Rafflesiaceae, 313, 321, 325
Ramalina
  maciformis, 62
  montagnei, 54
  obtusata, 45
  polymorpha, 45
  siliquosa, 52, 53
  subfarinacea, 52
Ranales, 298
Ranunculaceae, 252, 323, 324
Ranunculus, 198, 246, 324
Raphanobrassica, 302
Raphanus, 302
Relics, 45, 224
Resedaceae, 308
Rhacomitrium, 35
Rheum, 307
Rhinantheae, 325
Rhingia rostrata, 246
Rhizobium, 227
Rhizocarpon, 43, 61
  geographicum, 59
  obscuratum, 50
  oederi, 49
  sphaericum, 60
Rhizopogon roseolus, 165
Rhus typhina, 60
Rorippa, 311
Rosa, 114, 132

Rosaceae, 197, 253, 310, 321, 323
Rosales, 198
Rosidae, 320
Rubus, 132
Rumex, 114, 115, 132, 307
Ruspolia, 193
Rutaceae, 304
Rutales, 304

S

Sagina abyssinica, 86
S. afroalpina, **78**, 79, 86
Salicaceae, 21, 303, 322
Salix, 303, 324
Sampling, 11, 12, 14, 342
Salt spray, 285
Salvadoraceae, 308
Santalaceae, 308, 313
Santalales, 311, 325
Sapindales, 304
Sapotaceae, 298
Saururus, 198, 199
Saxifragaceae, 253
Schismatomma decolorans, 60
Schistosoma
  capense, 177
  haematobium, 173–186
  japonicum, 173
  mansoni, 173
Schizandra, 198, 199
Schizandraceae, 197
Sclerophyton circumscriptum, 60
Scolicidae, 292
Scrophulariaceae, 245, 296, 297, 305, 313, 325
Seasonality, 72, 81, 204, 236, 285
Seeds, 93–117
Selection, 16, 19, 71, 87, 177, 216, 236, 257, 286, 295, 313, 324
  natural, 15, 320
Self-fertilization, 176, 194, 286
Senecio, 83
  erici-rosenii, **75**
  hoffmannii, 261
  keniodendron, **73**, 74, **75**, 81

Sheep, 228
Silaum silaus, 314
Silene
  alba, 102, 105–110, 114–116
  colorata, 101
  cretica, 103–105, 113
  dichotoma, 114
  dioica, 105–113, 115–117
  gallica, 105, 114
  italica, 103, 104
  noctiflora, 103, 104
  nocturna, 101
  nutans, 306
  pendula, 101
  perfoliata, 101
  viscosa, 103, 104
  vulgaris, 99, 100, 114, 115
Silenoideae, 97–117
Sinapis, 302
Sitona, 228
Smilax, 304
Soil moisture, 221
  pH, 6–8, 235
  type, 39
Solanaceae, 292, 297, 298, 305
Soldanella, 71
Sorbus, 61
Sparaganothis xanthoides, 228
Spartina, 303
Speciation, 1–24, 31–64, 291, 297, 300, 310, 312
Species-richness, 201–210
Spergula, 307
  arvensis, 114
Spergularia, 307
Sphegidae, 292
Sphinctrina, 61
Sphingidae, 298, 300
Spiraeoideae, 323
Stachytarphata, 269
Statistics, 143
Stemphylium loti, 227
Stenocybe pullatula, 60
S. septata, 61
Sterculiaceae, 205
Stereocaulon nanodes, 60

Sticta, 34
  filix, 34
Striga, 313
Strigula elegans, 60
Stoebe, 81
Succession, 142
Swertia subnivalis, 86
S. volkensii, 86
Symbiosis, 151–168, 324
Symphytum, 302
Symplocarpus foetidus, 225
Syntaxonomy, 141–150
Syrphidae, 260

T

Tableau d'association, 132
Tegeticula, 246
Teloschistes flavicans, 49
Temperature, 268
  (*see also* Climate)
Tenthredo aceurima, 228
Teracolinae, 308
Tetrapanax papyrifer, 313
Theba pisana, 227
Thelephora terrestris, 153
Thelocarpon, 61
Thysanoptera, 227
Toninia coeruleonigricans, 60
Transplant experiments, 32
Trematoda, 173–186
Trentepohlia, 33
Trifolium, 79
  acaule, 86
  elgonense, **80**, 86
  parviflorum, 165
  repens, 213–237
Tripleurospermum, 301
Trisetaria pumila, 302
Trisetum, 302
  pumilum, 303
Triticeae, 302
Triticum, 302
Triuridaceae, 152
Trochilus polytmus, 267, 275, 277
Trochodendron, 197–199

Tropaeolaceae, 300, 308, 309
Tropaeolum, 300
  majus, 299
Tropics, 71–90, 129, 201–210, 319
Tsuga canadensis, 111
Tubiflorae, 305
Types, 144
Typha, 111

U

Umbelliferae, 259, 300, 304, 314, 336, 340
Uredinales, 302
Urginea maritima, 284, 285
Usnea, 64
  fragilescens, 47
  glabrescens, 47
  subfloridana, 47
Ustilago bromivora, 302
  bullata, 302

V

Variation, 10
  continuous, 14
  genetic, 1–24
  geographical, 142
Velacumantus australis, 178
Verbenaceae, 296
Vernonia, 293

Veronica, 79
  glandulosa, 86
  gunae, 86
Verrucaria, 44
  striatula, 44
Vicariance, 84, 87, 127, 128, 143, 312
Vicia faba, 344
  sativa, 236
Virus, 215
Viscaceae, 325
Vitaceae, 313

W

Wahlenbergia, 86
  pusilla, 79
Wind, 284

X

Xanthoria parietina, 38, 49, 50

Z

Zauschneria, 259
  californica, 259
Zenilla longicauda, 229
Zonation, 72
Zoogeography, 182
Zygaena, 226, 228, 229
  filipendula, 225
  lonicera, 225
Zygaenidae, 226, 257